DIFFERENCE EQUATIONS

An Introduction with Applications

DIFFERENCE EQUATIONS
An Introduction with Applications

Walter G. Kelley
Department of Mathematics
University of Oklahoma
Norman, Oklahoma

Allan C. Peterson
Department of Mathematics and Statistics
University of Nebraska
Lincoln, Nebraska

ACADEMIC PRESS, INC.
Harcourt Brace Jovanovich, Publishers

Boston   San Diego   New York
London   Sydney   Tokyo   Toronto

This book is printed on acid-free paper. ∞

Copyright © 1991 by Academic Press, Inc.
All rights reserved.
No part of this publication may be reproduced or
transmitted in any form or by any means, electronic
or mechanical, including photocopy, recording, or
any information storage and retrieval system, without
permission in writing from the publisher.

ACADEMIC PRESS, INC.
1250 Sixth Avenue, San Diego, CA 92101

United Kingdom Edition published by
ACADEMIC PRESS LIMITED
24–28 Oval Road, London NW1 7DX

Library of Congress Cataloging-in-Publication Data:

Kelley, Walter G.
   Difference equations: an introduction with applications
/ Walter G. Kelley. Allan C. Peterson.
     p. cm.
   Includes bibliographical references and index.
   ISBN 0-12-403325-3 (alk. paper)
   1. Difference equations. I. Peterson, Allan C. II. Title.
QA431.K44 1991
515' .625 — dc20   90-14437
                                                                         CIP

Printed in the United States of America
91 92 93 94   9 8 7 6 5 4 3 2 1

To our wives, Marilyn and Tina

Table of Contents

Preface	ix
Chapter 1 Introduction	1
Chapter 2 The Difference Calculus	15
2.1 The Difference Operator	15
2.2 Summation	24
2.3 Generating Functions and Approximate Summation	36
Chapter 3 Linear Difference Equations	51
3.1 First Order Equations	51
3.2 General Results for Linear Equations	60
3.3 Equations with Constant Coefficients	70
3.4 Applications	78
3.5 Equations with Variable Coefficients	91
3.6 Nonlinear Equations That Can Be Linearized	100
3.7 The z-Transform	107
Chapter 4 Stability Theory	151
4.1 Initial Value Problems for Linear Systems	151
4.2 Stability of Linear Systems	160
4.3 Stability of Nonlinear Systems	171
4.4 Chaotic Behavior	183
Chapter 5 Asymptotic Methods	207
5.1 Introduction	207
5.2 Asymptotic Analysis of Sums	212
5.3 Linear Equations	221
5.4 Nonlinear Equations	234
Chapter 6 The Self-Adjoint Second Order Linear Equation	249
6.1 Introduction	249
6.2 Sturmian Theory	260

6.3 Green's Functions	266
6.4 Disconjugacy	274
6.5 The Riccati Equation	284
6.6 Oscillation	293
Chapter 7 The Sturm-Liouville Problem	**311**
7.1 Introduction	311
7.2 Finite Fourier Analysis	314
7.3 A Nonhomogeneous Problem	324
Chapter 8 Discrete Calculus of Variations	**339**
8.1 Introduction	339
8.2 Necessary Conditions	340
8.3 Sufficient Conditions and Disconjugacy	354
Chapter 9 Boundary Value Problems for Nonlinear Equations	**371**
9.1 Introduction	371
9.2 The Lipschitz Case	374
9.3 Existence of Solutions	382
9.4 Boundary Value Problems for Differential Equations	388
Chapter 10 Partial Difference Equations	**399**
10.1 Discretization of Partial Differential Equations	399
10.2 Solutions of Partial Difference Equations	409
Answers to Selected Problems	**423**
References	**433**
Index	**453**

Preface

This book uses elementary analysis and linear algebra to investigate solutions to difference equations. We expect that the reader will have encountered difference equations in one or more of the following contexts: the approximation of solutions of equations by Newton's method, the discretization of differential equations, the computation of special functions, the counting of elements in a defined set (combinatorics), and the discrete modelling of economic or biological phenomena. In this book, we give examples of how difference equations arise in each of these areas, as well as examples of numerous applications to other subjects.

Our goal is to present an overview of the various facets of difference equations that can be studied by elementary mathematical methods. We hope to convince the reader that difference equations is a rich field, both interesting and useful. The reader will not find here a text on numerical analysis (plenty of good ones already exist). Although much of the contents of this book is closely related to the techniques of numerical analysis, we have, except in a few places, omitted discussion of the concerns of computation by computer.

We assume that the reader has no background in difference equations. The first three chapters provide an elementary introduction to the subject. A good course in calculus should suffice as a preliminary to reading this material. Chapter 1 gives seven elementary examples, including the definition of the Gamma function, which will be important in later chapters. Chapter 2 surveys briefly the fundamentals of difference calculus. In Chapter 3, the basic theory for linear difference equations is developed, and several methods are given for finding closed form solutions, including annihilators, generating functions and z-transforms. Also included are sections on applications and on transforming nonlinear equations into linear ones.

Chapter 4, which is essentially independent from the earlier chapters, is concerned mainly with stability theory for autonomous systems of equations. The Putzer algorithm for computing A^t, where A is an n by n matrix, is presented, leading to the solution of autonomous linear systems with constant coefficients. The chapter covers many of the fundamental stability results for linear and nonlinear systems, using eigenvalue criteria, stairstep diagrams, Liapunov functions and linearization. The last section is a brief introduction to chaotic behavior.

Approximations of solutions to difference equations for large values of the independent variable are studied in Chapter 5. This chapter is mostly independent of Chapter 4, but does use some of the results from Chapters 2 and 3. Here, one will find the asymptotic analysis of sums, the theorems of Poincaré and Perron on asymptotic behavior of solutions to linear equations, and the asymptotic behavior of solutions to nonlinear autonomous equations, with applications to Newton's method and the modified Newton's method.

Chapters 6 through 9 develop a wide variety of distinct but related topics involving second order difference equations from the theory given in Chapter 3. Chapter 6 contains a detailed study of the self-adjoint equation. This chapter includes generalized zeros, interlacing of zeros of independent solutions, disconjugacy, Green's functions, boundary value problems for linear equations, Riccati equations, and oscillation of solutions. Sturm-Liouville problems for difference equations are considered in Chapter 7. These problems lead to a consideration of finite Fourier series, properties of eigenpairs for self-adjoint Sturm-Liouville problems, nonhomogeneous problems, and a Rayleigh inequality for finding upper bounds on the smallest eigenvalue. Chapter 8 treats the discrete calculus of variations for sums, including the Euler-Lagrange difference equation, transversality conditions, the Legendre necessary condition for a local extremum, and some sufficient conditions. Disconjugacy plays an important role here and, indeed, the methods in this chapter are used to sharpen some of the results from Chapter 6. In Chapter 9, several existence and uniqueness results for nonlinear boundary value problems are proved, using the contraction mapping theorem and Brouwer fixed point theorems in Euclidean space. A final section relates these results to similar theorems for differential equations.

The last chapter takes a brief look at partial difference equations. It is shown how these arise from the discretization of partial differential equations. Computational molecules are introduced in order to determine what sort of initial and boundary conditions are needed to produce unique solutions of partial difference equations. Some special methods for finding explicit solutions are summarized.

This book can be used as a textbook at a variety of different levels ranging from middle undergraduate to beginning graduate, depending on the choice of topics. There are many exercises of varying degrees of difficulty (120 just in Chapter 3 alone). Answers to selected problems can be found near the end of the book. There is also a large bibliography of books and papers on difference equations for further study.

Preliminary portions of this book have been used by the authors in courses at the University of Oklahoma and the University of Nebraska. We are indebted to D. Hankerson and J. Hooker, who have taught courses from segments of the book at Auburn University and Southern Illinois University at Carbondale, respectively, and have offered helpful suggestions. We would also like to thank the following individuals who have influenced the book directly or indirectly: C. Ahlbrandt, G. Diaz, S. Elaydi, P. Eloe, L. Erbe, B. Harris, J. Henderson, L. Hall, L. Jackson, G. Ladas, R. Nau, W. Patula, T. Peil, J. Ridenhour, J. Schneider, and D. Smith. Shireen Ray deserves a special word of thanks for expertly "TeX-ing" the manuscript. Finally, Walter Kelley is grateful to the Graduate College and the Research Council at the University of Oklahoma for travel support during the writing of the book.

Chapter 1
Introduction

Mathematical computations frequently are based on equations that allow us to compute the value of a function recursively from a given set of values. Such an equation is called a "difference equation" or "recurrence equation." These equations occur in numerous settings and forms, both in mathematics itself and in its applications to statistics, computing, electrical circuit analysis, dynamical systems, economics, biology, and other fields.

The following elementary examples have been chosen to illustrate something of the diversity of the uses of difference equations and of the types of these equations that arise. Many more examples will appear later in the book.

Example 1.1. In 1626, Peter Minuit purchased Manhattan Island for goods worth \$24. If the \$24 could have been invested at an annual interest rate of 7% compounded quarterly, what would it have been worth in 1986?

Let $y(t)$ be the value of the investment after t quarters of a year. Then $y(0) = 24$. Since the interest rate is 1.75% per quarter, $y(t)$ satisfies the difference equation

$$y(t+1) = y(t) + .0175 y(t)$$
$$= (1.0175) y(t)$$

for $t = 0, 1, 2, \cdots$. Computing y recursively, we have

$$y(1) = 24(1.0175),$$
$$y(2) = 24(1.0175)^2,$$
$$\vdots$$
$$y(t) = 24(1.0175)^t.$$

After 360 years, or 1440 quarters, the value of the investment is

$$y(1440) = 24(1.0175)^{1440}$$
$$\simeq 1.697 \times 10^{12}$$

(about 1.7 trillion dollars!).

Example 1.2. It is observed that the decrease in the mass of a radioactive substance over a fixed time period is proportional to the mass that was present at the beginning of the time period. If the half life of radium is 1600 years, find a formula for its mass as a function of time.

Let $m(t)$ represent the mass of the radium after t years. Then

$$m(t+1) - m(t) = -km(t),$$

where k is a positive constant. Then

$$m(t+1) = (1-k)m(t)$$

for $t = 0, 1, 2, \cdots$. Using iteration as in the preceding example, we find

$$m(t) = m(0)(1-k)^t.$$

Since the half life is 1600,

$$m(1600) = m(0)(1-k)^{1600} = \frac{1}{2}m(0),$$

so

$$1 - k = \left(\frac{1}{2}\right)^{\frac{1}{1600}},$$

and we have finally that

$$m(t) = m(0)\left(\frac{1}{2}\right)^{\frac{t}{1600}}.$$

Introduction

This problem is traditionally solved in calculus and physics textbooks by setting up and integrating the differential equation $m'(t) = -km(t)$. However, the solution presented here using a difference equation is somewhat shorter and employs only elementary algebra.

Example 1.3. (The Tower of Hanoi Problem) The problem is to find the minimum number of moves $y(t)$ required to move t rings from the first peg in Figure 1.1 to the third peg. A move consists of transferring a single ring from one peg to another with the restriction that a larger ring may not be placed on a smaller ring. The reader should find $y(t)$ for some small values of t before reading further.

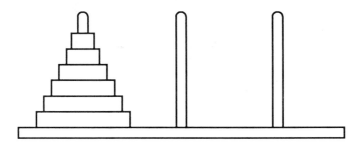

Fig. 1.1 Initial position of the rings

We can find the solution of this problem by finding a relationship between $y(t+1)$ and $y(t)$. Suppose there are $t+1$ rings to be moved. An essential intermediate stage in a successful solution is shown in Fig. 1.2. Note that exactly $y(t)$ moves are required to obtain this arrangement since the minimum number of moves needed to move t rings from peg 1 to peg 2 is the same as the minimum number of moves to move t rings from peg 1 to peg 3. Now a single move places the largest ring on peg 3, and $y(t)$ additional moves are needed to move the other t rings from peg 2 to peg 3. We are led to the difference equation

$$y(t+1) = y(t) + 1 + y(t),$$

or

$$y(t+1) - 2y(t) = 1.$$

The solution which satisfies $y(1) = 1$ is

$$y(t) = 2^t - 1.$$

(See Exercise 1.7.) Check the answers you got for $t = 2$ and $t = 3$.

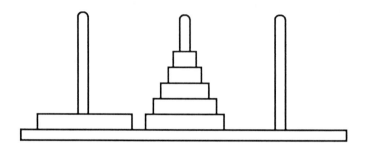

Fig. 1.2 An intermediate position

Example 1.4. (Airy equation) Suppose we wish to solve the differential equation

$$y''(x) = x\, y(x).$$

The Airy equation appears in many calculations in applied mathematics, e.g., in the study of nearly discontinuous periodic flow of electric current and in the description of the motion of particles governed by the Schrodinger equation in quantum mechanics. One approach is to seek power series solutions of the form

$$y(x) = \sum_{k=0}^{\infty} a_k x^k.$$

Substitution of the series into the differential equation yields

$$\sum_{k=2}^{\infty} a_k k(k-1) x^{k-2} = \sum_{k=0}^{\infty} a_k x^{k+1}.$$

The change of index $k \to k+3$ in the series on the left side of the equation gives us

$$\sum_{k=-1}^{\infty} a_{k+3}(k+3)(k+2)x^{k+1} = \sum_{k=0}^{\infty} a_k x^{k+1}.$$

In order that these series be equal for an interval of x values, the coefficients of x^{k+1} must be the same for all $k = -1, 0, \cdots$. For $k = -1$, we have

$$a_2(2)(1) = 0,$$

so $a_2 = 0$. For $k = 0, 1, 2, \cdots$,

$$a_{k+3}(k+3)(k+2) = a_k$$

or

$$a_{k+3} = \frac{a_k}{(k+3)(k+2)}.$$

The last equation is a difference equation that allows us to compute (in principle) all coefficients a_k in terms of the coefficients a_0 and a_1. Note that $a_{3n+2} = 0$ for $n = 0, 1, 2, \cdots$ since $a_2 = 0$.

Treating a_0 and a_1 as arbitrary constants we obtain the general solution of the Airy equation expressed as a power series:

$$y(x) = a_0 \left[1 + \frac{x^3}{3 \cdot 2} + \frac{x^6}{6 \cdot 5 \cdot 3 \cdot 2} + \cdots \right] + a_1 \left[x + \frac{x^4}{4 \cdot 3} + \frac{x^7}{7 \cdot 6 \cdot 4 \cdot 3} + \cdots \right].$$

Returning to the difference equation, we have

$$\frac{a_{k+3}}{a_k} = \frac{1}{(k+3)(k+2)} \to 0 \quad \text{as } k \to \infty,$$

and the ratio test implies that the power series converges for all values of x.

Example 1.5. Suppose a sack contains r red marbles and g green marbles. The following procedure is repeated n times: a marble is drawn at random from the sack, its color is noted and it is replaced. We want to compute the number $W(n,k)$ of ways of obtaining exactly k red marbles among the n draws.

We will be taking the order in which marbles are drawn into account here. For example, if the sack contains two red marbles R_1, R_2 and one green marble G, then the possible outcomes with $n=2$ draws are GG, GR_1, GR_2, R_1R_1, R_1R_2, R_1G, R_2R_1, R_2R_2 and R_2G, so $W(2,0)=1, W(2,1)=4$ and $W(2,2)=4$.

There are two cases. In the first case, the kth red marble is drawn on the nth draw. Since there are $W(n-1,k-1)$ ways of drawing $k-1$ red marbles on the first $n-1$ draws, the total number of ways that this case can occur is $rW(n-1,k-1)$.

In the second case, a green marble is drawn on the nth draw. The k red marbles were drawn on the first $n-1$ draws, so in this case the total is $gW(n-1,k)$.

Since these two cases are exhaustive and mutually exclusive, we have

$$W(n,k) = rW(n-1,k-1) + gW(n-1,k),$$

which is a difference equation in two variables, sometimes called a "partial difference equation." Mathematical induction can be used to verify the formula

$$W(n,k) = \binom{n}{k} r^k g^{n-k},$$

where $k = 0, 1, \cdots, n$ and $n = 1, 2, 3, \cdots$. The notation $\binom{n}{k}$ represents the binomial coefficient $\frac{n!}{k!(n-k)!}$.

From the Binomial Theorem, the total number of possible outcomes is

$$\sum_{k=0}^{n} \binom{n}{k} r^k g^{n-k} = (r+g)^n,$$

Introduction

so the probability of drawing exactly k red marbles is

$$\frac{\binom{n}{k}r^k g^{n-k}}{(r+g)^n} = \binom{n}{k}\left(\frac{r}{r+g}\right)^k \left(\frac{g}{r+g}\right)^{n-k},$$

a fundamental formula in probability theory.

Example 1.6. Perhaps the most useful of the higher transcendental functions is the gamma function $\Gamma(z)$, which is defined by

$$\Gamma(z) = \int_0^\infty e^{-t} t^{z-1} dt$$

if the real part of z is positive. Formally applying integration by parts, we have

$$\begin{aligned}\Gamma(z+1) &= \int_0^\infty e^{-t} t^z \, dt \\ &= [-e^{-t} t^z]_0^\infty - \int_0^\infty (-e^{-t}) z \; t^{z-1} dt \\ &= z \int_0^\infty e^{-t} t^{z-1} dt,\end{aligned}$$

so that Γ satisfies the difference equation

$$\Gamma(z+1) = z\Gamma(z).$$

Note that here, as in Example 1.2, the independent variable is not restricted to discrete values. If the value of $\Gamma(z)$ is known for some z whose real part belongs to $(0,1)$, then we can compute $\Gamma(z+1), \Gamma(z+2), \cdots$ recursively. Furthermore, if we write the difference equation in the form

$$\Gamma(z) = \frac{\Gamma(z+1)}{z},$$

then $\Gamma(z)$ can be given a useful meaning for all z with the real part less than or equal to zero except $z = 0, -1, -2, \cdots$.

Now

$$\Gamma(1) = \int_0^\infty e^{-t} dt = 1,$$

$$\Gamma(2) = \Gamma(1+1) = 1\Gamma(1) = 1,$$
$$\Gamma(3) = \Gamma(2+1) = 2\Gamma(2) = 2,$$
$$\Gamma(4) = \Gamma(3+1) = 3\Gamma(3) = 3\cdot 2,$$
$$\vdots$$
$$\Gamma(n+1) = n\Gamma(n) = n(n-1)! = n!.$$

We see that the gamma function extends the factorial to most of the complex plane. In Fig. 1.3, $\Gamma(x)$ is graphed for real values of x in the interval $(-4, 4)$, $x \neq 0, -1, -2, -3$.

Many other higher transcendental functions also satisfy difference equations (see Exercise 1.15). A good reference, which contains graphs and properties of the special functions discussed in this book, is Spanier and Oldham [235].

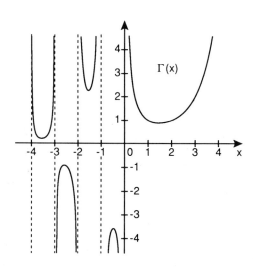

Fig. 1.3 Graph of $\Gamma(x)$ *for* $-4 < x < 4$

Introduction

Example 1.7. Euler's method for approximating the solution of the initial value problem

$$x'(t) = f(t, x(t)),$$
$$x(t_0) = x_0,$$

is obtained by replacing $x'(t)$ by the difference quotient $\frac{x(t+h)-x(t)}{h}$ for some small displacement h. We have

$$\frac{x(t+h) - x(t)}{h} = f(t, x(t))$$

or

$$x(t+h) = x(t) + h\, f(t, x(t)).$$

In order to change this difference equation to a more conventional form, let $x_n = x(t_0 + nh)$ for $n = 0, 1, 2, \cdots$. Then

$$x_{n+1} = x_n + h\, f(t_0 + nh, x_n) \qquad (n = 0, 1, \cdots),$$

where x_0 is given. The approximating values x_n can now be computed recursively; however, the approximations may be useful only for restricted values of n. For example, if $f(t, x) = x^2$ and $t_0 = 0$, then the initial value problem has the solution

$$x(t) = \frac{x_0}{1 - x_0 t},$$

which "blows up" for $t = \frac{1}{x_0}$ ($x_0 \neq 0$), while the solution of the corresponding difference equation $x_{n+1} = x_n + h x_n^2$ with x_0 given exists for all n. This is our first example of a nonlinear difference equation.

Exercises

1.1 We invest $500 in a bank where interest is compounded monthly at a rate of 6% a year. How much do we have in the bank after 10 years? How long does it take for our money to double?

1.2 The population of the city of Sludge Falls at the end of each year is proportional to the population at the beginning of that year. If the population increases from 50,000 in 1970 to 75,000 in 1980, what will the population be in the year 2000?

1.3 Suppose the population of bacteria in a culture is observed to increase over a fixed time period by an amount proportional to the population at the beginning of that period. If the initial population was 10 thousand and the population after two hours is 100 thousand, find a formula for the population as a function of time.

1.4 The amount of the radioactive isotope lead Pb-209 at the end of each hour is proportional to the amount present at the beginning of the hour. If the half life of Pb-209 is 3.3 hours, how long does it take for 80% of a certain amount of Pb-209 to decay?

1.5 In 1517, the King of France, Francis I, bought Leonardo da Vinci's painting, the "Mona Lisa," for his bathroom for 4000 gold florins (492 ounces of gold). If the gold had been invested at an annual rate of 3% (paid in gold), then how many ounces of gold would have accumulated by the end of this year?

1.6 A body of temperature 80°F is placed at time $t = 0$ in a large body of water with constant temperature 50°F. After 20 minutes the temperature of the body is 70°F. Experiments indicate that at the end of each minute the difference in temperature between the body and the water is proportional to the difference at the beginning of that minute. What was the temperature of the body after 10 minutes? When will the temperature be 60°?

1.7 In each of the following, show that $y(t)$ is a solution of the difference equation:
 (a) $y(t+1) - 2y(t) = 1$, $y(t) = A2^t - 1$.
 (b) $y(t+1) - y(t) = t + 1$, $y(t) = \frac{1}{2}t^2 + \frac{1}{2}t + A$.
 (c) $y(t+2) + y(t) = 0$, $y(t) = A\cos\frac{\pi}{2}t + B\sin\frac{\pi}{2}t$.
 (d) $y(t+2) - 4y(t+1) + 4y(t) = 0$, $y(t) = A2^t + Bt2^t$.
Here A and B are constants.

1.8 Let $R(t)$ denote the number of regions into which t lines divide the plane if no two lines are parallel and no three lines intersect at a single point. For example, $R(3) = 7$ (see Fig. 1.4).
  (a) Show that $R(t)$ satisfies the difference equation
$$R(t+1) = R(t) + t + 1.$$
  (b) Use Exercise 1.7 to find $R(t)$.

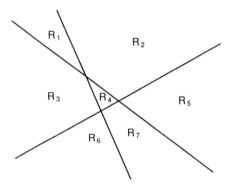

Fig. 1.4 Partition of the plane by lines

1.9 (a) Using substitution, find the difference equation satisfied by the coefficients a_k of the series $y(x) = \sum_{k=0}^{\infty} a_k x^k$ in order that $y(x)$ satisfies the differential equation $y''(x) = xy'(x)$.
  (b) Compute $\{a_k\}_{k=2}^{\infty}$ in terms of a_0 and a_1 and show that the infinite series converges for all x.

1.10 Verify that $W(n,k) = \binom{n}{k} r^k g^{n-k}$ satisfies the difference equation
$$W(n,k) = r\, W(n-1, k-1) + g\, W(n-1, k).$$

1.11 Let $D(n,k)$ be the number of ways that n distinct pieces of candy can be distributed among k identical boxes so that there is some candy in each box. Show that

$$D(n,k) = D(n-1, k-1) + k\, D(n-1, k).$$

(Hint: for a specific piece of candy consider two cases: (1) it is alone in a box or (2) it is with at least one other piece.)

1.12 Suppose that n men, having had too much to drink, choose among n checked hats at random.

(a) If $W(n)$ is the number of ways that no man gets his own hat, show that

$$W(n) = (n-1)(W(n-1) + W(n-2)).$$

(b) The probability that no man gets his own hat is given by $P(n) = \frac{1}{n!}W(n)$. Show that

$$P(n) = P(n-1) + \frac{1}{n}(P(n-2) - P(n-1)).$$

(c) Show

$$P(n) = \frac{1}{2!} - \frac{1}{3!} + \ldots + \frac{(-1)^n}{n!}.$$

1.13
(a) Show that

$$\left[\Gamma(\tfrac{1}{2})\right]^2 = 4\int_0^\infty \int_0^\infty e^{-(x^2+y^2)}\,dx\,dy.$$

(b) Use part *a* to show $\Gamma(\tfrac{1}{2}) = \sqrt{\pi}$.

(c) What are the values of $\Gamma\left(\tfrac{5}{2}\right)$ and $\Gamma\left(\tfrac{-3}{2}\right)$?

1.14 Verify

$$2^n \frac{\Gamma\left(n + \frac{1}{2}\right)}{\Gamma\left(\frac{1}{2}\right)} = (2n-1)(2n-3)\cdots(3)(1),$$

where n is an integer greater than zero.

1.15 The exponential integral $E_n(x)$ is defined by

$$E_n(x) = \int_1^\infty \frac{e^{-xt}}{t^n}\, dt \qquad (x > 0),$$

where n is a positive integer. Show that $E_n(x)$ satisfies the difference equation

$$E_{n+1}(x) = \frac{1}{n}\left[e^{-x} - x\, E_n(x)\right].$$

1.16 Use Euler's method to obtain a difference equation which approximates the logistic differential equation

$$x'(t) = ax(t)(1 - x(t)),$$

if a is a constant.

Chapter 2
The Difference Calculus

2.1 The Difference Operator

In Chapter 3, we will begin a systematic study of difference equations. Many of the calculations involved in solving and analyzing these equations can be simplified by use of the *difference calculus*, a collection of mathematical tools quite similar to the *differential calculus*. The present chapter briefly surveys the most important aspects of the difference calculus. It is not essential to memorize all the formulas presented here, but it is useful to have an overview of the available techniques and to observe the differences and similarities between the difference and differential calculus.

Just as the differential operator plays the central role in the differential calculus, the difference operator is the basic component of calculations involving finite differences.

Definition 2.1. Let $y(t)$ be a function of a real or complex variable t. The "difference operator" Δ is defined by

$$\Delta y(t) = y(t+1) - y(t).$$

For the most part, we will take the domain of y to be a set of consecutive integers such as the natural numbers $N = \{1, 2, 3, \cdots\}$. However, sometimes it is useful to choose a continuous set of t values such as the interval $[0, \infty)$ or the complex plane as the domain.

The step size of one unit used in the definition is not really a restriction. Consider a difference operation with a step size $h > 0$, say $z(s+h) - z(s)$. Let $y(t) = z(th)$. Then

$$\begin{aligned} z(s+h) - z(s) &= z(th+h) - z(th) \\ &= y(t+1) - y(t) \\ &= \Delta y(t). \end{aligned}$$

(See also Example 1.7.)

Occasionally we will apply the difference operator to a function of two or more variables. In this case, a subscript will be used to indicate which variable is to be shifted by one unit. For example,

$$\Delta_t \, te^n = (t+1)e^n - te^n = e^n,$$

while

$$\Delta_n \, te^n = te^{n+1} - te^n = te^n(e-1).$$

Higher order differences are defined by comparing the difference operator with itself. The second order difference is

$$\Delta^2 y(t) = \Delta(\Delta y(t))$$

$$= \Delta(y(t+1) - y(t))$$

$$= (y(t+2) - y(t+1)) - (y(t+1) - y(t))$$

$$= y(t+2) - 2y(t+1) + y(t).$$

The following formula for the nth order difference is readily verified by induction:

$$\Delta^n y(t) = y(t+n) - ny(t+n-1) + \frac{n(n-1)}{2!} y(t+n-2)$$

$$+ \cdots + (-1)^n y(t) \qquad (2.1)$$

$$= \sum_{k=0}^{n} (-1)^k \binom{n}{k} y(t+n-k).$$

An elementary operator that is often used in conjunction with the difference operator is the shift operator.

2.1 The Difference Operator

Definition 2.2. The "shift operator" E is defined by

$$Ey(t) = y(t+1).$$

If I denotes the identity operator, i.e., $Iy(t) = y(t)$, then we have

$$\Delta = E - I.$$

In fact, Eq. (2.1) is similar to the *binomial theorem* from algebra:

$$\begin{aligned}\Delta^n y(t) &= (E-I)^n y(t) \\ &= \sum_{k=0}^{n} \binom{n}{k}(-I)^k E^{n-k} y(t) \\ &= \sum_{k=0}^{n} \binom{n}{k}(-1)^k y(t+n-k).\end{aligned}$$

These calculations can be verified just as in algebra since the compostion of the operators I and E has the same properties as the multiplication of numbers. In much the same way, we have

$$E^n y(t) = \sum_{k=0}^{n} \binom{n}{k} \Delta^{n-k} y(t).$$

The fundamental properties of Δ are given in the following theorem.

Theorem 2.1.

a) $\Delta^m(\Delta^n y(t)) = \Delta^{m+n} y(t)$ for all positive integers m and n,

b) $\Delta(y(t) + z(t)) = \Delta y(t) + \Delta z(t)$,

c) $\Delta(Cy(t)) = C\Delta y(t)$ if C is a constant,

d) $\Delta(y(t)z(t)) = y(t)\Delta z(t) + Ez(t)\Delta y(t)$,

e) $\Delta\left(\frac{y(t)}{z(t)}\right) = \frac{z(t)\Delta y(t) - y(t)\Delta z(t)}{z(t)Ez(t)}$.

Proof. Consider the product rule (d).

$$\Delta(y(t)z(t)) = y(t+1)z(t+1) - y(t)z(t)$$

$$= y(t+1)z(t+1) - y(t)z(t+1)$$

$$+ y(t)z(t+1) - y(t)z(t)$$

$$= \Delta y(t) E z(t) + y(t)\Delta z(t).$$

The other parts are also straightforward. ∎

The formulas in Theorem 2.1 closely resemble the sum rule, the product rule, and the quotient rule from the differential calculus. However, note the appearance of the shift operator in parts (d) and (e).

In addition to the general formulas for computing differences, we will need a collection of formulas for differences of particular functions. Here is a list for some basic functions.

Theorem 2.2. Let a be a constant. Then

a) $\Delta a^t = (a-1)a^t$,

b) $\Delta \sin at = 2 \sin \frac{a}{2} \cos a(t + \frac{1}{2})$,

c) $\Delta \cos at = -2 \sin \frac{a}{2} \sin a(t + \frac{1}{2})$,

d) $\Delta \log at = \log(1 + \frac{1}{t})$,

e) $\Delta \log \Gamma(t) = \log t$.

(Here $\log t$ represents any logarithm of the positive number t.)

Proof. We leave the verification of parts (a)-(d) as exercises. For part (e),

$$\Delta \log \Gamma(t) = \log \Gamma(t+1) - \log \Gamma(t)$$
$$= \log \frac{\Gamma(t+1)}{\Gamma(t)}$$
$$= \log t \quad \text{(see Example 1.6)}. \quad \blacksquare$$

2.1 The Difference Operator

It is readily verified that all the formulas in Theorem 2.2 remain valid if a constant "shift" is introduced in the t variable. For example,

$$\Delta a^{t+k} = (a-1)a^{t+k}.$$

Now the formulas in Theorems 2.1 and 2.2 can be used in combination to find the differences of more complicated expressions. However, it may be just as easy (or easier!) to use the definition directly.

Example 2.1. Compute $\Delta \sec \pi t$.

First, we use Theorem 2.1(e) and Theorem 2.2(c):

$$\Delta \sec \pi t = \Delta \frac{1}{\cos \pi t}$$

$$= \frac{(\cos \pi t)(\Delta 1) - (1)(\Delta \cos \pi t)}{\cos \pi t \cos \pi(t+1)}$$

$$= \frac{2 \sin \frac{\pi}{2} \sin \pi(t + \frac{1}{2})}{\cos \pi t \cos \pi(t+1)}$$

$$= \frac{2(\sin \pi t \cos \frac{\pi}{2} + \cos \pi t \sin \frac{\pi}{2})}{\cos \pi t (\cos \pi t \cos \pi - \sin \pi t \sin \pi)}$$

$$= \frac{2 \cos \pi t}{(\cos \pi t)(-\cos \pi t)} = -2 \sec \pi t.$$

The definition of Δ can be used to obtain the same result more quickly:

$$\Delta \sec \pi t = \sec \pi(t+1) - \sec \pi t$$
$$= \frac{1}{\cos \pi(t+1)} - \frac{1}{\cos \pi t}$$
$$= \frac{1}{\cos \pi t \cos \pi - \sin \pi t \sin \pi} - \frac{1}{\cos \pi t}$$
$$= \frac{1}{-\cos \pi t} - \frac{1}{\cos \pi t}$$
$$= -2 \sec \pi t.$$

One of the most basic special formulas in the differential calculus is the power rule

$$\frac{d}{dt} t^n = n t^{n-1}.$$

Unfortunately, the difference of a power is complicated and, as a result, is not very useful:

$$\Delta_t t^n = (t+1)^n - t^n$$
$$= \sum_{k=0}^{n} \binom{n}{k} t^k - t^n$$
$$= \sum_{k=0}^{n-1} \binom{n}{k} t^k.$$

Our next definition introduces a function that will satisfy a version of the power rule for finite differences.

Definition 2.3. The "factorial function" $t^{(r)}$ is defined as follows, according to the value of r:

a) if $r = 1, 2, 3, \cdots$, then $t^{(r)} = t(t-1)(t-2) \cdots (t-r+1)$,

b) if $r = 0$, then $t^{(0)} = 1$,

c) if $r = -1, -2, -3, \cdots$, then $t^{(r)} = \frac{1}{(t+1)(t+2)\cdots(t-r)}$,

2.1 The Difference Operator

d) if r is not an integer, then

$$t^{(r)} = \frac{\Gamma(t+1)}{\Gamma(t-r+1)}.$$

It is understood that the definition of $t^{(r)}$ is given only for those values of t and r that make the formulas meaningful. For example, $(-2)^{(-3)}$ is not defined since the expression in part (c) would involve division by zero, and $(\frac{1}{2})^{(\frac{3}{2})}$ is meaningless since $\Gamma(0)$ is undefined. (Some books use the convention that $\Gamma(x)/\Gamma(y) = 0$ if $\Gamma(x)$ is defined and $y = 0, -1, -2, \cdots$.)

The expression for $t^{(r)}$ in part (d) can be shown to agree with the simpler expressions in parts (a), (b), and (c) if r is an integer, except for certain discrete values of t that make the gamma function undefined. Let r be a positive integer. Then

$$\frac{\Gamma(t+1)}{\Gamma(t-r+1)} = \frac{t\Gamma(t)}{\Gamma(t-r+1)} = \frac{t(t-1)\Gamma(t-1)}{\Gamma(t-r+1)}$$
$$= \cdots = \frac{t(t-1)\cdots(t-r+1)\Gamma(t-r+1)}{\Gamma(t-r+1)}$$
$$= t(t-1) \cdots (t-r+1),$$

so (a) is a special case of (d). In a similar way, (b) and (c) are particular cases of (d).

Recall that the binomial coefficients are defined by

$$\binom{n}{k} = \frac{n(n-1)\cdots(n-k+1)}{k!},$$

so it follows immediately from Definition 2.3(a) that

$$\binom{n}{k} = \frac{n^{(k)}}{\Gamma(k+1)}$$

if n and k are positive integers with $n \geq k$. This relationship between the binomial coefficients and the factorial function suggests the following definition of an extended binomial coefficient:

Definition 2.4. The "binomial coefficient" $\binom{t}{r}$ is defined by

$$\binom{t}{r} = \frac{t^{(r)}}{\Gamma(r+1)}.$$

Graphs of some binomial coefficients are given in Figs. 2.1 and 2.2.

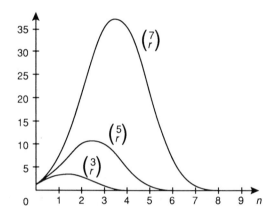

Fig. 2.1 Binomial coefficients as functions of r

With these definitions, we have the following result, which contains a power rule for differences and a closely related formula for the difference of a binomial coefficient.

Theorem 2.3. Whenever $t^{(r)}$ is defined, we have

a) $\Delta_t \, t^{(r)} = r t^{(r-1)}$,
b) $\Delta_t \binom{t}{r} = \binom{t}{r-1}$ $(r \neq 0)$.

Proof. Before we consider the general case, let's prove (a) for a positive integer r.

$$\begin{aligned}
\Delta_t \, t^{(r)} &= (t+1)^{(r)} - t^{(r)} \\
&= (t+1)(t)\cdots(t-r+2) - t(t-1)\cdots(t-r+1) \\
&= t(t-1)\cdots(t-r+2)[(t+1) - (t-r+1)] \\
&= r t^{(r-1)}.
\end{aligned}$$

2.1 The Difference Operator

Now let r be arbitrary. From (d) of Definition 2.3, we have

$$\Delta_t \, t^{(r)} = \Delta_t \frac{\Gamma(t+1)}{\Gamma(t-r+1)}$$

$$= \frac{\Gamma(t+2)}{\Gamma(t-r+2)} - \frac{\Gamma(t+1)}{\Gamma(t-r+1)}$$

$$= \frac{(t+1)\Gamma(t+1)}{\Gamma(t-r+2)} - \frac{(t-r+1)\Gamma(t+1)}{\Gamma(t-r+2)}$$

$$= r \frac{\Gamma(t+1)}{\Gamma(t-r+2)} = rt^{(r-1)}.$$

Part (b) follows easily:

$$\Delta_t \binom{t}{r} = \Delta_t \frac{t^{(r)}}{\Gamma(r+1)} = \frac{rt^{(r-1)}}{\Gamma(r+1)}$$

$$= \frac{t^{(r-1)}}{\Gamma(r)} = \binom{t}{r-1}. \qquad \blacksquare$$

Example 2.2. Find a solution to the difference equation

$$y(t+2) - 2y(t+1) + y(t) = t(t-1).$$

The difference equation can be written in the form

$$\Delta^2 y(t) = t^{(2)}.$$

From Theorem 2.3, $\Delta^2 t^{(4)} = \Delta 4t^{(3)} = 12t^{(2)}$, so $y(t) = \frac{t^{(4)}}{12}$ is a solution of the difference equation.

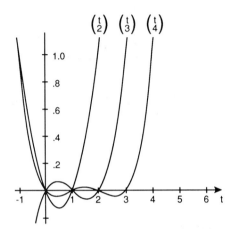

Fig. 2.2 Binomial coefficients as functions of t

2.2 Summation

In order to make effective use of the difference operator, we introduce in this section its right inverse operator, which is sometimes called the "indefinite sum."

Definition 2.5. An "indefinite sum" (or "antidifference") of $y(t)$, denoted $\sum y(t)$, is any function so that

$$\Delta(\sum y(t)) = y(t)$$

for all t in the domain of y.

The reader will recall that the indefinite integral plays a similar role in the differential calculus:

$$\frac{d}{dt}(\int y(t)\, dt) = y(t).$$

2.2 Summation

The indefinite integral is not unique, for example

$$\int \cos t \, dt = \sin t + C,$$

where C is any constant. The indefinite sum is also not unique, as we see in the next example.

Example 2.3. Compute the indefinite sum $\sum 6^t$.

From Theorem 2.2(a), $\Delta 6^t = 5 \cdot 6^t$, so we have $\Delta \frac{6^t}{5} = 6^t$. It follows that $\frac{6^t}{5}$ is an indefinite sum of 6^t. What are the others?

Let $C(t)$ be a function with the same domain as 6^t so that $\Delta C(t) = 0$. Then

$$\Delta\left(\frac{6^t}{5} + C(t)\right) = \Delta\left(\frac{6^t}{5}\right) = 6^t,$$

so $\frac{6^t}{5} + C(t)$ is an indefinite sum of 6^t. Further, if $f(t)$ is any indefinite sum of 6^t, then

$$\Delta\left(f(t) - \frac{6^t}{5}\right) = \Delta f(t) - \Delta \frac{6^t}{5} = 6^t - 6^t = 0,$$

so $f(t) = \frac{6^t}{5} + C(t)$ for some $C(t)$ with $\Delta C(t) = 0$. It follows that we have found all indefinite sums of 6^t, and we write

$$\sum 6^t = \frac{6^t}{5} + C(t),$$

where $C(t)$ is any function with the same domain as 6^t and $\Delta C(t) = 0$.

In a similar way, one can prove Theorem 2.4.

Theorem 2.4. If $z(t)$ is an indefinite sum of $y(t)$, then every indefinite sum of $y(t)$ is given by

$$\sum y(t) = z(t) + C(t),$$

where $C(t)$ has the same domain as y and $\Delta C(t) = 0$.

Example 2.3. (continued)

What sort of function must $C(t)$ be? The answer depends on the domain of $y(t)$. Let's consider first the most common case where the domain is a set of integers, say the natural numbers $N = \{1, 2, \cdots\}$. Then

$$\Delta C(t) = C(t+1) - C(t) = 0$$

for $t = 1, 2, 3, \cdots$, i.e., $C(1) = C(2) = C(3) = \cdots$, so $C(t)$ is a constant function! In this case, we simply write

$$\sum 6^t = \frac{6^t}{5} + C,$$

where C is any constant.

On the other hand, if the domain of y is the set of all real numbers, then the equation

$$\Delta C(t) = C(t+1) - C(t) = 0$$

says that $C(t+1) = C(t)$ for all real t, which means that C can be any periodic function having period one. For example, we could choose $C(t) = 2\sin 2\pi t$, or $C(t) = -5\cos 4\pi(t-\pi)$, in Theorem 2.4 and obtain an indefinite sum.

Since the discrete case will be the most important case in the remainder of this book, we state the following corollary.

Corollary 2.1. Let $y(t)$ be defined on a set of the type $\{a, a+1, a+2, \cdots\}$, where a is any real number, and let $z(t)$ be an indefinite sum of $y(t)$. Then every indefinite sum of $y(t)$ is given by

$$\sum y(t) = z(t) + C,$$

where C is an arbitrary constant.

Theorems 2.2 and 2.3 provide us with a useful collection of indefinite sums.

Theorem 2.5. Let a be a constant. Then

a) $\sum a^t = \frac{a^t}{a-1} + C(t), \quad (a \neq 1)$,

2.2 Summation

b) $\sum \sin at = -\frac{\cos a(t-\frac{1}{2})}{2\sin \frac{a}{2}} + C(t)$, $(a \neq 2n\pi)$,

c) $\sum \cos at = \frac{\sin a(t-\frac{1}{2})}{2\sin \frac{a}{2}} + C(t)$, $(a \neq 2n\pi)$,

d) $\sum \log t = \log \Gamma(t) + C(t)$, $(t > 0)$,

e) $\sum t^{(a)} = \frac{t^{(a+1)}}{a+1} + C(t)$, $(a \neq -1)$,

f) $\sum \binom{t}{a} = \binom{t}{a+1} + C(t)$,

where $\Delta C(t) = 0$.

Proof. Consider (b). From Theorem 2.2(c),

$$\Delta \cos a(t - \frac{1}{2}) = -2\sin \frac{a}{2} \sin at,$$

so by Theorem 2.4,

$$\sum \sin at = \frac{-\cos a(t-\frac{1}{2})}{2\sin \frac{a}{2}} + C(t).$$

The other parts are similar. ∎

As in Theorem 2.2, the preceding formulas can be generalized somewhat by introducing a constant shift in the t variable.

Example 2.2. (continued) Find the solution of

$$y(t+2) - 2y(t+1) + y(t) = t^{(2)}, \quad (t = 0, 1, 2, \cdots),$$

so that $y(0) = -1$, $y(1) = 3$.

Since $\Delta^2 y(t) = t^{(2)}$, by Corollary 2.1 and Theorem 2.5(a),

$$\Delta y(t) = \frac{t^{(3)}}{3} + C$$

and

$$y(t) = \frac{t^{(4)}}{12} + Ct + D,$$

where C and D are constants. Using the values of y at $t = 0$ and $t = 1$, we find $D = -1$ and $C = 4$, so the unique solution is

$$y(t) = \frac{t^{(4)}}{12} + 4t - 1.$$

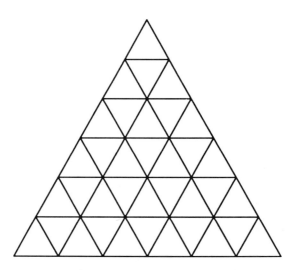

Fig. 2.3 Partition into equilateral triangles

Example 2.4. Suppose that each side of an upward pointing equilateral triangle is divided into n equal parts. These are then used to partition the original triangle into smaller equilateral triangles as shown in Fig. 2.3. How many upward pointing triangles $y(n)$ of all sizes are there?

First note that $y(1) = 1$, $y(2) = 4$, $y(3) = 10$, and so forth. Consider the case that each side of the original triangle has been divided into $n + 1$ equal parts. If we ignore the last row of small triangles, then the remaining portion has $y(n)$ upward pointing triangles. Now taking into account the last row, there are $(n + 1) + n + (n - 1) + \cdots + 1$ additional upward pointing triangles. Thus,

$$y(n + 1) = y(n) + (n + 1) + \cdots + 1$$

2.2 Summation

or

$$\Delta y(n) = \frac{(n+1)(n+2)}{2} = \frac{1}{2}(n+2)^{(2)},$$

so

$$y(n) = \frac{1}{6}(n+2)^{(3)} + C.$$

Since $y(1) = 1$, $C = 0$. Hence there are

$$y(n) = \frac{1}{6}(n+2)^{(3)}$$

upward pointing triangles.

We can derive a number of general properties of indefinite sums from Theorem 2.1.

Theorem 2.6.

a) $\sum (y(t) + z(t)) = \sum y(t) + \sum z(t)$,

b) $\sum Dy(t) = D \sum y(t)$ if D is constant,

c) $\sum (y(t)\Delta z(t)) = y(t)z(t) - \sum Ez(t)\Delta y(t)$,

d) $\sum (Ey(t)\Delta z(t)) = y(t)z(t) - \sum z(t)\Delta y(t)$.

Remark. Parts (c) and (d) of Theorem 2.6 are known as "summation by parts" formulas.

Proof. Parts (a) and (b) are immediate from Theorem 2.1. To prove (c), start with

$$\Delta(y(t)z(t)) = y(t)\Delta z(t) + Ez(t)\Delta y(t).$$

By Theorem 2.4,

$$\sum (y(t)\Delta z(t) + Ez(t)\Delta y(t)) = y(t)z(t) + C(t).$$

Then (c) follows from (a) and rearrangement. Finally, (d) is just a rearrangement and relabeling of (c). ∎

The summation by parts formulas can be used to compute certain indefinite sums much as the integration by parts formula is used to compute integrals. Moreover, these formulas turn out to be of fundamental importance in the analysis of difference equations, as we will see later.

Example 2.5. Compute $\sum ta^t$ $(a \neq 1)$.

In Theorem 2.6(c), choose $y(t) = t$ and $\Delta z(t) = a^t$; then $z(t) = \frac{a^t}{a-1}$. We have

$$\sum ta^t = t\frac{a^t}{a-1} - \sum \frac{a^{t+1}}{a-1} \Delta t + C(t)$$
$$= \frac{ta^t}{a-1} - \frac{a}{a-1} \sum a^t + C(t)$$
$$= \frac{ta^t}{a-1} - \frac{a}{(a-1)^2} a^t + C(t),$$

where $\Delta C(t) = 0$, and we have made use of Theorem 2.5(a).

Example 2.6. Compute $\sum \binom{t}{5}\binom{t}{2}$.

Let $y(t) = \binom{t}{2}$ and $\Delta z(t) = \binom{t}{5}$ in Theorem 2.6(c). By Theorem 2.5(f), we can take $z(t) = \binom{t}{6}$. Then

$$\sum \binom{t}{5}\binom{t}{2} = \binom{t}{6}\binom{t}{2} - \sum \binom{t+1}{6}\binom{t}{1} + C(t).$$

Now we apply summation by parts to the last sum with $y_1(t) = \binom{t}{1}$, $\Delta z_1(t) = \binom{t+1}{6}$ and $z_1(t) = \binom{t+1}{7}$:

$$\sum \binom{t}{5}\binom{t}{2} = \binom{t}{6}\binom{t}{2} - [\binom{t+1}{7}\binom{t}{1} - \sum \binom{t+2}{7}] + C(t)$$
$$= \binom{t}{6}\binom{t}{2} - \binom{t+1}{7}t + \binom{t+2}{8} + C(t),$$

where $\Delta C(t) = 0$.

For the remainder of this section, we will assume that the domain of $y(t)$ is a set of consecutive integers, which for the sake of being specific we

2.2 Summation

will take to be the natural numbers $N = \{1, 2, 3, \cdots\}$. Sequence notation will be used for the function $y(t)$:

$$y(t) \leftrightarrow \{y_n\},$$

where $n \in N$. In the later chapters, both functional and sequence notation will be utilized.

In what follows it will be convenient to use the convention

$$\sum_{k=a}^{b} y_k = 0$$

whenever $a > b$. Observe that for m fixed and $n \geq m$,

$$\Delta_n \Big(\sum_{k=m}^{n-1} y_k \Big) = y_n,$$

and for p fixed and $p \geq n$,

$$\Delta_n \Big(\sum_{k=n}^{p} y_k \Big) = -y_n.$$

Corollary 2.1 tells us that

$$\sum y_n = \sum_{k=m}^{n-1} y_k + C \quad (m \leq n) \tag{2.2}$$

for some constant C and, alternatively, that

$$\sum y_n = -\sum_{k=n}^{p} y_k + D \quad (p \geq n) \tag{2.3}$$

for some constant D. Equations (2.2) and (2.3) give us a way of relating indefinite sums to definite sums.

Example 2.7. Compute the definite sum $\sum_{k=1}^{n-1}(\frac{2}{3})^k$.

By Eq. (2.2) and Theorem 2.5(a),

$$\sum_{k=1}^{n-1}(\frac{2}{3})^k = \sum(\frac{2}{3})^n + C$$

$$= \frac{(\frac{2}{3})^n}{\frac{2}{3} - 1} + C$$

$$= -3(\frac{2}{3})^n + C \quad (n = 2, 3, \cdots).$$

To evaluate C, let $n = 2$:

$$\frac{2}{3} = -3(\frac{2}{3})^2 + C,$$
$$2 = C,$$

so

$$\sum_{k=1}^{n-1}(\frac{2}{3})^k = 2 - 3(\frac{2}{3})^n \quad (n = 2, 3, \cdots).$$

There is a useful formula for computing definite sums, which is analogous to the *fundamental theorem of calculus*:

Theorem 2.7. If z_n is an indefinite sum of y_n, then

$$\sum_{k=m}^{n-1} y_k = [z_k]_m^n = z_n - z_m.$$

Theorem 2.7 is an immediate consequence of Eq. (2.2) (see Exercise 2.26).

2.2 Summation

Example 2.8. Compute $\sum_{k=1}^{l} k^2$.

Recall that $k^{(1)} = k$ and $k^{(2)} = k(k-1)$. Then $k^2 = k^{(1)} + k^{(2)}$, so

$$\sum k^2 = \sum k^{(1)} + \sum k^{(2)}$$
$$= \frac{k^{(2)}}{2} + \frac{k^{(3)}}{3} + C$$

by Theorem 2.5(e).

From Theorem 2.7, we have

$$\sum_{k=1}^{l} k^2 = \left[\frac{k^{(2)}}{2} + \frac{k^{(3)}}{3}\right]_1^{l+1}$$
$$= \frac{(l+1)^{(2)}}{2} + \frac{(l+1)^{(3)}}{3} - \frac{1^{(2)}}{2} - \frac{1^{(3)}}{3}$$
$$= \frac{(l+1)l}{2} + \frac{(l+1)l(l-1)}{3}$$
$$= \frac{l(l+1)(2l+1)}{6}.$$

The next theorem gives two versions of the summation by parts method for definite sums.

Theorem 2.8.

a) If $m < n$, then

$$\sum_{k=m}^{n-1} a_k b_k = b_n \sum_{k=m}^{n-1} a_k - \sum_{k=m}^{n-1} \left(\sum_{i=m}^{k} a_i\right) \Delta b_k.$$

b) If $p \geq n$, then

$$\sum_{k=n}^{p} a_k b_k = b_{n-1} \sum_{k=n}^{p} a_k + \sum_{k=n}^{p} \left(\sum_{i=k}^{p} a_i\right) \Delta b_{k-1}.$$

Proof. To prove (a), choose $y(n) = b_n$ and $z(n) = \sum_{k=m}^{n-1} a_k$ in Theorem 2.6(c):

$$\sum a_n b_n = b_n \sum_{k=m}^{n-1} a_k - \sum (\sum_{i=m}^{n} a_i)\Delta b_n.$$

From (2.2),

$$\sum_{k=m}^{n-1} a_k b_k = b_n \sum_{k=m}^{n-1} a_k - \sum_{k=m}^{n-1} (\sum_{i=m}^{k} a_i)\Delta b_k + C.$$

With $n = m + 1$, we have

$$a_m b_m = b_{m+1} a_m - a_m \Delta b_m + C,$$

so $C = 0$, and we obtain (a).

The expression in (b) is obtained from Theorem 2.6(d) with $y = b_{n-1}$ and $z = \sum_{k=n}^{p} a_k$:

$$\sum b_n(-a_n) = b_{n-1} \sum_{k=n}^{p} a_k - \sum (\sum_{i=n}^{p} a_i)\Delta b_{n-1},$$

so by Eq. (2.3)

$$\sum_{k=n}^{p} a_k b_k = b_{n-1} \sum_{k=n}^{p} a_k + \sum_{k=n}^{p} (\sum_{i=k}^{p} a_i)\Delta b_{k-1} + C_1.$$

It is easy to check that $C_1 = 0$. ∎

Remark. The equation in Theorem 2.8(a) is known as *Abel's summation formula*.

2.2 Summation

Example 2.9. Compute $\sum_{k=1}^{n-1} k2^k$.

By Theorem 2.8(a) with $a_k = 2^k$ and $b_k = k$,

$$\sum_{k=1}^{n-1} k2^k = n\sum_{k=1}^{n-1} 2^k - \sum_{k=1}^{n-1}\left(\sum_{i=1}^{k} 2^i\right).$$

From Theorem 2.7 and Theorem 2.5(a),

$$\sum_{k=1}^{n-1} 2^k = 2^n - 2^1.$$

Returning to our calculation, we have

$$\begin{aligned}\sum_{k=1}^{n-1} k2^k &= n(2^n - 2) - \sum_{k=1}^{n-1}(2^{k+1} - 2) \\ &= n2^n - 2n - 2(2^n - 2) + 2(n-1) \\ &= n2^n - 2^{n+1} + 2.\end{aligned}$$

The same result can also be obtained from the calculation in Example 2.5:

$$\sum n2^n = n2^n - 2^{n+1} + C.$$

Then

$$\sum_{k=1}^{n-1} k2^k = n2^n - 2^{n+1} - (2^1 - 2^2) = n2^n - 2^{n+1} + 2.$$

The methods used in Example 2.9 allow us to compute any definite sum of sequences of the form $p(n)a^n$, $p(n)\sin an$, $p(n)\cos an$ and $p(n)\binom{n}{a}$, where $p(n)$ is a polynomial in n. However, we must have as many repetitions of summation by parts as the degree of p.

There is a special method of summation that is based on Eq. (2.1) for the nth difference of a function:

$$\Delta^n y(0) = \sum_{k=0}^{n} (-1)^k \binom{n}{k} y(n-k)$$

$$= \sum_{i=0}^{n} (-1)^{n-i} \binom{n}{i} y(i),$$

where we have used the change of index $i = n - k$ and the fact that $\binom{n}{n-i} = \binom{n}{i}$. It follows that

$$\sum_{i=0}^{n} (-1)^i \binom{n}{i} y(i) = (-1)^n \Delta^n y(0). \qquad (2.4)$$

Example 2.10. Compute $\sum_{i=0}^{n}(-1)^i \binom{n}{i}\binom{i+a}{m}$.

Let $y(i) = \binom{i+a}{m}$ in Eq. (2.4). From Theorem 2.3(b), $\Delta^n \binom{i+a}{m} = \binom{i+a}{m-n}$, so Eq. (2.4) gives immediately

$$\sum_{i=0}^{n} (-1)^i \binom{n}{i}\binom{i+a}{m} = (-1)^n \binom{a}{m-n}.$$

Other examples of indefinite and definite sums are contained in the exercises.

2.3. Generating Functions and Approximate Summation

In Section 2.2, we discussed a number of methods by which finite sums can be computed. However, most sums, like most integrals, cannot be expressed in terms of the elementary functions of calculus. There are functions such as $y(t) = \frac{1}{t}$ which can be integrated exactly,

$$\int_a^b \frac{1}{t} dt = \log \frac{b}{a}, \quad (b > a > 0),$$

2.3 Generating Functions and Approximate Summation

but for which there is no elementary formula for the corresponding sum:

$$\sum_{k=1}^{n} \frac{1}{k}.$$

The main result of this section, called the *Euler summation formula*, will give us a technique for approximating a sum if the corresponding integral can be computed. In order to formulate this result, we will use the concept of a generating function, which is itself important in the analysis of difference equations, and a family of special functions called the Bernoulli polynomials.

Definition 2.6. Let $\{y_k(t)\}$ be a sequence of (possibly constant) functions. Suppose there is a function $g(t,x)$ so that

$$g(t,x) = \sum_{k=0}^{\infty} y_k(t) x^k$$

for all x in an open interval about 0. Then g is called the "generating function" for $\{y_k(t)\}$.

In other words, for each t, $y_k(t)$ is the kth coefficient in the power series for $g(t,x)$ with respect to x at $x = 0$. Recall that these coefficients can be computed with the formula

$$y_k(t) = \frac{1}{k!} \frac{\partial^k}{\partial x^k} g(t,0), \tag{2.5}$$

but that there may be easier ways to calculate the $y_k(t)$.

One relationship between generating functions and difference equations is illustrated by Example 1.4, in which the solutions of the Airy equation are generating functions for the sequences $\{a_k\}$ that satisfy the difference equation

$$a_{k+3} = \frac{a_k}{(k+3)(k+2)}.$$

This association of differential equations and difference equations will be utilized in Chapter 3 to solve certain difference equations.

Example 2.11. Let $y_k(t) = (f(t))^k$, for some function $f(t)$. To compute the generating function for $y_k(t)$, we must sum the series

$$\sum_{k=0}^{\infty}(f(t))^k x^k = \sum_{k=0}^{\infty}(f(t)x)^k.$$

By using the methods of the previous section, or by simply recognizing this to be a geometric series, we obtain the sum

$$\frac{1}{1-f(t)x} = g(t,x)$$

for $|f(t)x| < 1$. This result can be used to find other generating functions. Using differentiation,

$$\frac{\partial}{\partial x}\left(\frac{1}{1-f(t)x}\right) = \frac{\partial}{\partial x}\sum_{k=0}^{\infty}(f(t))^k x^k,$$

$$\frac{f(t)}{(1-f(t)x)^2} = \sum_{k=0}^{\infty} k(f(t))^k x^{k-1},$$

$$\frac{xf(t)}{(1-f(t)x)^2} = \sum_{k=0}^{\infty} k(f(t))^k x^k,$$

so $\frac{xf(t)}{(1-f(t)x)^2}$ is the generating function for the sequence $\{k(f(t))^k\}$.

Definition 2.7. The "Bernoulli polynomials" $B_k(t)$ are defined by the equation

$$\frac{xe^{tx}}{e^x - 1} = \sum_{k=0}^{\infty} \frac{B_k(t)}{k!} x^k;$$

in other words, $\frac{xe^{tx}}{e^x-1}$ is the generating function for the sequence $\frac{B_k(t)}{k!}$.

Definition 2.8. The "Bernoulli numbers" B_k are given by $B_k = B_k(0)$, the value of the kth Bernoulli polynomial at $t = 0$.

We could use Eq. (2.5) to compute the first few Bernoulli polynomials, but it is easier to use the equation in Definition 2.7 directly. First,

2.3 Generating Functions and Approximate Summation

multiply both sides of the equation by $\frac{e^x-1}{x}$:

$$e^{tx} = \frac{e^x - 1}{x} \sum_{k=0}^{\infty} \frac{B_k(t)}{k!} x^k.$$

Then expand the exponential functions on each side in their Taylor series about zero and collect terms containing the same power of x:

$$1 + tx + \frac{t^2 x^2}{2!} + \frac{t^3 x^3}{3!} + \cdots$$

$$= (1 + \frac{x}{2!} + \frac{x^2}{3!} + \cdots)(B_0(t) + \frac{B_1(t)}{1!} x + \frac{B_2(t)}{2!} x^2 + \cdots)$$

$$= B_0(t) + (\frac{B_1(t)}{1!} + \frac{B_0(t)}{2!})x + (\frac{B_2(t)}{2!} + \frac{B_1(t)}{2!1!} + \frac{B_0(t)}{3!})x^2 + \cdots.$$

Equating coefficients of like powers of x, we have

$$B_0(t) = 1, \quad B_1(t) + \frac{B_0(t)}{2} = t, \quad \frac{B_2(t)}{2} + \frac{B_1(t)}{2} + \frac{B_0(t)}{6} = \frac{t^2}{2},$$

and so forth. The first few Bernoulli polynomials are given by

$$B_0(t) = 1, \quad B_1(t) = t - \frac{1}{2}, \quad B_2(t) = t^2 - t + \frac{1}{6}, \tag{2.6}$$

$$B_3(t) = t^3 - \frac{3}{2}t^2 + \frac{1}{2}t, \quad \cdots.$$

Then the first four Bernoulli numbers are

$$B_0 = 1, \quad B_1 = -\frac{1}{2}, \quad B_2 = \frac{1}{6}, \quad B_3 = 0. \tag{2.7}$$

Here are several properties of these polynomials.

Theorem 2.9.

a) $B'_k(t) = kB_{k-1}(t)$ $(k \geq 1)$,

b) $\Delta_t B_k(t) = kt^{k-1}$ $(k \geq 0)$,

c) $B_k = B_k(0) = B_k(1)$ $(k \neq 1)$,

d) $B_{2m+1} = 0$ $(m \geq 1)$.

Proof. To prove (a), we apply differentiation with respect to t to both sides of the equation in Definition 2.7:

$$\frac{x^2 e^{tx}}{e^x - 1} = \sum_{k=0}^{\infty} \frac{B'_k(t)}{k!} x^k,$$

or

$$\sum_{k=0}^{\infty} \frac{B_k(t)}{k!} x^{k+1} = \sum_{k=0}^{\infty} \frac{B'_k(t)}{k!} x^k.$$

Now make the change of index $k \to k-1$ in the left-hand sum:

$$\sum_{k=1}^{\infty} \frac{B_{k-1}(t)}{(k-1)!} x^k = \sum_{k=0}^{\infty} \frac{B'_k(t)}{k!} x^k.$$

Equating coefficients, we obtain (a).

Next, take the difference of both sides of the equation in Definition 2.7:

$$\sum_{k=0}^{\infty} \frac{\Delta_t B_k(t)}{k!} x^k = \frac{x}{e^x - 1}(e^{(t+1)x} - e^{tx})$$

$$= xe^{tx}$$

$$= \sum_{k=0}^{\infty} \frac{t^k x^{k+1}}{k!}$$

$$= \sum_{k=1}^{\infty} \frac{t^{k-1} x^k}{(k-1)!}.$$

2.3 Generating Functions and Approximate Summation

Then (b) follows immediately by equating coefficients.
Parts (c) and (d) are left as exercises. ∎

Note that Theorem 2.9(a) implies that each $B_k(t)$ is a polynomial of degree k. Also, from part (b) we have an additional summation formula.

Corollary 2.2. If $k = 0, 1, 2, \cdots$, then

$$\sum t^k = \frac{1}{k+1} B_{k+1}(t) + C(t),$$

when $\Delta C(t) = 0$.

Additional information about Bernoulli polynomials and numbers is contained in the exercises.

The next theorem is the Euler summation formula.

Theorem 2.10. Suppose that the $2m^{\text{th}}$ derivative of $y(t)$, $y^{(2m)}(t)$, is continuous on $[1, n]$ for some integers $m \geq 1$ and $n \geq 2$. Then

$$\sum_{k=1}^{n} y(k) = \int_{1}^{n} y(t)\, dt + \frac{y(n) + y(1)}{2} + \sum_{i=1}^{m} \frac{B_{2i}}{(2i)!} [y^{(2i-1)}(n) - y^{(2i-1)}(1)]$$
$$- \frac{1}{(2m)!} \int_{1}^{n} y^{(2m)}(t) B_{2m}(t - [t])\, dt,$$

where $[t]$ = the greatest integer less than or equal to t.
The graph of $t - [t]$ is sketched in Fig. 2.4.

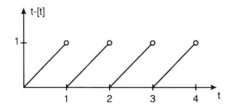

Fig. 2.4 Graph of $t - [t]$.

Example 2.12. Approximate $\sum_{k=1}^{n} k^{\frac{1}{2}}$.

Let $y(t) = t^{\frac{1}{2}}$ and $m = 1$. From Theorem 2.10 and (2.7),

$$\sum_{k=1}^{n} k^{\frac{1}{2}} = \int_{1}^{n} t^{\frac{1}{2}}\, dt + \frac{n^{\frac{1}{2}}+1}{2} + \frac{1}{24}[n^{-\frac{1}{2}} - 1]$$

$$- \frac{1}{2}\int_{1}^{n} (-\frac{1}{4t^{\frac{3}{2}}})B_2(t - [t])\, dt$$

$$= \frac{2}{3}n^{\frac{3}{2}} + \frac{1}{2}n^{\frac{1}{2}} + \frac{1}{24}n^{-\frac{1}{2}} - \frac{5}{24}$$

$$+ \frac{1}{8}\int_{1}^{n} t^{-\frac{3}{2}}B_2(t - [t])\, dt.$$

Now $B_2(x) = x^2 - x + \frac{1}{6}$, by Eq. (2.6), and an easy max-min argument shows that

$$-\frac{1}{12} \leq B_2(x) \leq \frac{1}{6}$$

for $0 \leq x \leq 1$. Since $0 \leq t - [t] \leq 1$ for all t, we have

$$-\frac{1}{96}\int_{1}^{n} t^{-\frac{3}{2}}\, dt \leq \frac{1}{8}\int_{1}^{n} t^{-\frac{3}{2}} B_2(t - [t])\, dt \leq \frac{1}{48}\int_{1}^{n} t^{-\frac{3}{2}}\, dt,$$

or

$$-\frac{1}{96}(2 - 2n^{-\frac{1}{2}}) \leq \frac{1}{8}\int_{1}^{n} t^{-\frac{3}{2}} B_2(t - [t])\, dt \leq \frac{1}{48}(2 - 2n^{-\frac{1}{2}}).$$

Using these inequalities in the earlier calculation, we finally arrive at the estimate

$$\frac{2}{3}n^{\frac{3}{2}} + \frac{1}{2}n^{\frac{1}{2}} + \frac{1}{16}n^{-\frac{1}{2}} - \frac{11}{48} \leq \sum_{k=1}^{n} k^{\frac{1}{2}} \leq \frac{2}{3}n^{\frac{3}{2}} + \frac{1}{2}n^{\frac{1}{2}} - \frac{1}{6}.$$

2.3 Generating Functions and Approximate Summation

Proof of Theorem 2.10. Integration by parts gives for each k,

$$\int_k^{k+1} (t - [t] - \frac{1}{2})y'(t)\, dt = \int_k^{k+1} (t - k - \frac{1}{2})y'(t)\, dt$$

$$= \frac{y(k+1) + y(k)}{2} - \int_k^{k+1} y(t)\, dt. \quad (2.8)$$

Note that in the first integral in Eq. (2.8),

$$t - [t] - \frac{1}{2} = B_1(t - [t]).$$

Similarly, we have by Theorem 2.9 (a) and (c),

$$\int_k^{k+1} B_i(t - [t])y^{(i)}(t)\, dt = \int_k^{k+1} B_i(t - k)y^{(i)}(t)\, dt$$

$$= \frac{B_{i+1}}{i+1}[y^{(i)}(k+1) - y^{(i)}(k)]$$

$$- \frac{1}{i+1}\int_k^{k+1} B_{i+1}(t - [t])y^{(i+1)}(t)\, dt,$$

$$(2.9)$$

for $i = 1, \cdots, 2m - 1$.

Summing Eq. (2.8) and Eq. (2.9) as k goes from 1 to $n - 1$, we have, respectively,

$$\int_1^n B_1(t - [t])y'(t)\, dt = \sum_{k=1}^n y(k) - \frac{1}{2}[y(1) + y(n)] \quad (2.10)$$

$$- \int_1^n y(t)\, dt,$$

$$\int_1^n B_i(t - [t])y^{(i)}(t)\, dt = \frac{B_{i+1}}{i+1}[y^{(i)}(n) - y^{(i)}(1)] \quad (2.11)$$

$$- \frac{1}{i+1}\int_1^n B_{i+1}(t - [t])y^{(i+1)}(t)\, dt.$$

Finally, we begin with Eq. (2.10) and use Eq. (2.11) repeatedly to obtain

$$\sum_{k=1}^{n} y(k) - \frac{1}{2}(y(1) + y(n)) - \int_{1}^{n} y(t)\, dt$$
$$= \int_{1}^{n} B_1(t - [t])y'(t)\, dt$$
$$= \frac{B_2}{2}[y'(n) - y'(1)] - \frac{1}{2}\int_{1}^{n} B_2(t - [t])y^{(2)}(t)\, dt$$
$$= \cdots$$
$$= \sum_{i=1}^{m} \frac{B_{2i}}{(2i)!}[y^{(2i-1)}(n) - y^{(2i-1)}(1)] - \frac{1}{(2m)!}\int_{1}^{n} B_{2m}(t - [t])y^{(2m)}(t)\, dt,$$

where we have used Theorem 2.9(d). Rearrangement yields the Euler summation formula. ∎

Theorem 2.10 will be fundamental to our discussion of asymptotic analysis of sums in Chapter 5.

Exercises

Section 2.1.

2.1 Show that Δ and E commute, i.e. $\Delta E y(t) = E \Delta y(t)$ for all $y(t)$.

2.2 Prove the quotient rule for differences (Theorem 2.1(e)).

2.3 Derive the formula

$$\Delta[x(t)y(t)z(t)] = \Delta x(t)Ey(t)Ez(t) + x(t)\Delta y(t)Ez(t) + x(t)y(t)\Delta z(t).$$

Write down five other formulas of this type.

2.4 Show that
(a) $\Delta a^t = (a-1)a^t$, if a is a constant,
(b) $\Delta e^{ct} = (e^c - 1)e^{ct}$, if c is a constant.

2.5 Show that for any constant a, $\Delta \sin at = 2\sin\frac{a}{2}\cos a(t+\frac{1}{2})$.

2.6 Verify the formula $\Delta \sinh at = 2\sinh\frac{a}{2}\cosh a(t+\frac{1}{2})$, where a is a constant.

2.7 Show that
(a) $\Delta \tan t = \sec^2 t \frac{\tan 1}{1-\tan 1 \tan t}$,
(b) $\Delta \tan^{-1} t = \tan^{-1}\left(\frac{1}{t^2+t+1}\right)$.

2.8 Compute $\Delta(3^t \cos t)$ by two methods:
(a) using Theorem 2.1(d) and Theorem 2.2(a),(c),
(b) directly from the definition of Δ.

2.9 Compute $\Delta^n t^{(3)}$ and $\Delta^n t^3$ for $n = 1, 2, 3, \cdots$.

2.10 Does the factorial function satisfy $t^{(r)}t^{(s)} = t^{(r+s)}$?

2.11 If $r = -1, -2, -3, \cdots$, show that

$$\frac{1}{(t+1)(t+2)\cdots(t-r)} = \frac{\Gamma(t+1)}{\Gamma(t-r+1)},$$

so that (c) and (d) of Definition 2.3 agree in this case.

2.12 Use the formula $t^{(r)} = \frac{1}{(t+1)(t+2)\cdots(t-r)}$, which is valid for $r = -1, -2, \cdots$, to show that $\Delta t^{(r)} = rt^{(r-1)}$ for those values of r.

2.13 Find a solution of the following difference equations:
 (a) $y(t+1) - y(t) = t^{(3)} + 3^t$,
 (b) $y(t+2) - 2y(t+1) + y(t) = \binom{t}{5}$.

2.14 Let n be a positive integer.
 (a) Show that $\binom{-t}{n} = (-1)^n \binom{t+n-1}{n}$,
 (b) Show $\Delta_t \binom{-t}{n} = -\binom{-t-1}{n-1}$.

2.15 If $f(t)$ is a polynomial of degree n, show that

$$f(t) = f(0) + \frac{\Delta f(0)}{1!} t^{(1)} + \cdots + \frac{\Delta^n f(0)}{n!} t^{(n)}.$$

2.16 Use the formula in 2.14 to write t^3 in terms of $t^{(1)}$, $t^{(2)}$, and $t^{(3)}$.

Section 2.2.

2.17 Show that
 (a) $\sum \cos at = \frac{\sin a(t-\frac{1}{2})}{2 \sin \frac{a}{2}} + C(t)$ $(a \neq 2n\pi)$,
 (b) $\sum \binom{t}{a} = \binom{t}{a+1} + C(t)$, where $\Delta C(t) = 0$.

2.18 Let $y(t)$ be the maximum number of points of intersection of t lines in the plane. Find a difference equation that $y(t)$ satisfies and use it to find $y(t)$.

2.19 Suppose that t points are chosen on the perimeter of a disc and all line segments joining these points are drawn. Let $z(t)$ be the maximum number of regions into which the disc can be divided by such line segments. Given that $z(t)$ satisfies the equation $\Delta^4 z(t) = 1$, find a formula for $z(t)$.

2.20 Find the total number of downward pointing triangles of all sizes in Example 2.4.

2.21 Use summation by parts to compute $\sum t \sin t$.

2.22 Use summation by parts to compute

$$\sum \frac{t}{(t+1)(t+2)(t+3)}.$$

2.23 Use the summation by parts formula to evaluate each of the following:
 (a) $\sum t^2 3^t$,

(b) $\sum \binom{t}{2} \cdot \binom{t}{7}$,

(c) $\sum \binom{t}{2}^2$.

2.24 Use the Abel summation formula to evaluate $\sum_{k=1}^{n-1} k 3^k$.

2.25 Show that $\sum \binom{t}{m}\binom{t}{n} = \sum_{k=0}^{n} (-1)^k \binom{t+k}{m+1+k}\binom{t}{n-k} + C(t)$, where $\Delta C(t) = 0$ and n is a positive integer. (Hint: use summation by parts.)

2.26 Let $z_n = \sum y_n$. Show that

$$\sum_{k=m}^{n-1} y_k = z_n - z_m.$$

2.27 Use Exercise 2.26 to show

$$\sum_{k=m}^{n-1} \binom{k+b}{a} = \binom{n+b}{a+1} - \binom{m+b}{a+1}.$$

2.28 Show that

$$\frac{1}{2} + \sum_{k=1}^{n-1} \cos ak = \frac{\sin(n-\frac{1}{2})a}{2\sin\frac{a}{2}}.$$

2.29 Compute $\sum_{k=1}^{8} \frac{1}{(k+1)(k+2)(k+3)}$.

2.30 Compute $\sum_{k=1}^{n-1} \frac{k}{2^k}$ using summation by parts.

2.31 Use Theorem 2.8(a) to compute $\sum_{k=2}^{n-1} k^2(k-1)$.

2.32 If $p > n$, prove that

$$\sum_{k=n}^{p} a_k b_k = b_n \sum_{k=n}^{p} a_k + \sum_{k=n+1}^{p} \left(\sum_{i=k}^{p} a_i\right)(b_k - b_{k-1}).$$

(Hint: start with Theorem 2.8(b).)

2.33 Use Eq. (2.4) to show that $\sum_{i=0}^{n} (-1)^i \binom{n}{i} = 0$ if $n \geq 2$.

2.34 Use Eq. (2.4) to compute $\sum_{i=0}^{n} \frac{(-1)^i}{1+i} \binom{n}{i}$.

2.35 The Sterling numbers s_k^n (of the second kind) are defined to be the solution of the partial difference equation

$$s_k^{n+1} = s_{k-1}^n + k s_k^n$$

with $s_n^n = s_1^n = 1$ for each n. Show that

$$x^n = \sum_{k=1}^{n} s_k^n x^{(k)}.$$

(Hint: use induction.)

2.36 Use the result of Exercise 2.35 to compute $\sum_{k=1}^{n-1} k^3$.

Section 2.3.

2.37 What are the generating functions of the sequences?
(a) $y_k = \frac{1}{k!}$.
(b) $\begin{cases} y_{2l} = \frac{(-1)^l}{(2l)!}, \\ y_{2l+1} = 0. \end{cases}$
(c) $y_0 = 0$, $y_k = \frac{2^k}{k}$ for $k \geq 1$.
(d) $y_k = k 2^k$.

2.38 Given that

$$g(t, x) = (1-x)^{-1} \exp\left(\frac{-xt}{1-x}\right) = \sum_{n=0}^{\infty} L_n(t) x^n$$

is the generating function for the Laguerre polynomials $L_n(t)$, find $L_n(t)$ for $0 \leq n \leq 3$.

2.39 Given that $g(t, x) = (1 - 2tx + x^2)^{-\frac{1}{2}}$ is the generating function for the Legendre polynomials $P_n(t)$, find $P_n(t)$ for $0 \leq n \leq 3$.

2.40 The function $g(t, x) = \exp(2tx - x^2)$ is the generating function for the Hermite polynomials $H_n(t)$. Compute $H_n(t)$ for $0 \leq n \leq 3$.

Exercises

2.41 Find the generating function for $y_k(t) = \cos kt$. (Hint: one way to do this is to write $\cos kt = Re(e^{ikt})$ and use Example 2.11.)

2.42 What is the generating function for
(a) $y_k = \binom{n}{k}$, $(k = 0, \cdots, n)$, $y_k = 0$, $(k > n)$?
(b) Use part (a) and differentiation to compute the sum $\sum_{k=0}^{n} k\binom{n}{k}$.

2.43 Show that $B_0(t) = 1$ using Eq. (2.5).

2.44 Show that $B_3(t) = t^3 - \frac{3}{2}t^2 + \frac{1}{6}t$.

2.45 Show that $B_k(0) = B_k(1)$ for $k \neq 1$.

2.46 Show $B_{2i+1} = 0$, $i \geq 1$.

2.47 Prove $\int_0^1 B_k(t)\,dt = 0$ for $k \geq 1$.

2.48 Use Corollary 2.2 to show ($k \geq 1$)

$$\sum_{i=1}^{n-1} i^k = \frac{1}{k+1}[B_{k+1}(n) - B_{k+1}].$$

2.49 Use the result of Exercise 2.48 to compute $\sum_{i=1}^{n-1} i^2$.

2.50 Prove $B_k(t) = (-1)^k B_k(1-t)$ for all k and all t.

2.51 Give an estimate for $\sum_{k=1}^{400} k^{\frac{1}{2}}$.

2.52 Use Theorem 2.10 with $m = 1$ to obtain the estimate

$$\sum_{k=1}^{n} \frac{1}{k} \simeq \log n + \frac{1}{2n} - \frac{1}{12n^2} + \frac{7}{12},$$

and show that the error is less than $\frac{1}{12}$.

2.53 Use Theorem 2.10 with $m = 1$ to derive the trapezoidal rule for approximating integrals:

$$\int_{x_1}^{x_n} f(x)\,dx = h\left[\frac{f(x_1)}{2} + f(x_2) + \cdots + f(x_{n-1}) + \frac{f(x_n)}{2}\right] + E(h),$$

where $h = x_k - x_{k-1}$ for $k = 2, \cdots, n$, and $|E(h)| \leq Ch^2$ for some constant C.

Chapter 3
Linear Difference Equations

In this rather long chapter, we examine a special class of difference equations, the so-called linear equations. The use of the words linear and nonlinear here are completely analogous to their usage in the field of differential equations. Furthermore we restrict most of our discussion to equations that involve a single independent and a single dependent variable. Equations with several independent or dependent variables will receive more attention in later chapters.

The study of linear equations is important for a number of reasons. Many types of problems are naturally formulated as linear equations (see Chapter 1, the fourth section of the present chapter, and the related exercises). Also, certain subclasses of the class of linear equations, such as first order equations and the equations with constant coefficients, represent large families of equations that can be solved explicitly. The class of linear equations has nice algebraic properties that permit the use of matrix methods, operational methods, transforms, generating functions and other special techniques. Finally, certain methods of analysis for nonlinear equations, such as establishing stability by linearization, depend on the properties of associated linear equations.

The sixth section of this chapter discusses circumstances under which nonlinear equations can be transformed into linear equations.

3.1 First Order Equations

Let $p(t)$ and $r(t)$ be given functions with $p(t) \neq 0$ for all t. The first order linear difference equation is

$$y(t+1) - p(t)y(t) = r(t). \qquad (3.1)$$

Equation (3.1) is said to be of first order because it involves the values of y at t and $t+1$ only, as in the first order difference operator $\Delta y(t) = y(t+1) - y(t)$. If $p(t) = 1$ for all t, then Eq. (3.1) is simply

$$\Delta y(t) = r(t),$$

Chapter 3. Linear Difference Equations

so from Chapter 2, the solution is

$$y(t) = \sum r(t) + C(t),$$

where $\Delta C(t) = 0$.

For simplicity, let's assume the domain of interest is a discrete set $t = a, a+1, a+2, \cdots$. Consider first the "homogeneous" equation

$$u(t+1) = p(t)u(t). \qquad (3.1)'$$

This equation is easily solved by iteration:

$$u(a+1) = p(a)u(a),$$
$$u(a+2) = p(a+1)p(a)u(a),$$
$$\vdots$$
$$u(a+n) = u(a) \prod_{k=0}^{n-1} p(a+k).$$

We can write the solution in the more convenient form

$$u(t) = u(a) \prod_{s=a}^{t-1} p(s) \quad (t = a, a+1, \cdots),$$

where it is understood that $\prod_{s=a}^{a-1} p(s) \equiv 1$ and for $t \geq a+1$, the product is taken over $a, a+1, \cdots, t-1$.

Now Eq. (3.1) can be solved by substituting $y(t) = u(t)v(t)$ into Eq. (3.1), where v is to be determined:

$$u(t+1)v(t+1) - p(t)u(t)v(t) = r(t)$$

or

$$v(t) = \sum \frac{r(t)}{Eu(t)} + C,$$

3.1 First Order Equations

so

$$y(t) = u(t)\left[\sum \frac{r(t)}{Eu(t)} + C\right].$$

The last equations with C an arbitrary constant gives us a representation of all solutions of Eq. (3.1) provided $u(t)$ is any nontrivial (i.e., nonzero) solution of Eq. (3.1)'. Let's summarize these results in a theorem.

Theorem 3.1. Let $p(t) \neq 0$ and $r(t)$ be given for $t = a, a+1, \cdots$. Then
a) the solutions of Eq. (3.1)' are

$$u(t) = u(a) \prod_{s=a}^{t-1} p(s), \quad (t = a+1, a+2, \cdots),$$

b) all solutions of Eq. (3.1) are given by

$$y(t) = u(t)\left[\sum \frac{r(t)}{Eu(t)} + C\right],$$

where C is a constant and $u(t)$ is any nonzero function from part (a).

Remark. The method we used to solve Eq. (3.1) is a special case of the method of "variation of parameters," which will be described later in this chapter.

Example 3.1. Find the solution $y(t)$ of

$$y(t+1) - ty(t) = (t+1)!, \quad (t = 1, 2, \cdots),$$

so that $y(1) = 5$.

First, note that the solutions of $u(t+1) - tu(t) = 0$ are

$$u(t) = u(1) \prod_{s=1}^{t-1} s = u(1)(t-1)!.$$

We can take $u(1) = 1$. Then

$$\begin{aligned} y(t) &= (t-1)! \left[\sum \frac{(t+1)!}{t!} + C \right] \\ &= (t-1)! \left[\sum (t+1) + C \right] \\ &= (t-1)! \left[\frac{B_2(t+1)}{2} + C \right] \quad \text{(from Corollary 2.2)} \\ &= \frac{(t+1)!}{2} + D(t-1)!. \end{aligned}$$

(We could have used the factorial function here.) To evaluate D, let $t = 1$:

$$5 = y(1) = \frac{2!}{2} + D \cdot 0!,$$

so $D = 4$. The solution is

$$y(t) = \frac{(t+1)!}{2} + 4(t-1)! \quad (t = 1, 2, \cdots).$$

It's a good idea to check these answers:

$$\begin{aligned} y(t+1) - ty(t) &= \frac{(t+2)!}{2} + 4t! - t\frac{(t+1)!}{2} - 4t! \\ &= \frac{(t+1)!}{2}[t+2-t] = (t+1)!. \end{aligned}$$

Example 3.2. Suppose we deposit \$2000 at the beginning of each year in an IRA that pays an annual interest rate of 8%. How much will we have in the IRA at the end of the tth year?

3.1 First Order Equations

Let $y(t)$ be the amount of money in the IRA at the end of the tth year. Then

$$y(t+1) = y(t) + (y(t) + 2000)(0.08) + 2000$$
$$= 1.08 y(t) + 2160.$$

A solution of the homogeneous equation $u(t+1) = 1.08 u(t)$ is $u(t) = (1.08)^t$. Then

$$y(t) = (1.08)^t \left[\sum \frac{2160}{(1.08)^{t+1}} + C \right]$$

$$= (1.08)^t \left[\frac{2160}{1.08} \sum \left(\frac{1}{1.08} \right)^t + C \right]$$

$$= (1.08)^t \left[\frac{2160}{1.08} \frac{\left(\frac{1}{1.08}\right)^t}{\frac{1}{1.08} - 1} + C \right]$$

(from Theorem 2.5(a))

$$= -27000 + C(1.08)^t.$$

Since $y(0) = 0$, we have $C = 27000$, so

$$y(t) = 27000[(1.08)^t - 1].$$

For example, at the end of twenty years we would have

$$y(20) = 27000[(1.08)^{20} - 1]$$
$$\simeq \$98,845.84.$$

(See Fig. 3.1.)

56   Chapter 3.   Linear Difference Equations

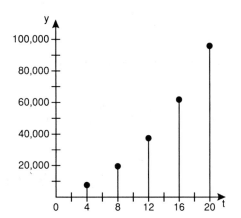

Fig. 3.1 Amount of money after t years

Of course, it is always possible to compute solutions of difference equations by direct step by step computation from the difference equation itself. However, roundoff error can be a serious problem. A simple but dramatic illustration of the possible effect of roundoff error is given by the following example due to Gautschi [84]:

$$y(t+1) - ty(t) = 1, \quad y(1) = 1 - e.$$

By Exercise 3.8 the exact solution is

$$y(t) = (t-1)! \left[1 - e + \sum_{k=1}^{t-1} \frac{1}{k!} \right].$$

Since

$$\sum_{k=1}^{\infty} \frac{1}{k!} = e - 1,$$

we must have $y(t) < 0$ for all t. Now let us attempt to compute $y(8)$ directly,

3.1 First Order Equations

starting with the approximate initial value $y(1) = -1.718$:

$$y(2) = y(1) + 1 = -.718$$
$$y(3) = 2y(2) + 1 = -.436$$
$$y(4) = 3y(3) + 1 = -.308$$
$$y(5) = 4y(4) + 1 = -.232$$
$$y(6) = 5y(5) + 1 = -.16$$
$$y(7) = 6y(6) + 1 = .04$$
$$y(8) = 7y(7) + 1 = 1.28.$$

At this point in the computation it is clear that the computed values of $y(t)$ are not close to the actual values and that the situation will deteriorate if we continue. Note that the only roundoff error occurs in the initial approximation since all of the other calculations are exact. For the actual behavior of the solution of this problem for large t see Exercise 5.12.

Now let's solve Eq. (3.1)$'$ and Eq. (3.1) for t in a discrete or continuous domain. For simplicity, we assume $p(t) > 0$. Apply the natural logarithm to both sides of Eq. (3.1)$'$:

$$\log|u(t+1)| = \log|u(t)| + \log p(t),$$
$$\Delta \log|u(t)| = \log p(t),$$
$$\log|u(t)| = \sum \log p(t) + D(t),$$

where $\Delta D(t) = 0$. Then

$$|u(t)| = e^{D(t)} e^{\sum \log p(t)},$$
$$u(t) = C(t) e^{\sum \log p(t)}, \qquad (3.2)$$

where $\Delta C(t) = 0$. Equation (3.2) is useful because it gives the solution of Eq. (3.1)$'$ in terms of the indefinite sum. Once $u(t)$ is found, then the solution $y(t)$ of Eq. (3.1) can be computed using Theorem 3.1(b) with the constant C replaced by an arbitrary function $C(t)$ so that $\Delta C(t) = 0$.

Chapter 3. Linear Difference Equations

Example 3.3. Solve the equation

$$u(t+1) = a\frac{(t-r_1)\cdots(t-r_n)}{(t-s_1)\cdots(t-s_m)}u(t),$$

where a, $r_1, \cdots, r_n$, $s_1, \cdots, s_m$ are constants.

For the moment, assume that all factors in the preceding expression are positive. Then

$$u(t) = C(t)e^{\sum[\log a + \log(t-r_1)+\cdots+\log(t-r_n)-\log(t-s_1)-\cdots-\log(t-s_m)]}$$
$$= C(t)e^{[t\log a+\log\Gamma(t-r_1)+\log\Gamma(t-r_n)-\log\Gamma(t-s_1)-\cdots-\log\Gamma(t-s_m)]}$$

by Theorem 2.5(d), so

$$u(t) = C(t)a^t\frac{\Gamma(t-r_1)\cdots\Gamma(t-r_n)}{\Gamma(t-s_1)\cdots\Gamma(t-s_m)},$$

where $\Delta C(t) = 0$. By direct substitution, we can show that this expression for $u(t)$ solves the difference equation for all values of t where the various gamma functions are defined (see Exercise 3.18). We can conclude that Eq. (3.1)' is solvable in terms of gamma functions if $p(t)$ is a rational function.

For example, consider

$$u(t+1) = \frac{t}{2t^2+3t+1}u(t).$$

The coefficient function factors as follows:

$$\frac{t}{2t^2+3t+1} = \frac{1}{2}\frac{t}{(t+1)(t+\frac{1}{2})},$$

so by the previous calculation, the solution is

$$u(t) = C(t)(\frac{1}{2})^t\frac{\Gamma(t)}{\Gamma(t+1)\Gamma(t+\frac{1}{2})}$$
$$= C(t)(\frac{1}{2})^t\frac{1}{t\Gamma(t+\frac{1}{2})}.$$

3.1 First Order Equations

There is an interesting relationship between Eq. (3.1) and ascending continued fractions. Let's rewrite Eq. (3.1) in the fractional form

$$y(t) = \frac{-r(t) + y(t+1)}{p(t)}.$$

Then

$$y(t+1) = \frac{-r(t+1) + y(t+2)}{p(t+1)}.$$

Substituting,

$$y(t) = \frac{-r(t) + \frac{-r(t+1)+y(t+2)}{p(t+1)}}{p(t)}.$$

Continuing in this way, we obtain the continued fraction

$$y(t) = \cfrac{-r(t) + \cfrac{-r(t+1) + \cfrac{-r(t+2) + \cfrac{-r(t+3)+\cdots}{p(t+3)}}{p(t+2)}}{p(t+1)}}{p(t)}.$$

If we formally divide out the continued fraction, we arrive at the infinite series

$$y(t) = \frac{-r(t)}{p(t)} + \frac{-r(t+1)}{p(t)p(t+1)} + \cdots,$$

or

$$y(t) = \sum_{k=0}^{\infty} \frac{-r(t+k)}{p(t)\cdots p(t+k)}. \tag{3.3}$$

When this series converges, its sum must be a solution of Eq. (3.1), as can be verified by substitution.

Example 3.4. For the equation

$$y(t+1) - ty(t) = -3^t,$$

Eq. (3.3) is

$$y(t) = \sum_{k=0}^{\infty} \frac{3^{t+k}}{t(t+1)\cdots(t+k)} = \frac{3^t}{t}\sum_{k=0}^{\infty} 3^k t^{(-k)},$$

a "factorial series." The ratio test shows that this series converges for all $t \neq 0, -1, -2, \cdots$, so the series represents one solution of the difference equation.

Factorial series will be considered again later in this chapter.

3.2 General Results for Linear Equations

The linear equation of the nth order is

$$p_n(t)y(t+n) + \cdots + p_0(t)y(t) = r(t), \tag{3.4}$$

where $p_0(t), \cdots, p_n(t)$ and $r(t)$ are assumed to be known and $p_0(t) \neq 0$, $p_n(t) \neq 0$ for all t. If $r(t) \not\equiv 0$, then we say Eq. (3.4) is "nonhomogeneous." As in Section 3.1, we will study Eq. (3.4) in association with the corresponding homogeneous equation

$$p_n(t)u(t+n) + \cdots + p_0(t)u(t) = 0. \tag{3.4}'$$

Note that Eq. (3.4) can also be written using the shift operator as

$$\left(p_n(t)E^n + \cdots + p_0(t)E^0\right)y(t) = r(t),$$

where $E^0 = I$. Since $E = \Delta + I$, it is also possible to write Eq. (3.4) in terms of the difference operator. However, the following example shows that the order of the equation is not apparent in that case.

3.2 General Results for Linear Equations

Example 3.5. What is the order of the equation

$$\Delta^3 y(t) + 3\Delta^2 y(t) + \Delta y(t) - y(t) = r(t)?$$

Let $\Delta = E - I$ and expand the powers of Δ:

$$(E^3 - 3E^2 + 3E - I)y(t) + 3(E^2 - 2E + I)y(t) \\ + (E - I)y(t) - y(t) = r(t)$$

or

$$y(t+3) - 2y(t+1) = r(t),$$

which is of order two in an appropriate domain.

Let's begin by observing the elementary fact that "initial value problems" for Eq. (3.4) have exactly one solution.

Theorem 3.2. Assume that $p_0(t), \cdots, p_n(t)$, and $r(t)$ are defined for $t = a, a+1, \cdots$ and $p_0(t) \neq 0$, $p_n(t) \neq 0$ for all t. Then for any t_0 in $\{a, a+1, \cdots\}$ and any numbers $y_0, \cdots, y_{n-1}$, there is exactly one $y(t)$ that satisfies Eq. (3.4) for $t = a, a+1, \cdots$ and $y(t_0 + k) = y_k$ for $k = 0, \cdots, n-1$.

Proof. The proof follows from iteration. For example,

$$y(t_0 + n) = \frac{r(t_0) - p_{n-1}(t_0)y_{n-1} - \cdots - p_0(t_0)y_0}{p_n(t_0)}$$

since $p_n(t_0) \neq 0$. Similarly, we can solve Eq. (3.4) for $y(t)$ when $t > t_0 + n$ in terms of the n preceding values of y. Since $p_0(t)$ is never 0, we can also solve for $y(t)$ when $t < t_0$. ∎

We will characterize the general solution of Eq. (3.4) through a sequence of theorems, beginning with the following basic result.

Theorem 3.3.
  a) If $u_1(t)$ and $u_2(t)$ solve Eq. (3.4)', then so does $Cu_1(t) + Du_2(t)$ for any constants C and D.
  b) If $u(t)$ solves Eq. (3.4)' and $y(t)$ solves Eq. (3.4), then $u(t) + y(t)$ solves Eq. (3.4).
  c) If $y_1(t)$ and $y_2(t)$ solve Eq. (3.4), then $y_1(t) - y_2(t)$ solves Eq. (3.4)'.

Proof. All parts can be proved by direct substitution. ∎

Corollary 3.1. *If $z(t)$ is a solution of Eq. (3.4) then every solution $y(t)$ of Eq. (3.4) takes the form*

$$y(t) = z(t) + u(t),$$

where $u(t)$ is some solution of Eq. (3.4)'.

Proof. This is just a restatement of Theorem 3.3(c). ∎

As a result of Corollary 3.1 the problem of finding all solutions of Eq.(3.4) reduces to two smaller problems:
 (1) find all solutions of Eq. (3.4)',
 (2) find one solution of Eq. (3.4).
This simplification is identical to that for linear differential equations. In order to analyze the first problem, we need some definitions.

Definition 3.1. The set of functions $\{u_1(t), \cdots, u_m(t)\}$ is "linearly dependent" on the set $t = a, a+1, \cdots$, if there are constants $C_1, \cdots, C_m$, not all zero, so that

$$C_1 u_1(t) + C_2 u_2(t) + \cdots + C_m u_m(t) = 0$$

for $t = a, a+1, \cdots$. Otherwise, the set is said to be "linearly independent."

Example 3.6. The functions $2^t, t2^t, t^2 2^t$ are linearly independent on every set $t = a, a+1, \cdots$, for if

$$C_1 2^t + C_2 t 2^t + C_3 t^2 2^t = 0, \quad (t = a, a+1, \cdots),$$

then

$$C_1 + C_2 t + C_3 t^2 = 0, \quad (t = a, a+1, \cdots),$$

but this equation can have an infinite number of roots only if $C_1 = C_2 = C_3 = 0$.

On the other hand, the functions $u_1(t) = 2$, $u_2(t) = 1 + \cos \pi t$ are linearly independent on the set $t = 1, 2, 3, \cdots$, but are linearly dependent for $t = \frac{1}{2}, \frac{3}{2}, \frac{5}{2}, \cdots$ since $u_1(t) - 2u_2(t) = 0$ for all such t.

3.2 General Results for Linear Equations

We will now define a matrix which is extremely useful in the study of linear equations.

Definition 3.2. The *matrix of Casorati* is given by

$$W(t) = \begin{bmatrix} u_1(t) & u_2(t) & \cdots & u_n(t) \\ u_1(t+1) & u_2(t+1) & \cdots & u_n(t+1) \\ \vdots & \vdots & \ddots & \vdots \\ u_1(t+n-1) & \cdots & \cdots & u_n(t+n-1) \end{bmatrix},$$

where $u_1, \cdots, u_n$ are given functions. The determinant

$$w(t) = \det W(t)$$

is called the "Casoratian."

It is not difficult to check that the Casoratian satisfies the equation

$$w(t) = \det \begin{bmatrix} u_1(t) & u_2(t) & \cdots & u_n(t) \\ \Delta u_1(t) & \Delta u_2(t) & \cdots & \Delta u_n(t) \\ \vdots & \vdots & \ddots & \vdots \\ \Delta^{n-1} u_1(t) & \cdots & \cdots & \Delta^{n-1} u_n(t) \end{bmatrix}. \qquad (3.5)$$

(See Exercise 3.23.) The Casoratian plays a role in the study of linear difference equations similar to that played by the Wronskian for linear differential equations. For example, we have the following characterization of dependence.

Theorem 3.4. Let $u_1(t), \cdots, u_n(t)$ be solutions of Eq. (3.4)' for $t = a, a+1, \cdots$. Then the following statements are equivalent:

a) The set $\{u_1(t), \cdots, u_n(t)\}$ is linearly dependent for $t = a, a+1, \cdots$,

b) $w(t) = 0$ for some t,

c) $w(t) = 0$ for all t.

Proof. First suppose $u_1(t), u_2(t), \cdots, u_n(t)$ are linearly dependent. Then there are constants $C_1, C_2, \cdots, C_n$ not all zero so that

$$C_1 u_1(t) + C_2 u_2(t) + \cdots + C_n u_n(t) = 0,$$
$$C_1 u_1(t+1) + C_2 u_2(t+1) + \cdots + C_n u_n(t+1) = 0,$$
$$\vdots$$
$$C_1 u_1(t+n-1) + C_2 u_2(t+n-1) + \cdots + C_n u_n(t+n-1) = 0,$$

for $t = a, a+1, \cdots$. Since this homogeneous system has a nontrivial solution $C_1, C_2, \cdots, C_n$, the determinant of the matrix of coefficients $w(t)$ is zero for $t = a, a+1, \cdots$.

Conversely, suppose $w(t_0) = 0$. Then there are constants $C_1, C_2, \cdots, C_n$, not all zero, so that

$$C_1 u_1(t_0) + C_2 u_2(t_0) + \cdots + C_n u_n(t_0) = 0$$
$$C_1 u_1(t_0+1) + C_2 u_2(t_0+1) + \cdots + C_n u_n(t_0+1) = 0$$
$$\vdots$$
$$C_1 u_1(t_0+n-1) + C_2 u_2(t_0+n-1) + \cdots + C_n u_n(t_0+n-1) = 0.$$

Let

$$u(t) = C_1 u_1(t) + C_2 u_2(t) + \cdots + C_n u_n(t).$$

Then u is a solution of Eq. (3.4)' and

$$u(t_0) = u(t_0+1) = \cdots = u(t_0+n-1) = 0.$$

It follows immediately from Theorem 3.2 that $u(t) = 0$ for all t, so the set $\{u_1, u_2, \cdots, u_n\}$ is linearly dependent. ∎

The importance of linear independence of solutions of Eq. (3.4)' is a consequence of the next theorem.

Theorem 3.5. If $u_1(t), \cdots, u_n(t)$ are independent solutions of Eq. (3.4)', then every solution $u(t)$ of Eq. (3.4)' can be written in the form

$$u(t) = C_1 u_1(t) + \cdots + C_n u_n(t)$$

for some constants $C_1, \cdots, C_n$.

3.2 General Results for Linear Equations

Proof. Let $u(t)$ be a solution of Eq. (3.4)'. Since $w(t) \neq 0$ for $t = a, a+1, \cdots$, the system of equations

$$C_1 u_1(a) + \cdots + C_n u_n(a) = u(a),$$
$$\vdots$$
$$C_1 u_1(a+n-1) + \cdots + C_n u_n(a+n-1) = u(a+n-1)$$

has a unique solution $C_1, \cdots, C_n$. Recall that a solution of Eq. (3.4)' is uniquely determined by its values at $t = a, a+1, \cdots, a+n-1$, so we must have

$$u(t) = C_1 u_1(t) + \cdots + C_n u_n(t)$$

for all t. ∎

Example 3.7. The equation

$$u(t+3) - 6u(t+2) + 11u(t+1) - 6u(t) = 0$$

has solutions 2^t, 3^t, 1 for all values of t. Their Casoratian is from Eq. (3.5)

$$w(t) = \det \begin{bmatrix} 2^t & 3^t & 1 \\ 2^t & 2 \cdot 3^t & 0 \\ 2^t & 4 \cdot 3^t & 0 \end{bmatrix} = 2^{t+1} 3^t,$$

which does not vanish. Consequently, the set $\{2^t, 3^t, 1\}$ is linearly independent, and all solutions of the equation have the form

$$u(t) = C_1 2^t + C_2 3^t + C_3.$$

We now turn our attention to the second step in solving Eq. (3.4), namely the calculation of at least one solution of Eq. (3.4). If we assume that n linearly independent solutions of Eq. (3.4)' are known, then the method of variation of parameters yields all solutions of Eq. (3.4) in terms of n indefinite sums. We will carry out the calculation for $n = 2$ since this case is representative of the general method.

Let u_1, u_2 be independent solutions of Eq. (3.4)′ with $n=2$. We seek a solution of Eq. (3.4) of the form

$$y(t) = a_1(t)u_1(t) + a_2(t)u_2(t),$$

where a_1 and a_2 are to be determined. Then

$$\begin{aligned}y(t+1) &= a_1(t+1)u_1(t+1) + a_2(t+1)u_2(t+1)\\ &= a_1(t)u_1(t+1) + a_2(t)u_2(t+1)\\ &\quad + \Delta a_1(t)u_1(t+1) + \Delta a_2(t)u_2(t+1).\end{aligned}$$

We eliminate the third and fourth terms of the last expression by choosing a_1 and a_2 so that

$$\Delta a_1(t)u_1(t+1) + \Delta a_2(t)u_2(t+1) = 0. \tag{3.6}$$

Next, we have

$$\begin{aligned}y(t+2) &= a_1(t+1)u_1(t+2) + a_2(t+1)u_2(t+2)\\ &= a_1(t)u_1(t+2) + a_2(t)u_2(t+2)\\ &\quad + \Delta a_1(t)u_1(t+2) + \Delta a_2(t)u_2(t+2).\end{aligned}$$

Now substitute the above expressions for $y(t)$, $y(t+1)$ and $y(t+2)$ into Eq. (3.4) and collect terms involving $a_1(t)$ and terms involving $a_2(t)$ to obtain

$$\begin{aligned}&p_2(t)y(t+2) + p_1(t)y(t+1) + p_0(t)y(t)\\ &= a_1(t)[p_2(t)u_1(t+2) + p_1(t)u_1(t+1) + p_0(t)u_1(t)]\\ &\quad + a_2(t)[p_2(t)u_2(t+2) + p_1(t)u_2(t+1) + p_0(t)u_2(t)]\\ &\quad + p_2(t)[u_1(t+2)\Delta a_1(t) + u_2(t+2)\Delta a_2(t)].\end{aligned}$$

Since u_1 and u_2 satisfy Eq. (3.4)′, the first two bracketed expressions are zero. Then $y(t)$ satisfies Eq. (3.4) if

$$u_1(t+2)\Delta a_1(t) + u_2(t+2)\Delta a_2(t) = \frac{r(t)}{p_2(t)}. \tag{3.7}$$

3.2 General Results for Linear Equations

To sum up, $y(t) = a_1(t)u_1(t) + a_2(t)u_2(t)$ is a solution of Eq. (3.4) if $\Delta a_1(t)$, $\Delta a_2(t)$ satisfy the linear equations (3.6) and (3.7). This system of linear equations has a unique solution since the matrix of coefficients is $W(t+1)$, which has a nonzero determinant by Theorem 3.4.

The result for the nth order equation is as follows.

Theorem 3.6. Let $u_1(t), \cdots, u_n(t)$ be independent solutions of Eq. (3.4)'. Then

$$y(t) = a_1(t)u_1(t) + \cdots + a_n(t)u_n(t)$$

is a solution of Eq. (3.4), provided $a_1, \cdots, a_n$ satisfy the matrix equation

$$W(t+1) \begin{bmatrix} \Delta a_1(t) \\ \vdots \\ \Delta a_n(t) \end{bmatrix} = \begin{bmatrix} 0 \\ \vdots \\ \frac{r(t)}{p_n(t)} \end{bmatrix}.$$

Remark. The technique of Theorem 3.6 for finding a solution for the nonhomogeneous equation is called the method of variation of parameters.

Example 3.8. Find all solutions of

$$y(t+2) - 7y(t+1) + 6y(t) = t.$$

In the next section, we will show how to obtain two independent solutions of the associated homogeneous equation. Here we simply produce them, $u_1(t) = 1$ and $u_2(t) = 6^t$, and ask the reader to check them by substitution. Equations (3.6) and (3.7) are

$$\Delta a_1(t) + 6^{t+1}\Delta a_2(t) = 0,$$
$$\Delta a_1(t) + 6^{t+2}\Delta a_2(t) = t,$$

with solutions

$$\Delta a_1(t) = -\frac{t}{5}, \quad \Delta a_2(t) = \frac{t}{30}6^{-t}.$$

Then

$$a_1(t) = \sum \left(-\frac{t}{5}\right) + C,$$
$$= -\frac{t^{(2)}}{10} + C, \quad \text{(by Theorem 2.5(e))}$$
$$= -\frac{t(t-1)}{10} + C,$$

$$a_2(t) = \frac{1}{30}\sum t\left(\frac{1}{6}\right)^t + D$$
$$= \frac{1}{30}[t(-\frac{6}{5})(\frac{1}{6})^t - \sum(-\frac{6}{5})(\frac{1}{6})^{t+1}] + D, \quad \text{(by Theorem 2.6(c))}$$
$$= \frac{1}{30}[-\frac{6}{5}t(\frac{1}{6})^t + (\frac{6}{5})\frac{1}{6}(-\frac{6}{5})(\frac{1}{6})^t] + D.$$
$$= -\frac{t}{25}(\frac{1}{6})^t - \frac{1}{125}(\frac{1}{6})^t + D.$$

Finally,

$$y(t) = a_1(t)(1) + a_2(t)6^t$$
$$= -\frac{t(t-1)}{10} + C - \frac{t}{25} - \frac{1}{125} + D6^t$$
$$= C + D6^t - \frac{t^2}{10} + \frac{3t}{50} - \frac{1}{125}$$

is the general solution of the difference equation. A quicker method for computing $y(t)$ will be given in the next section.

For the case $n = 2$, Theorem 3.6 can be used to obtain an explicit representation of the solution $y(t)$ of Eq. (3.4) with $y(a) = y(a+1) = 0$.

Corollary 3.2. The unique solution $y(t)$ of Eq. (3.4) with $n = 2$ satisfying $y(a) = y(a+1) = 0$ is given by the variation of parameters formula

$$y(t) = \sum_{k=a}^{t-1} \frac{u_1(k+1)u_2(t) - u_2(k+1)u_1(t)}{p_2(k)w(k+1)} r(k),$$

where u_1 and u_2 are independent solutions of Eq. (3.4)'.

3.2 General Results for Linear Equations

Note. In order to obtain $y(a) = 0$ from the preceding expression for $y(t)$, we need the convention that a sum where the lower limit of summation is larger than the upper limit of summation is understood to be zero. This convention will be used frequently thoughout the book.

For the proof of Corollary 3.2, see Exercise 3.33. Exercise 3.34 contains additional information.

The results of this section can be extended to linear difference equations on a general t domain. The definition of linear dependence can be modified as follows.

Definition 3.1'. The set of functions $\{u_1(t), \cdots, u_m(t)\}$ is "linearly dependent" on a set S if there are functions $C_1(t), \cdots, C_m(t)$ on S so that $\Delta C_k(t) = 0$ $(k = 1, \cdots, m)$, $C_k(t) \neq 0$ for some k and some t, and

$$C_1(t)u_1(t) + \cdots + C_m(t)u_m(t) = 0$$

for all t in S. Otherwise, the set is said to be "linearly independent."

Note. Some books give a different definition of linear independence; see Miller [183].

For example, the set $\{1, t^2\}$ is linearly independent on the real line. Let $C_1(t)$, $C_2(t)$ be functions of period one so that

$$C_1(t) + C_2(t)t^2 = 0$$

for all t. If there is a value $t = t_0$ so that $C_2(t_0) \neq 0$, then

$$t^2 = -\frac{C_1(t)}{C_2(t)}$$

for $t = t_0, t_0 + 1, \cdots$. Since the righthand side of this equation is constant for these t values, we have a contradiction. Consequently, $C_2(t) = 0$ for all t and also $C_1(t) = 0$ for all t, so $\{1, t^2\}$ is independent.

If (c) is eliminated from Theorem 3.4 and the constants $C_1, \cdots, C_n$ in Theorem 3.5 are replaced by functions $C_1(t), \cdots, C_n(t)$ so that $\Delta C_k(t) = 0$ $(k = 1, \cdots, n)$, then everything we have proved is also true for general domains. See Milne-Thomson [184] for further details.

3.3 Equations with Constant Coefficients

We now turn to the problem of finding n linearly independent solutions of Eq. (3.4)′ in the case that the coefficient functions are all constants. Since $p_n \neq 0$, we can divide through both sides of Eq. (3.4)′ by p_n and relabel the resulting equation to obtain

$$u(t+n) + p_{n-1}u(t+n-1) + \cdots + p_0 u(t) = 0, \qquad (3.8)$$

where $p_0, \cdots, p_{n-1}$ are constants and $p_0 \neq 0$.

Definition 3.3.

a) The polynomial $\lambda^n + p_{n-1}\lambda^{n-1} + \cdots + p_0$ is called the "characteristic polynomial" for Eq. (3.8).

b) The equation $\lambda^n + \cdots + p_0 = 0$ is the "characteristic equation" for Eq. (3.8).

c) The solutions $\lambda_1, \cdots, \lambda_k$ of the characteristic equation are the "characteristic roots."

If we introduce the shift operator E into Eq. (3.8), then it takes on the form of its characteristic equation and has similar factors:

$$(E^n + p_{n-1}E^{n-1} + \cdots + p_0)u(t) = 0,$$

or

$$(E - \lambda_1)^{\alpha_1} \cdots (E - \lambda_k)^{\alpha_k} u(t) = 0, \qquad (3.9)$$

where $\alpha_1 + \cdots + \alpha_k = n$ and the order of the factors is immaterial. Note that each characteristic root is nonzero since $p_0 \neq 0$.

Let's solve the equation

$$(E - \lambda_1)^{\alpha_1} u(t) = 0. \qquad (3.10)$$

Certainly, any solution of Eq. (3.10) will also be a solution of Eq. (3.9). If $\alpha_1 = 1$, then Eq. (3.10) is simply $u(t+1) = \lambda_1 u(t)$, which has as a solution

3.3 Equations with Constant Coefficients

$u(t) = \lambda_1^t$. If $\alpha_1 > 1$, let $u(t) = \lambda_1^t v(t)$ in Eq. (3.10):

$$(E - \lambda_1)^{\alpha_1} \lambda_1^t v(t)$$
$$= \sum_{i=0}^{\alpha_1} \binom{\alpha_1}{i} (-\lambda_1)^{\alpha_1 - i} E^i \lambda_1^t v(t),$$
$$= \sum_{i=0}^{\alpha_1} \binom{\alpha_1}{i} (-\lambda_1)^{\alpha_1 - i} \lambda_1^{t+i} E^i v(t),$$
$$= \lambda_1^{\alpha_1 + t} \sum_{i=0}^{\alpha_1} \binom{\alpha_1}{i} (-1)^{\alpha_1 - i} E^i v(t),$$
$$= \lambda_1^{\alpha_1 + t} (E - 1)^{\alpha_1} v(t),$$
$$= \lambda_1^{\alpha_1 + t} \Delta^{\alpha_1} v(t) = 0,$$

if $v(t) = 1, t, t^2, \cdots, t^{\alpha_1 - 1}$. Consequently, Eq. (3.10) has α_1 solutions $\lambda_1^t, t\lambda_1^t, \cdots, t^{\alpha_1 - 1}\lambda_1^t$. As in Example 3.6, these are easily seen to be linearly independent. By applying this argument to each factor of Eq. (3.9), we obtain n solutions of Eq. (3.8), which are linearly independent. (See Exercise 3.31 for the verification of independence in the case of distinct characteristic roots.)

Theorem 3.7. Suppose that Eq. (3.8) has characteristic roots $\lambda_1, \cdots, \lambda_k$ with multiplicities $\alpha_1, \cdots, \alpha_k$, respectively. Then Eq. (3.8) has the n independent solutions $\lambda_1^t, \cdots, t^{\alpha_1 - 1}\lambda_1^t, \lambda_2^t, \cdots, t^{\alpha_2 - 1}\lambda_2^t, \cdots, \lambda_k^t, \cdots, t^{\alpha_k - 1}\lambda_k^t$.

Example 3.9. Find all solutions of

$$u(t + 3) - 7u(t + 2) + 16u(t + 1) - 12u(t) = 0, \quad (t = a, a + 1, \cdots).$$

The characteristic equation is

$$\lambda^3 - 7\lambda^2 + 16\lambda - 12 = 0$$

or

$$(\lambda - 2)^2 (\lambda - 3) = 0.$$

Then Theorem 3.7 tells us that three independent solutions of the difference equations are

$$u_1(t) = 2^t, \quad u_2(t) = t2^t, \quad u_3(t) = 3^t.$$

Let's verify independence:

$$w(t) = \det \begin{bmatrix} 2^t & t2^t & 3^t \\ 2^{t+1} & (t+1)2^{t+1} & 3^{t+1} \\ 2^{t+2} & (t+2)2^{t+2} & 3^{t+2} \end{bmatrix} = 3^t 2^{2t} \det \begin{bmatrix} 1 & t & 1 \\ 2 & 2(t+1) & 3 \\ 4 & 4(t+2) & 9 \end{bmatrix}$$
$$= 3^t 2^{2t+1} \neq 0.$$

The general solution of the difference equation is

$$u(t) = C_1 2^t + C_2 t 2^t + C_3 3^t,$$

where C_1, C_2, C_3 are arbitrary constants.

If the characteristic roots include a complex pair $\lambda = a \pm ib$, then real-valued solutions of Eq. (3.8) can be found by using polar form

$$\lambda = re^{\pm i\theta} = r(\cos\theta \pm i\sin\theta),$$

where $a^2 + b^2 = r^2$ and $\tan\theta = b/a$. Then

$$\lambda^t = r^t e^{\pm i\theta t} = r^t(\cos\theta t \pm i\sin\theta t).$$

Since linear combinations of solutions are also solutions, we obtain the independent real solutions $r^t \cos\theta t$ and $r^t \sin\theta t$. Repeated complex roots are handled in a similar way.

Example 3.10. Find independent real solutions of

$$u(t+2) - 2u(t+1) + 4u(t) = 0.$$

From the characteristic equation $\lambda^2 - 2\lambda + 4 = 0$, we have $\lambda = 1 \pm \sqrt{3}i$.

3.3 Equations with Constant Coefficients

The polar coordinates are $r = \sqrt{1+3} = 2$ and $\theta = \tan^{-1}\sqrt{3} = \frac{\pi}{3}$, so two real solutions are

$$u_1(t) = 2^t \cos \frac{\pi}{3} t, \quad u_2(t) = 2^t \sin \frac{\pi}{3} t.$$

These are independent since $w(t) = \sqrt{3} \cdot 4^t \neq 0$.

The general equation with constant coefficients,

$$y(t+n) + p_{n-1} y(t+n-1) + \cdots + p_0 y(t) = r(t) \qquad (3.11)$$

can be solved by the "annihilator method" if $r(t)$ is a solution of some homogeneous equation with constant coefficients. The central idea is contained in the following simple result.

Theorem 3.8. Suppose that $y(t)$ solves Eq. (3.11), i.e.,

$$(E^n + p_{n-1} E^{n-1} + \cdots + p_0) y(t) = r(t),$$

and that $r(t)$ satisfies

$$(E^m + q_{m-1} E^{m-1} + \cdots + q_0) r(t) = 0.$$

Then $y(t)$ satisfies

$$(E^m + \cdots + q_0)(E^n + \cdots + p_0) y(t) = 0.$$

Proof. Simply apply the operator $E^m + \cdots + q_0$ to both sides of Equation (3.11). ∎

We illustrate the annihilator method by solving again the equation in Example 3.8, which was solved in the last section using variation of parameters.

Example 3.11. $y(t+2) - 7y(t+1) + 6y(t) = t$.

First, rewrite the equation in operator form:

$$(E^2 - 7E + 6) y(t) = t$$

or

$$(E-1)(E-6)y(t) = t.$$

Now t satisfies the homogeneous equation

$$(E-1)^2 t = \Delta^2 t = 0.$$

By Theorem 3.7, $y(t)$ satisfies

$$(E-1)^3(E-6)y(t) = 0.$$

(Here $(E-1)^2$ is the "annihilator," which eliminates the nonzero function from the righthand side of the equation.)

From our discussion of homogeneous equations, we have

$$y(t) = C_1 6^t + C_2 + C_3 t + C_4 t^2.$$

The next step is to substitute this expression for $y(t)$ into the original equation to determine the coefficients. Note that $C_1 6^t + C_2$ satisifies the homogeneous portion of that equation, so it is sufficient to substitute $y(t) = C_3 t + C_4 t^2$:

$$C_3(t+2) + C_4(t+2)^2 - 7C_3(t+1) - 7C_4(t+1)^2 + 6C_3 t + 6C_4 t^2 = t$$

or

$$t^2[C_4 - 7C_4 + 6C_4] + t[4C_4 + C_3 - 14C_4 - 7C_3 + 6C_3]$$
$$+ [4C_4 + 2C_3 - 7C_4 - 7C_3] = t.$$

Equating coefficients, we have

$$-10C_4 = 1,$$
$$-5C_3 - 3C_4 = 0,$$

3.3 Equations with Constant Coefficients

so $C_4 = -\frac{1}{10}$ and $C_3 = \frac{3}{50}$. Then

$$y(t) = C_1 6^t + C_2 + \frac{3}{50}t - \frac{1}{10}t^2,$$

which agrees with the answer in Example 3.8.

Example 3.12. Solve $\Delta y(t) = 3^t \sin \frac{\pi}{2}t$ $(t = a, a+1, \cdots)$.

The function $3^t \sin \frac{\pi}{2}t$ must satisfy an equation with complex characteristic roots. From the discussion preceding Example 3.10, we see that the polar coordinates of these roots are $r = 3$, $\theta = \pm\frac{\pi}{2}$, so $\lambda = 3e^{\pm\frac{\pi}{2}i} = \pm 3i$. Then $3^t \sin \frac{\pi}{2}t$ satisfies

$$(E - 3i)(E + 3i)u(t) = (E^2 + 9)u(t) = 0,$$

so $y(t)$ satisfies

$$(E^2 + 9)(E - 1)y(t) = 0,$$

which has general solution

$$y(t) = C_1 + C_2 3^t \sin \frac{\pi}{2}t + C_3 3^t \cos \frac{\pi}{2}t.$$

By substituting this expression into the original equation, we find

$$C_2 3^t \left(3\cos \frac{\pi}{2}t - \sin \frac{\pi}{2}t\right) + C_3 3^t \left(-3\sin \frac{\pi}{2}t - \cos \frac{\pi}{2}t\right) = 3^t \sin \frac{\pi}{2}t.$$

Then $C_2 = -\frac{1}{10}$, $C_3 = -\frac{3}{10}$, and finally

$$y(t) = C_1 - \frac{3^t}{10}\left(\sin \frac{\pi}{2}t + 3\cos \frac{\pi}{2}t\right),$$

where C_1 is arbitrary.

Systems of linear difference equations with constant coefficients can be solved by the methods of this section. For example, suppose we have two equations in two unknowns

$$L(E)y(t) + M(E)z(t) = r(t),$$
$$P(E)y(t) + Q(E)z(t) = s(t),$$

where $y(t)$ and $z(t)$ are the unknowns and L, M, P, and Q are polynomials. Simply apply $Q(E)$ to the first equation and $M(E)$ to the second equation and subtract to obtain

$$(Q(E)L(E) - M(E)P(E))\,y(t) = Q(E)r(t) - M(E)s(t),$$

which is a single linear equation with constant coefficients. Once $y(t)$ is found, it can be substituted into one of the original equations to produce an equation for $z(t)$.

Example 3.13. Solve the system

$$y(t+2) - 3y(t) + z(t+1) - z(t) = 5^t,$$
$$y(t+1) - 3y(t) + z(t+1) - 3z(t) = 2 \cdot 5^t.$$

First, write the system in operator form

$$(E^2 - 3)y(t) + (E - 1)z(t) = 5^t,$$
$$(E - 3)y(t) + (E - 3)z(t) = 2 \cdot 5^t.$$

Apply $(E - 3)$ to the first equation and $(E - 1)$ to the second equation and subtract:

$$[(E^2 - 3)(E - 3) - (E - 3)(E - 1)]\,y(t) = (E - 3)5^t - (E - 1)2 \cdot 5^t$$

or

$$(E - 3)(E - 2)(E + 1)y(t) = -6 \cdot 5^t.$$

By the annihilator method, an appropriate trial solution is $y(t) = C5^t$.

3.3 Equations with Constant Coefficients

Substitution yields $C = -\frac{1}{6}$, so

$$y(t) = C_1 3^t + C_2 2^t + C_3(-1)^t - \frac{5^t}{6}.$$

If we substitute this expression for y into the second of the original equations, we have

$$(E-3)z(t) = C_2 2^t + 4C_3(-1)^t + \frac{7}{3}5^t.$$

The annihilator method gives us

$$z(t) = C_4 3^t - C_2 2^t - C_3(-1)^t + \frac{7}{6}5^t.$$

Finally, substituting the expressions for y and z into the first of the original equations, we have

$$(E^2 - 3)y(t) + (E-1)z(t)$$
$$= \left(6C_1 3^t + C_2 2^t - 2C_3(-1)^t - \frac{11}{3}5^t\right)$$
$$+ \left(2C_4 3^t - C_2 2^t + 2C_3(-1)^t + \frac{14}{3}5^t\right)$$
$$= (6C_1 + 2C_4)3^t + 5^t$$
$$= 5^t,$$

so $C_4 = -3C_1$. The general solutions are

$$y(t) = C_1 3^t + C_2 2^t + C_3(-1)^t - \frac{5^t}{6},$$
$$z(t) = -3C_1 3^t - C_2 2^t - C_3(-1)^t + \frac{7}{6}5^t.$$

An alternate method for solving homogeneous systems with constant coefficients will be presented in Chapter 4.

3.4 Applications

Even though linear equations with constant coefficients represent a very restrictive class of difference equations, they appear in a variety of applications. The following eight examples are fairly representative. Example 3.20 introduces constant coefficient equations of a different type, the so-called "convolution equations." Additional details of those and other applications are contained in the exercises.

Example 3.14. (The Fibonacci sequence)

The Fibonacci sequence is

$$1,\ 1,\ 2,\ 3,\ 5,\ 8,\ 13,\ 21,\ \cdots,$$

where each integer after the first two is the sum of the two integers immediately preceding it. Certain natural phenomena, such as the spiral patterns on the surfaces of pine cones, appear to be governed by this sequence. These numbers also occur in the analysis of algorithms and are of sufficient interest to mathematicians that a journal is devoted to the study of their properties.

If $u(t)$ denotes the tth term in the Fibonacci sequence for $t = 1, 2, \cdots$, then $u(t)$ satisfies the initial value problem

$$\begin{aligned} u(t+2) - u(t+1) - u(t) &= 0, \quad (t = 1, 2, \cdots) \\ u(1) &= 1, \quad u(2) = 1. \end{aligned}$$

The characteristic equation is $\lambda^2 - \lambda - 1 = 0$, so $\lambda = \frac{1 \pm \sqrt{5}}{2}$. Then the general solution of the difference equation is

$$C_1 \left(\frac{1+\sqrt{5}}{2} \right)^t + C_2 \left(\frac{1-\sqrt{5}}{2} \right)^t.$$

By using the initial conditions, we find $C_1 = -C_2 = \frac{1}{\sqrt{5}}$, so

$$u(t) = \frac{1}{\sqrt{5}} \left(\frac{1+\sqrt{5}}{2} \right)^t - \frac{1}{\sqrt{5}} \left(\frac{1-\sqrt{5}}{2} \right)^t.$$

3.4 Applications

for $t = 1, 2, \cdots$. Although $\sqrt{5}$ is predominant in this formula, all these numbers must be integers!

Note that

$$\frac{u(t+1)}{u(t)} = \frac{\left(\frac{1+\sqrt{5}}{2}\right) - \left(\frac{1-\sqrt{5}}{2}\right)\left(\frac{1-\sqrt{5}}{1+\sqrt{5}}\right)^t}{1 - \left(\frac{1-\sqrt{5}}{1+\sqrt{5}}\right)^t} \to \frac{1+\sqrt{5}}{2}$$

as $t \to \infty$. The ratio $\frac{1+\sqrt{5}}{2}$ is known as the "golden section," and it was considered by the ancient Greeks to be the most aesthetically pleasing ratio for the length of a rectangle to its width.

Example 3.15. (A crystal lattice)

A crystal lattice is sometimes modeled mathematically by viewing it as an infinite collection of objects connected by springs. We consider vibrations along a fixed direction (see Fig. 3.2). Let v_n be the displacement of the nth object from equilibrium. Then the equations of motion are

$$m_{n+1}\frac{d^2 v_{n+1}}{dt^2} = k_{n+1}(v_{n+2} - v_{n+1}) + k_n(v_n - v_{n+1}),$$

where m_n is the mass of the nth object and the k_n's are spring constants.

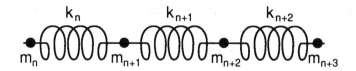

Fig. 3.2 *Masses connected by springs*

By using the substitution

$$v_n = u_n e^{-i\omega t},$$

we obtain a difference equation for the quantities u_n:

$$-m_{n+1}\omega^2 u_{n+1} = k_{n+1}(u_{n+2} - u_{n+1}) + k_n(u_n - u_{n+1}).$$

In an ideal crystal, we may assume the coefficients are independent of n, say $k_n = k$ and $m_n = m$ for all n. The difference equation becomes

$$u_{n+2} + \left(\frac{\omega^2 m}{k} - 2\right) u_{n+1} + u_n = 0,$$

with characteristic roots

$$\lambda = 1 - \frac{\omega^2 m}{2k} \pm \omega \sqrt{\frac{m}{k}} \sqrt{\frac{\omega^2 m}{4k} - 1}.$$

We consider only the case $\frac{\omega^2 m}{4k} < 1$. Then these roots are a complex conjugate pair

$$\lambda = 1 - \frac{\omega^2 m}{k} \pm i\omega \sqrt{\frac{m}{k}} \sqrt{1 - \frac{\omega^2 m}{4k}}.$$

Now

$$|\lambda|^2 = \left(1 - \frac{\omega^2 m}{2k}\right)^2 + \omega^2 \frac{m}{k}\left(1 - \frac{\omega^2 m}{4k}\right)$$
$$= 1,$$

so $\lambda = e^{\pm i\theta}$, for some θ. The general solution in complex form is then

$$u_n = Ce^{i\theta n} + De^{-i\theta n},$$

and

$$v_n(t) = Ce^{i(n\theta - \omega t)} + De^{i(-n\theta - \omega t)}$$

represents a linear combination of a wave moving to the right and a wave moving to the left.

3.4 Applications

Example 3.16. (Predator-prey model) Let $x(t)$ and $y(t)$ denote predator and prey populations, respectively, of two species after t years. Suppose that the changes in the populations are given by the equations

$$\Delta x(t) = -.1x(t) + .2y(t),$$
$$\Delta y(t) = -.1x(t) - .4y(t).$$

Let the initial populations be $x(0) = y(0) = 5,000$.

The above system has operator form

$$(E - .9)x(t) - .2y(t) = 0,$$
$$.1x(t) + (E - .6)y(t) = 0.$$

Using the methods of Section 3.3, we find that $y(t)$ satisfies

$$(E^2 - 1.5E + .56)y(t) = 0$$

or

$$(E - .7)(E - .8)y(t) = 0.$$

Then

$$y(t) = A(.7)^t + B(.8)^t.$$

Since

$$x(t) = (-10E + 6)y(t),$$

we have that

$$x(t) = -A(.7)^t - 2B(.8)^t.$$

The initial conditions lead us to $A = 15,000$ and $B = -10,000$.

The populations after t years are then

$$x(t) = -15,000(.7)^t + 20,000(.8)^t,$$
$$y(t) = 15,000(.7)^t - 10,000(.8)^t.$$

It is interesting to observe that the prey population becomes extinct in finite time, so the equation for the predator population is independent of $y(t)$ from that time. (See Fig. 3.3.) Of course, the predator population dies out somewhat later.

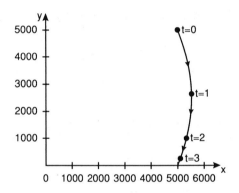

Fig. 3.3 Predator and prey populations

Example 3.17. (The Chebyshev polynomials) The nth Chebyshev polynomial of the first kind is

$$T_n(x) = \cos(n \cos^{-1} x) \quad (n \geq 0).$$

Note that $T_0(x) = 1$ and $T_1(x) = x$. Letting $\theta = \cos^{-1}(x)$, we have

$$\begin{aligned} T_{n+2}(x) - 2xT_{n+1}(x) + T_n(x) &= \cos(n+2)\theta - 2\cos\theta\cos(n+1)\theta + \cos n\theta \\ &= \cos n\theta \cos 2\theta - \sin\theta \sin 2\theta \\ &\quad - 2\cos n\theta \cos^2\theta + 2\sin n\theta \cos\theta \sin\theta \\ &\quad + \cos n\theta = 0, \quad (n \geq 0), \end{aligned}$$

3.4 Applications

since $\cos 2\theta = 2\cos^2\theta - 1$ and $\sin 2\theta = 2\cos\theta\sin\theta$. Consequently, $T_n(x)$ (as a function of n with x fixed) satisfies a homogeneous linear difference equation with constant coefficients.

The Chebyshev polynomials can be computed recursively from this equation,

$$\begin{aligned} T_2(x) &= 2xT_1(x) - T_0(x), \\ &= 2x^2 - 1, \\ T_3(x) &= 2xT_2(x) - T_1(x) \\ &= 4x^3 - 3x, \end{aligned}$$

and so forth. A simple induction argument shows that $T_n(x)$ is a polynomial of degree n.

Let

$$w(x) = \frac{1}{\sqrt{1-x^2}}.$$

With the change of variable $\theta = \cos^{-1} x$, we have

$$\begin{aligned} \int_{-1}^{1} T_n(x)T_m(x)w(x)\,dx &= \int_{-1}^{1} \cos(n\cos^{-1}x)\cos(m\cos^{-1}x)\frac{dx}{\sqrt{1-x^2}}, \\ &= \int_0^{\pi} \cos n\theta \cos m\theta\,d\theta, \\ &= \int_0^{\pi} \frac{\cos(n+m)\theta + \cos(n-m)\theta}{2}\,d\theta, \\ &= 0, \end{aligned}$$

if $m \neq n$. We say that the Chebyshev polynomials are "orthogonal" on $[-1,1]$ with "weight function" $w(x)$. Because of this and several other nice properties, they are of fundamental importance in the branch of approximation theory that involves the approximation of continuous functions by polynomials.

More generally, it can be shown (see Atkinson [20]) that every family $\{\varphi_n\}$ of orthogonal polynomials satisfies a second order homogeneous linear difference equation of the form

$$\varphi_{n+1}(x) + (a_n x + b_n)\varphi_{n+1}(x) + c_n \varphi_n(x) = 0,$$

so these equations are of general utility in the computation of orthogonal polynomials.

Example 3.18. (Water rationing) Due to water rationing, Al can water his lawn only during the hours from 9PM to 9AM. Suppose he can add a quantity q of water to the topsoil during this period, but that half of the total amount of water in the topsoil is lost through evaporation or absorption during the period from 9AM to 9PM.

Assume that the topsoil contains an initial quantity I of water at 9PM on the first day of rationing. Let $y(t)$ be the amount of water in the soil at the end of the tth 12 hour period thereafter.

Now if t is odd,

$$y(t+2) = \frac{1}{2}y(t) + q,$$

and if t is even,

$$y(t+2) = \frac{1}{2}y(t) + \frac{q}{2},$$

so in general

$$y(t+2) - \frac{1}{2}y(t) = \frac{q}{4}\left(3 - (-1)^t\right).$$

Since the homogeneous portion of this equation has characteristic roots $\lambda = \pm\frac{1}{\sqrt{2}}$, it has solutions

$$C(\sqrt{2})^{-t} + D(-\sqrt{2})^{-t}.$$

By the annihilator method, a trial solution for the nonhomogeneous equation is $A + B(-1)^t$.

3.4 Applications

We find easily that $A = \frac{3q}{2}$, $B = -\frac{q}{2}$, so the general solution is

$$y(t) = C(\sqrt{2})^{-t} + D(-\sqrt{2})^{-t} + \frac{q}{2}\left[3 - (-1)^t\right].$$

Finally, by using the initial values $y(0) = I$ and $y(1) = I + q$, we have

$$y(t) = \frac{I-q}{2}(\sqrt{2})^{-t}\left\{\sqrt{2}\left[1 - (-1)^t\right] + \left[1 + (-1)^t\right]\right\}$$
$$+ \frac{q}{2}\left[3 - (-1)^t\right],$$

where $t = 0, 1, 2, \cdots$. Note that for large values of t, $y(t)$ essentially oscillates betwen q and $2q$.

Example 3.19. (A tridiagonal determinant) Let D_n be the value of the determinant of the following n by n matrix:

$$\begin{vmatrix} a & b & 0 & 0 & \cdots & 0 & 0 & 0 \\ c & a & b & 0 & \cdots & 0 & 0 & 0 \\ 0 & c & a & b & \cdots & 0 & 0 & 0 \\ 0 & 0 & c & a & \cdots & 0 & 0 & 0 \\ \vdots & \vdots & \vdots & \vdots & \ddots & \vdots & \vdots & \vdots \\ 0 & 0 & 0 & 0 & \cdots & a & b & 0 \\ 0 & 0 & 0 & 0 & \cdots & c & a & b \\ 0 & 0 & 0 & 0 & \cdots & 0 & c & a \end{vmatrix}$$

If we expand the corresponding $n + 2$ by $n + 2$ determinant by the first row, we obtain

$$D_{n+2} = aD_{n+1} - bcD_n.$$

This homogeneous difference equation has characteristic roots

$$\lambda = \frac{a \pm \sqrt{a^2 - 4bc}}{2}.$$

Chapter 3. Linear Difference Equations

We consider here only the case $a^2 - 4bc < 0$. Then

$$\lambda = \frac{a}{2} \pm i \frac{\sqrt{4bc - a^2}}{2}.$$

The polar coordinate r for these complex roots is

$$r = \frac{\sqrt{a^2 + (4bc - a^2)}}{4} = \sqrt{bc}.$$

Choose θ so that

$$\cos\theta = \frac{a}{2\sqrt{bc}},$$

$$\sin\theta = \frac{\sqrt{4bc - a^2}}{2\sqrt{bc}},$$

so

$$\lambda = \sqrt{bc}(\cos\theta \pm i\sin\theta).$$

The general solution is

$$D_n = (C_1 \cos n\theta + C_2 \sin n\theta)(bc)^{\frac{n}{2}}.$$

Since $D_1 = a$ and $D_2 = a^2 - bc$, the constants C_1 and C_2 can be computed, and we find

$$D_n = (bc)^{\frac{n}{2}} (\cos n\theta + \cot\theta \sin n\theta)$$
$$= (bc)^{\frac{n}{2}} \frac{\sin(n+1)\theta}{\sin\theta}, \quad (n \geq 1).$$

Note that in certain cases, the values of D_n are periodic. For example, if $a = b = c = 1$ then

$$D_n = \frac{2}{\sqrt{3}} \sin(n+1)\frac{\pi}{3}, \quad (n \geq 1),$$

3.4 Applications

which yields the sequence $1, 0, -1, -1, 0, 1, 1, 0, -1, -1, \cdots$.

Example 3.20. (Epidemiology) If x_n denotes the fraction of susceptible individuals in a certain population during the nth day of an epidemic, then the following equation represents one possible model (see Lauwerier [162, Chapter 8]) of the spread of the illness:

$$\log \frac{1}{x_{n+1}} = \sum_{k=0}^{n} (1 + \varepsilon - x_{n-k}) A_k, \qquad (n \geq 0),$$

where A_k is a measure of how infectious the ill individuals are during the kth day and ε is a small positive constant. If we let $x_n = e^{-z_n}$, then the equation for z_n is

$$z_{n+1} = \sum_{k=0}^{n} (1 + \varepsilon - e^{-z_{n-k}}) A_k.$$

This is a nonlinear equation; however, note that during the early stages of the epidemic x_n is near 1, so z_n is near 0. Replacing $e^{-z_{n-k}}$ by the approximation $1 - z_{n-k}$, we obtain the linearized equation

$$y_{n+1} = \sum_{k=0}^{n} (\varepsilon + y_{n-k}) A_k, \qquad (n \geq 0),$$

with $y_0 = 0$.

Even though this is a linear equation with constant coefficients, it is not of the type studied previously since each y_{n+1} depends on all of the preceding members of the sequence $y_0, \cdots, y_n$. However, the method of generating functions is useful here because of the special form of the sum $\sum_{k=0}^{n} y_{n-k} A_k$, which is called a sum of "convolution type."

We seek a generating function $Y(t)$ for $\{y_n\}$,

$$Y(t) = \sum_{n=0}^{\infty} y_n t^n,$$

and also set

$$A(t) = \sum_{n=0}^{\infty} A_n t^{n+1}.$$

By the usual procedure for multiplying power series (the "Cauchy" product)

$$\begin{aligned} A(t)Y(t) &= y_0 A_0 t + (y_1 A_0 + y_0 A_1)t^2 + \cdots \\ &= \sum_{n=0}^{\infty} \left(\sum_{k=0}^{n} y_{n-k} A_k \right) t^{n+1}. \end{aligned}$$

Now multiply both sides of the difference equation for y_n by t^{n+1} and sum to obtain

$$\sum_{n=0}^{\infty} y_{n+1} t^{n+1} = \varepsilon \sum_{n=0}^{\infty} \left(\sum_{k=0}^{n} A_k \right) t^{n+1} + \sum_{n=0}^{\infty} \left(\sum_{k=0}^{n} y_{n-k} A_k \right) t^{n+1}.$$

The sum on the left is simply $Y(t)$ since $y_0 = 0$ and the last sum is $A(t)Y(t)$. The first sum on the right is the product of the series $\sum_{n=0}^{\infty} A_n t^{n+1}$ and $\sum_{n=0}^{\infty} t^n = \frac{1}{1-t}$. We have that

$$Y(t) = \varepsilon A(t) \frac{1}{1-t} + A(t) Y(t)$$

or

$$Y(t) = \frac{\varepsilon A(t)}{(1-t)(1-A(t))}$$

is the generating function for $\{y_n\}$.

In a few special cases, the sequence $\{y_n\}$ can be computed explicitly. For example, if $A_k = c\alpha^k$, $0 < \alpha < 1$, then

$$A(t) = \frac{ct}{1-\alpha t} \quad \text{and} \quad Y(t) = \frac{\varepsilon ct}{(1-t)(1-\alpha t - ct)}.$$

3.4 Applications

By partial fractions,

$$\frac{\varepsilon ct}{(1-t)(1-(\alpha+c)t)} = \frac{\varepsilon c}{1-(\alpha+c)}\left[\frac{1}{1-t} - \frac{1}{1-(\alpha+c)t}\right],$$

so

$$Y(t) = \frac{\varepsilon c}{1-(\alpha+c)}\left[\sum_{n=0}^{\infty} t^n - \sum_{n=0}^{\infty}(\alpha+c)^n t^n\right],$$

and finally

$$y_n = \frac{\varepsilon c}{1-(\alpha+c)}\left[1-(\alpha+c)^n\right].$$

If $\alpha + c < 1$, then y_n will remain small for all n, so the outbreak does not reach epidemic proportions in that case.

Example 3.21. (A tiling problem) In how many ways can the floor of a hallway that is three units wide and n units long be tiled with tiles, each of which is two units by one unit? We assume that n is even so that the tiling can be done without breaking any tiles.

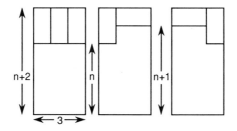

Fig. 3.4 Three initial tiling patterns

Let $y(n)$ be the number of different arrangements of the tiles that will accomplish the tiling. There are three different ways we can start to tile an $n+2$ by 3 hallway (see Fig. 3.4). There are $y(n)$ ways to complete the first hallway in the figure. Let $z(n+2)$ be the number of ways to finish the second hallway. By symmetry, there are also $z(n+2)$ ways to finish the third hallway. It follows that

$$y(n+2) = y(n) + 2z(n+2).$$

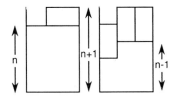

Fig. 3.5 *Two secondary patterns*

From Fig. 3.5, we see that there are two different ways to begin tiling the remainder of the second hallway in Fig. 3.4. Now there are $y(n)$ ways to complete the first of these and $z(n)$ ways to complete the second, so

$$z(n+2) = y(n) + z(n).$$

We need to solve the system

$$(E^2 - 1)y(n) - 2E^2 z(n) = 0$$
$$-y(n) + (E^2 - 1)z(n) = 0.$$

Eliminating $z(n)$ we obtain

$$(E^4 - 4E^2 + 1)y(n) = 0.$$

The characteristic roots of this equation are

$$\lambda = \pm(2+\sqrt{3})^{\frac{1}{2}}, \ \pm(2-\sqrt{3})^{\frac{1}{2}},$$

so since n is even

$$y(n) = A(2+\sqrt{3})^{\frac{n}{2}} + B(2-\sqrt{3})^{\frac{n}{2}}.$$

From the initial conditions $y(2) = 3$, $y(4) = 11$, we have

$$A(2+\sqrt{3}) + B(2-\sqrt{3}) = 3$$
$$A(2+\sqrt{3})^2 + B(2-\sqrt{3})^2 = 11.$$

The solutions of this system are

$$A = \frac{1+\sqrt{3}}{2\sqrt{3}}, \quad B = \frac{\sqrt{3}-1}{2\sqrt{3}},$$

so

$$y(n) = \frac{1+\sqrt{3}}{2\sqrt{3}}(2+\sqrt{3})^{\frac{n}{2}} + \frac{\sqrt{3}-1}{2\sqrt{3}}(2-\sqrt{3})^{\frac{n}{2}}.$$

For $n = 2, 4, \cdots$, the second term is positive and less than one. Then $y(n)$ is given by the integer part of the first term plus one. For example, there are 413,403 ways to tile a 20 by 3 hallway!

3.5 Equations with Variable Coefficients

Second and higher order linear equations with variable coefficients cannot be solved in closed form in most cases. Consequently, our discussion in this section will not be a general one but will present several methods that are often useful and can lead to explicit solutions in certain cases.

Recall that the nth order linear equation can be written in operator form as

$$(p_n(t)E^n + \cdots + p_0(t))y(t) = r(t).$$

If we are very lucky, the operator may factor into linear factors in a manner similar to the case of constant coefficients. Then the solutions can be found by solving a series of first order equations. The following example, which comes from probability theory (see Exercise 1.12), illustrates the procedure.

Example 3.22. Solve

$$\left(E^2 - (t+1)E - (t+1)\right)y(t) = 0.$$

The operator factors thus:

$$(E+1)(E-(t+1))y(t) = 0.$$

(Check this!) Consider the first order equation

$$(E+1)v(t) = 0.$$

The solution is $v(t) = (-1)^t C$. In order to solve the original equation, set

$$(E-(t+1))y(t) = (-1)^t C.$$

The homogeneous portion has general solution $D\Gamma(t+1)$, so from Theorem 3.1,

$$y(t) = D\Gamma(t+1) + C\Gamma(t+1)\sum \frac{(-1)^t}{\Gamma(t+2)}.$$

If t takes on discrete values $0, 1, 2, \cdots$, then

$$y(t) = Dt! + Ct!\sum_{k=0}^{t-1}\frac{(-1)^k}{(k+1)!},$$

where C and D are arbitrary constants.

Note that the factors in this example do not commute. In fact, the solutions of the equation

$$(E-(t+1))(E+1)y(t) = 0$$

are quite different (see Exercise 3.73).

It sometimes happens that we can find one nonzero solution of

3.5 Equations with Variable Coefficients

a homogeneous equation. In this case, the order of the equation can be reduced by one. For a second order equation, it is then possible to find a second solution that is independent of the first and, consequently, to generate the general solution.

As a first step, we show that the Casoratian satisfies a simple first order equation.

Lemma 3.1. Let $u_1(t), \cdots, u_n(t)$ be solutions of the equation

$$p_n(t)u(t+n) + \cdots + p_0(t)u(t) = 0,$$

and let $w(t)$ be the corresponding Casoratian. Then $w(t)$ satisfies

$$w(t+1) = (-1)^n \frac{p_0(t)}{p_n(t)} w(t). \tag{3.12}$$

Proof. The value of $w(t+1)$ is unchanged if we replace the last row by

$$(n^{\text{th}} \text{ row}) + \frac{p_1}{p_n} \times (1^{\text{st}} \text{ row}) + \cdots + \frac{p_{n-1}}{p_n} \times ((n-1)^{\text{st}} \text{ row}).$$

The difference equation can then be used to show that the new last row is

$$\left[-\frac{p_0}{p_n} u_1(t), \cdots, -\frac{p_0}{p_n} u_n(t) \right].$$

Then we have

$$w(t+1) = \det \begin{bmatrix} u_1(t+1) & \cdots & u_n(t+1) \\ \vdots & \ddots & \vdots \\ u_1(t+n-1) & \cdots & u_n(t+n-1) \\ -\frac{p_0}{p_n} u_1(t) & \cdots & -\frac{p_0}{p_n} u_n(t) \end{bmatrix}$$

$$= (-1)^n \frac{p_0(t)}{p_n(t)} w(t)$$

by rearrangement. ∎

Now assume $u_1(t)$ is a nonzero solution of

$$p_2(t)u(t+2) + p_1(t)u(t+1) + p_0(t)u(t) = 0, \qquad (3.13)$$

and let $u_2(t)$ denote another solution. Recall that

$$\Delta \frac{u_2(t)}{u_1(t)} = \frac{u_1(t)\Delta u_2(t) - u_2(t)\Delta u_1(t)}{u_1(t)u_1(t+1)}$$

$$= \frac{w(t)}{u_1(t)u_1(t+1)}.$$

Then

$$u_2(t) = u_1(t) \sum \frac{w(t)}{u_1(t)u_1(t+1)}, \qquad (3.14)$$

and we have the following theorem.

Theorem 3.9. If $u_1(t)$ is a solution of Eq. (3.13) that is never zero and $p_0(t)$ and $p_2(t)$ are not zero, then Eq. (3.14) yields an independent solution of Eq. (3.13), where $w(t)$ is a nonzero solution of Eq. (3.12).

Theorem 3.9 is known as the method of "reduction of order" for a second order equation. A technique for reducing the order of a higher order equation is outlined in Exercise 3.76.

Example 3.23. Solve the equation

$$y(t+2) - y(t+1) - \frac{1}{t+1}y(t) = 0.$$

By inspection, $u_1(t) = t+1$ is a solution. The Casoratian $w(t)$ satisfies

$$w(t+1) = -\frac{1}{t+1}w(t),$$

3.5 Equations with Variable Coefficients

so we can choose

$$w(t) = \frac{(-1)^t}{t!}.$$

Then

$$u_2(t) = (t+1)\sum_{k=0}^{t-1} \frac{(-1)^k}{(k+2)!}.$$

Example 3.24. Let a and b be constants in the equation

$$t(t+1)\Delta^2 u(t) + at\Delta u(t) + bu(t) = 0,$$

which is similar to the Cauchy-Euler differential equation. By substituting the trial solution $u(t) = (t+r-1)^{(r)}$, we have

$$t(t+1)r(r-1)(t+r-1)^{(r-2)} + atr(t+r-1)^{(r-1)} + b(t+r-1)^{(r)} = 0.$$

We need the following identities (see Exercise 3.79):

$$t(t+r-1)^{(r-1)} = (t+r-1)^{(r)}, \quad (3.15)$$
$$t(t+1)(t+r-1)^{(r-2)} = (t+r-1)^{(r)}. \quad (3.16)$$

Then we have

$$r(r-1)(t+r-1)^{(r)} + ar(t+r-1)^{(r)} + b(t+r-1)^{(r)} = 0,$$

or

$$r^2 + (a-1)r + b = 0. \quad (3.17)$$

If Eq. (3.17) has distinct real roots r_1, r_2, then the difference equation has independent solutions

$$u_i(t) = (t+r_i-1)^{(r_i)}, \quad (i=1,2).$$

In the case of repeated roots, Theorem 3.9 can be applied to obtain a second solution. Consider

$$t(t+1)\Delta^2 u(t) - 5t\Delta u(t) + 9u(t) = 0.$$

Here $r = 3$, so

$$u_1(t) = (t+2)^{(3)} = (t+2)(t+1)t$$

is a solution. Rewritten in standard form, the equation is

$$t(t+1)u(t+2) - (2t^2 + 7t)u(t+1) + (t+3)^2 u(t) = 0,$$

so for Eq. (3.12) we have

$$w(t+1) = \frac{(t+3)^2}{t(t+1)} w(t),$$

and we can take $w(t) = (t+2)^2(t+1)^2 t$. From Eq. (3.14),

$$u_2(t) = (t+2)^{(3)} \sum \frac{(t+2)^2(t+1)^2 t}{(t+2)^{(3)}(t+3)^{(3)}}$$

$$= (t+2)^{(3)} \sum \frac{1}{t+3}.$$

The general solution is

$$u(t) = (t+2)(t+1)t \left[C + D \sum \frac{1}{t+3} \right].$$

If the coefficients p_0, p_1, p_2 in Eq. (3.13) are polynomials, then a generating function for a solution of Eq. (3.13) can be shown to satisfy a differential equation, which may be soluable in terms of familiar functions. This is a reversal of the procedure for finding power series solutions of differential equations (see Example 1.4).

3.5 Equations with Variable Coefficients

Example 3.25. Solve

$$(n+2)u_{n+2} - (n+3)u_{n+1} + 2u_n = 0,$$

$n = 0, 1, 2, \cdots$.

Let a generating function be

$$g(x) = \sum_{n=0}^{\infty} u_n x^n.$$

First, multiply each term in the difference equation by x^n and sum as n goes from 0 to ∞:

$$\sum_{n=0}^{\infty}(n+2)u_{n+2}x^n - \sum_{n=0}^{\infty}(n+3)u_{n+1}x^n + 2\sum_{n=0}^{\infty}u_n x^n = 0.$$

Now make a change of index in the first two summations so that the index on u is n in each sum:

$$\sum_{n=2}^{\infty} n u_n x^{n-2} - \sum_{n=1}^{\infty}(n+2)u_n x^{n-1} + 2\sum_{n=0}^{\infty} u_n x^n = 0. \qquad (3.18)$$

Since $g'(x) = \sum_{n=1}^{\infty} n u_n x^{n-1}$, the first sum in Eq. (3.18) is

$$\sum_{n=2}^{\infty} n u_n x^{n-2} = \frac{1}{x}\left(g'(x) - u_1\right).$$

The second sum in Eq. (3.18) is

$$\sum_{n=1}^{\infty}(n+2)u_n x^{n-1} = \sum_{n=1}^{\infty} n u_n x^{n-1} + 2\sum_{n=1}^{\infty} u_n x^{n-1}$$

$$= g'(x) + \frac{2}{x}\left(g(x) - u_0\right).$$

98   Chapter 3.   Linear Difference Equations

Substituting these expressions into Eq. (3.18), we have

$$\frac{1}{x}(g'(x) - u_1) - g'(x) - \frac{2}{x}(g(x) - u_0) + 2g(x) = 0,$$

or

$$g'(x) - 2g(x) = \frac{u_1 - 2u_0}{1 - x}.$$

For $u_1 = 2u_0$, this last equation has the elementary solution

$$g(x) = e^{2x}$$
$$= \sum_{n=0}^{\infty} \frac{2^n}{n!} x^n,$$

so $u_n = \frac{2^n}{n!}$, $(n = 0, 1, 2, \cdots)$.

In this calculation, it was necessary to introduce only the first derivative of $g(x)$ since the coefficient functions are polynomials of degree at most one. More generally, the order of the differential equation will equal the degree of the polynomial of highest degree.

A second solution is easily found by using Theorem 3.9. By (3.12),

$$w(n+1) = \frac{2}{n+2} w(n),$$

so we can choose $w(n) = \frac{2^n}{(n+1)!}$. A second solution is

$$v_n = \frac{2^n}{n!} \sum \frac{\frac{2^n}{(n+1)!}}{\frac{2^n}{n!} \frac{2^{n+1}}{(n+1)!}}$$

$$= \frac{2^n}{n!} \sum_{k=0}^{n-1} \frac{k!}{2^{k+1}}.$$

In Example 3.4, we saw that solutions of certain first order equations can be expressed as factorial series. Higher order equations may also

3.5 Equations with Variable Coefficients

have such solutions. One approach is to substitute a trial series such as $u(t) = \sum_{k=0}^{\infty} a_k t^{(-k)}$ into the equation and try to determine the coefficients a_k. Of course, it might turn out that all the a_k's are zero, as we get only the zero solution for our efforts! In fact, the calculations are usually quite involved. See Milne-Thomson [184] for a thorough discussion of this topic. The following example is deceptively simple.

Example 3.26. Find a factorial series solution of

$$2u(t+2) + (t+2)(t+1)u(t+1) - (t+2)(t+1)u(t) = 0,$$

or

$$2u(t+2) + (t+2)(t+1)\Delta u(t) = 0.$$

Substitute $u(t) = \sum_{k=0}^{\infty} a_k t^{(-k)}$:

$$\sum_{k=0}^{\infty} 2a_k(t+2)^{(-k)} + (t+2)(t+1)\sum_{k=1}^{\infty} a_k(-k)t^{(-k-1)} = 0.$$

Since

$$(t+2)(t+1)t^{(-k-1)} = (t+2)^{(-k+1)},$$

we have

$$\sum_{k=0}^{\infty} 2a_k(t+2)^{(-k)} + \sum_{k=1}^{\infty} a_k(-k)(t+1)^{(-k+1)} = 0.$$

Make the change of index $k \to k+1$ in the second series and combine the series to obtain

$$\sum_{k=0}^{\infty} [2a_k - (k+1)a_{k+1}](t+2)^{(-k)} = 0.$$

Then a_0 is arbitrary and

$$a_{k+1} = \frac{2}{k+1} a_k \quad (k \geq 0),$$

so

$$a_k = \frac{2^k}{k!} a_0.$$

A factorial series solution is

$$u(t) = a_0 \sum_{k=0}^{\infty} \frac{2^k}{k!} t^{(-k)},$$

and the series converges for all t except the negative integers.

3.6 Nonlinear Equations That Can Be Linearized

As defined in Chapter 1, a difference equation is a relation

$$y(t+n) = f(t, y(t), \cdots, y(t+n-1)),$$

so that values of y can be computed recursively from known values. It is not to be expected that explicit formulas can be found for the solutions of these equations except in special cases. However, there are a number of important examples of nonlinear equations that can be transformed into equivalent linear equations by a change of dependent variable.

One class of equations for which this approach is successful is the Riccati equation

$$y(t+1)y(t) + p(t)y(t+1) + q(t)y(t) + r(t) = 0. \tag{3.19}$$

Let $y(t) = \frac{z(t+1)}{z(t)} - p(t)$. Direct substitution of this expression into Eq. (3.19) yields the linear equation

$$z(t+2) + [q(t) - p(t+1)] z(t+1) + [r(t) - p(t)q(t)] z(t) = 0, \tag{3.20}$$

3.6 Nonlinear Equations That Can Be Linearized

which may be solvable by one of the methods discussed earlier in the chapter. Then solutions of Eq. (3.19) are obtained from the relationship between y and z.

Example 3.27. $y(t+1)y(t) + 2y(t+1) + 4y(t) + 9 = 0$.

The change of variable $y(t) = \frac{z(t+1)}{z(t)} - 2$ gives us, from Eq. (3.20),

$$z(t+2) + 2z(t+1) + z(t) = 0,$$

which has general solution

$$z(t) = A(-1)^t + Bt(-1)^t.$$

The general solution of the Riccati equation is

$$\begin{aligned} y(t) &= \frac{A(-1)^{t+1} + B(t+1)(-1)^{t+1}}{A(-1)^t + Bt(-1)^t} - 2 \\ &= \frac{-1 - C(t+1)}{1 + Ct} - 2 \\ &= \frac{-3 - C(3t+1)}{1 + Ct}, \end{aligned}$$

where C is arbitrary. Actually, this last form of the solution is not quite general since it omits the solution $y(t) = -\frac{t+1}{t} - 2$, which results from $A = 0$.

Sometimes the structure of a difference equation will suggest a substitution that converts the equation into a linear one.

Example 3.28. $y(t+2) = \frac{y(t+1)}{y(t)}$.

In this case, the equation can be simplified by applying a logarithm:

$$\log y(t+2) = \log y(t+1) - \log y(t).$$

Let $z(t) = \log y(t)$ and rearrange to obtain

$$z(t+2) - z(t+1) + z(t) = 0.$$

By the methods of Section 3.3,

$$z(t) = A\cos\frac{\pi}{3}t + B\sin\frac{\pi}{3}t,$$

so

$$y(t) = C^{\cos\frac{\pi}{3}t} D^{\sin\frac{\pi}{3}t},$$

for some constants C and D. Note that all solutions have period 6! Check this conclusion by iteration of the difference equation.

Next, we consider a more systematic method for searching for a change of dependent variable that may linearize an equation. This technique is based on Lie's transformation group method (see Maeda [173]). To make the discussion as clear as possible, we restrict our attention to the equation

$$y(t+1) = f(y(t)), \qquad (3.21)$$

in which there is no explicit t dependence.

Let us begin by assuming that a solution $\xi(y)$ of the functional equation

$$D\xi(f(y)) = \xi(y)\frac{df}{dy}(y) \qquad (3.22)$$

is known for some constant D. Then define a new dependent variable z by

$$\frac{dz}{dy} = \frac{1}{\xi(y)} \qquad (3.23)$$

for y belonging to an open interval I in which $\xi(y)$ is different from 0.

3.6 Nonlinear Equations That Can Be Linearized

Using the chain rule, we have

$$\frac{d}{dy}z(f(y)) = \frac{dz}{dy}(f(y))\frac{df}{dy}(y)$$

$$= \frac{1}{\xi(f(y))}\frac{df}{dy}(y) \quad \text{(by Eq. (3.23))}$$

$$= \frac{D}{\xi(y)} \quad \text{(by Eq. (3.22))}$$

$$= D\frac{dz}{dy}(y) \quad \text{(by Eq. (3.23))}.$$

Now integrate to obtain

$$z(f(y)) = Dz(y) + C$$

or

$$z(y(t+1)) = Dz(y(t)) + C,$$

which is a linear equation of first order with constant coefficients!

In summary, we have shown that if $\xi(y)$ satisfies Eq. (3.22) for some constant D and if $\xi(y) \neq 0$ for y in an open interval I, then the change of variable given by Eq. (3.23) produces a first order linear equation which is equivalent to Eq. (3.21) as long as y is in I.

Example 3.29. $y(t+1) = ay(t)(1 - y(t))$.

In biology this equation is known as the discrete logistic model. Such models occur in the study of populations which reproduce at discrete intervals, such as once a year. Here a is a constant and $f(y) = ay(1-y)$, so Eq. (3.22) is

$$D\xi(ay(1-y)) = \xi(y)a(1-2y).$$

The form of this last equation suggests that we try a linear expression for ξ, say $\xi(y) = cy + d$. We obtain

$$-Dcay^2 + Dcay + Dd = -2acy^2 + (ac - 2ad)y + ad.$$

Equating coefficients leads to

$$D = a = 2 \text{ and } c = -2d.$$

Let $c = -1$ and $d = \frac{1}{2}$; from Eq. (3.23)

$$\frac{dz}{dy} = \frac{1}{-y + \frac{1}{2}},$$

so we take

$$z = -\log(\frac{1}{2} - y)$$

or

$$y = \frac{1}{2} - e^{-z}.$$

Now substitute the last expression into

$$y(t+1) = 2y(t)(1 - y(t))$$

to obtain

$$\frac{1}{2} - e^{-z(t+1)} = 2\left(\frac{1}{2} - e^{-z(t)}\right)\left(e^{-z(t)} + \frac{1}{2}\right)$$

or

$$e^{-z(t+1)} = 2e^{-2z(t)}$$

or

$$z(t+1) = 2z(t) - \ln 2.$$

3.6 Nonlinear Equations That Can Be Linearized

Then

$$z(t) = C \cdot 2^t + \ln 2,$$

and finally

$$y(t) = \frac{1}{2}(1 - A^{2^t}),$$

where A is arbitrary.

Note that the choice of a linear ξ leads to a general solution only for the case $a = 2$.

An alternate approach is to start with a particular ξ in Eq.(3.22) and to solve Eq. (3.22) for f to discover which equations can be linearized by the change of variable Eq. (3.23). In this way, it is possible to catalog many nonlinear equations that are equivalent to first order linear equations with constant coefficients.

Example 3.30. Choose $\xi(y) = \sqrt{y(1-y)}$ in Eq. (3.22):

$$D\sqrt{f(1-f)} = \sqrt{y(1-y)}\frac{df}{dy}.$$

This first order differential equation can be solved by separation of variables:

$$D\int \frac{dy}{\sqrt{y(1-y)}} = \int \frac{df}{\sqrt{f(1-f)}}$$

or

$$D \cdot 2\sin^{-1}\sqrt{y} + C^* = 2\sin^{-1}\sqrt{f}.$$

Then

$$f(y) = \sin^2\left(D\sin^{-1}\sqrt{y} + C\right),$$

where C is an arbitrary constant.

If $D = 1$, we obtain the family of functions

$$f(y) = \left(\sqrt{y}\cos C + \sqrt{1-y}\sin C\right)^2.$$

For $D = 2$, we have

$$f(y) = \left(2\sqrt{y}\sqrt{1-y}\cos C + (1-2y)\sin C\right)^2,$$

and the choice $C = 0$ gives

$$f(y) = 4y(1-y),$$

which is the function in Example 3.25 with $a = 4$! Other values of D lead to more complicated expressions.

Let's solve

$$y(t+1) = 4y(t)\left(1 - y(t)\right).$$

From Eq. (3.23), $z = 2\sin^{-1}\sqrt{y}$, so $y = \sin^2 \frac{z}{2}$. This change of variable in the difference equation results in

$$\sin^2 \frac{z(t+1)}{2} = 4\sin^2 \frac{z(t)}{2}\cos^2 \frac{z(t)}{2},$$
$$= \sin^2 z(t)$$

or

$$z(t+1) = 2z(t).$$

Then $z(t) = A \cdot 2^t$, so

$$y(t) = \sin^2 B \cdot 2^t,$$

where B is an arbitrary constant. Of course, this solution is valid only for $0 \leq y \leq 1$.

Additional examples using Eq. (3.22), as well as a generalization, are contained in the exercises.

3.7 The z-Transform

The z-transform is a mathematical device similar to a generating function which provides an alternate method for solving linear difference equations, as well as certain summation equations. In this section we will define the z-transform, derive several of its properties, and consider an application. The z-transform is very important in the analysis and design of digital control systems. Jury [140] is a good source of information on this topic.

The z-transform of a sequence $\{y_k\}$ is a function $Y(z)$ of a complex variable defined by

$$Y(z) = Z(y_k) = \sum_{k=0}^{\infty} \frac{y_k}{z^k}$$

for those values of z for which the series converges. Note that we should write $Z(\{y_k\})$, but we will use the abbreviated form $Z(y_k)$. Also note that the generating function for y_k can be obtained from the z-transform of y_k by the substitution $z = \frac{1}{x}$.

Example 3.31. Find the z-transform of the sequence $\{y_k = 1\}$.

$$\begin{aligned} Y(z) = Z(1) &= \sum_{k=0}^{\infty} \frac{1}{z^k} \\ &= \frac{1}{1 - z^{-1}} \\ &= \frac{z}{z-1}, \quad |z| > 1. \end{aligned}$$

Example 3.32. Find the z-transform of the sequence $\{u_k = a^k\}$.

$$\begin{aligned} U(z) = Z(a^k) &= \sum_{k=0}^{\infty} \frac{a^k}{z^k} \\ &= \sum_{k=0}^{\infty} \left(\frac{a}{z}\right)^k = \frac{1}{1 - \frac{a}{z}} \\ &= \frac{z}{z-a}, \quad |z| > |a|. \end{aligned}$$

Example 3.33. Find the z-transform of $\{v_k = k\}_{k=0}^{\infty}$.

$$\begin{aligned} V(z) = Z(k) &= \sum_{k=0}^{\infty} \frac{k}{z^k} \\ &= \sum_{k=0}^{\infty} \frac{k+1}{z^{k+1}} \\ &= \frac{1}{z} \sum_{k=0}^{\infty} \frac{k+1}{z^k} \\ &= \frac{1}{z} V(z) + \frac{1}{z} Z(1), \quad |z| > 1, \end{aligned}$$

so by rearrangement

$$\begin{aligned} \frac{z-1}{z} V(z) &= \frac{1}{z} \frac{z}{z-1}, \quad |z| > 1, \\ V(z) &= \frac{z}{(z-1)^2}, \quad |z| > 1. \end{aligned}$$

These formulas for z-transforms along with some others are collected in Table 1 at the end of this section. Of course, this table is easily converted into a table of generating functions by the substitution $z = \frac{1}{x}$.

Theorem 3.10. (Linearity Theorem) If a and b are constants, then

$$Z(au_k + bv_k) = aZ(u_k) + bZ(v_k)$$

for those z in the common domain of $U(z)$ and $V(z)$.

Proof. Simply compute

$$\begin{aligned} Z(au_k + bv_k) &= \sum_{k=0}^{\infty} \frac{au_k + bv_k}{z^k} \\ &= a \sum_{k=0}^{\infty} \frac{u_k}{z^k} + b \sum_{k=0}^{\infty} \frac{v_k}{z^k} \\ &= aZ(u_k) + bZ(v_k). \quad \blacksquare \end{aligned}$$

3.7 The z-Transform

Example 3.34. Find the z-transform of $\{\sin ak\}_{k=0}^{\infty}$.

The following calculation makes use of the Linearity Theorem:

$$\begin{aligned}
Z(\sin ak) &= Z\left(\frac{1}{2i}e^{iak} - \frac{1}{2i}e^{-iak}\right) \\
&= \frac{1}{2i}Z(e^{iak}) - \frac{1}{2i}Z(e^{-iak}) \\
&= \frac{1}{2i}\frac{z}{z - e^{ia}} - \frac{1}{2i}\frac{z}{z - e^{-ia}} \\
&= \frac{z^2 - ze^{-ia} - z^2 + ze^{ia}}{2i[z^2 - (e^{ia} + e^{-ia})z + 1]} \\
&= \frac{z \sin a}{z^2 - 2(\cos a)z + 1}.
\end{aligned}$$

Similarly, one can show

$$Z(\cos ak) = \frac{z^2 - z \cos a}{z^2 - 2z \cos a + 1}.$$

Theorem 3.11. If $Y(z) = Z(y_k)$ for $|z| > r$, then

$$Z\left((k+n-1)^{(n)} y_k\right) = (-1)^n z^n \frac{d^n Y}{dz^n}(z)$$

for $|z| > r$.

Proof.

By definition

$$Y(z) = \sum_{k=0}^{\infty} \frac{y_k}{z^k} = \sum_{k=0}^{\infty} y_k z^{-k}$$

for $|z| > r$. The nth derivative is

$$\frac{d^n Y}{dz^n}(z) = (-1)^n \sum_{k=0}^{\infty} k(k+1)\cdots(k+n-1) y_k z^{-k-n}$$

$$= \frac{(-1)^n}{z^n} \sum_{k=0}^{\infty} \frac{(k+n-1)^{(n)} y_k}{z^k}.$$

Hence

$$Z\left((k+n-1)^{(n)} y_k\right) = (-1)^n z^n \frac{d^n Y}{dz^n}(z). \qquad \blacksquare$$

For $n = 1$ in Theorem 3.11 we get the special case

$$Z(k y_k) = -z Y'(z).$$

Example 3.35. Find $Z(k a^k)$.

$$Z(k a^k) = -z \frac{d}{dz} Z(a^k)$$
$$= -z \frac{d}{dz} \left(\frac{z}{z-a}\right)$$
$$= \frac{az}{(z-a)^2}.$$

Example 3.36. Find $Z(k^2)$.

$$Z(k^2) = Z(k \cdot k)$$
$$= -z \frac{d}{dz} Z(k)$$
$$= -z \frac{d}{dz} \left[\frac{z}{(z-1)^2}\right]$$
$$= \frac{z(z+1)}{(z-1)^3}.$$

3.7 The z-Transform

Define the unit step sequence $u(n)$ by

$$u_k(n) = \begin{cases} 0, & 0 \leq k \leq n-1 \\ 1, & n \leq k. \end{cases}$$

Note that the unit step sequence has a single "step" of unit height located at $k = n$.

The following result is known as a "shifting theorem."

Theorem 3.12. For n a positive integer

$$Z(y_{k+n}) = z^n Z(y_k) - \sum_{m=0}^{n-1} y_m z^{n-m},$$

$$Z(y_{k-n} u_k(n)) = z^{-n} Z(y_k).$$

In Fig. 3.6 the various sequences used in this theorem are illustrated.

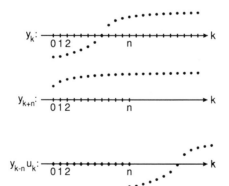

Fig. 3.6 Sequences used in Theorem 3.12

Proof. First observe that

$$Z(y_{k+n}) = \sum_{k=0}^{\infty} y_{k+n} z^{-k}$$

$$= \sum_{k=n}^{\infty} y_k z^{-k+n}$$

$$= z^n \left[\sum_{k=0}^{\infty} y_k z^{-k} - \sum_{m=0}^{n-1} y_m z^{-m} \right]$$

$$= z^n Z(y_k) - \sum_{m=0}^{n-1} y_m z^{n-m}.$$

For the second part, we have

$$Z(y_{n-k} u_k(n)) = \sum_{k=0}^{\infty} y_{k-n} u_k(n) z^{-k}$$

$$= \sum_{k=n}^{\infty} y_{k-n} z^{-k}$$

$$= \sum_{k=0}^{\infty} y_k z^{-k-n}$$

$$= z^{-n} Z(y_k). \blacksquare$$

Example 3.37. Find $Z(u_k(n))$.

Theorem 3.12 gives immediately that

$$Z(u_k(n)) = z^{-n} Z(1)$$

$$= \frac{z^{1-n}}{z-1}.$$

Example 3.38. Find $Z(y_k)$ if $y_k = 2$, $0 \le k \le 99$, $y_k = 5$, $100 \le k$.

3.7 The z-Transform

$$\begin{aligned} Z(y_k) &= Z\left(2 + 3u_k(100)\right) \\ &= \frac{2z}{z-1} + \frac{3z^{-99}}{z-1} \\ &= \frac{2z^{100}+3}{z^{99}(z-1)}. \end{aligned}$$

Theorem 3.13. For any integer $n \geq 0$

$$Z\left((k+n-1)^{(n)}\right) = \frac{n!z^n}{(z-1)^{n+1}}$$

$$Z\left(k^{(n)}\right) = \frac{n!z}{(z-1)^{n+1}}$$

for $|z| > 1$.

Proof. Letting $y_k = 1$ in Theorem 3.11 we have

$$\begin{aligned} Z\left((k+n-1)^{(n)}\right) &= (-1)^n z^n \frac{d^n}{dz^n} \frac{z}{(z-1)} \\ &= (-1)^n z^n \frac{(-1)^n n!}{(z-1)^{n+1}} \\ &= \frac{n!z^n}{(z-1)^{n+1}}, \end{aligned}$$

which is the first formula. Now using Theorem 3.12 we get

$$Z\left((k+n-1)^{(n)}\right) = z^{n-1} Z(k^{(n)}) + \sum_{m=0}^{n-2} m^{(n)} z^{n-1-m}.$$

Hence

$$Z\left(k^{(n)}\right) = \frac{1}{z^{n-1}} \left(\frac{n!z^n}{(z-1)^{n+1}}\right)$$

$$= \frac{n!z}{(z-1)^{n+1}}. \qquad \blacksquare$$

Theorem 3.14. (Initial value and final value theorem)

(a) If $Y(z)$ exists for $|z| > r$, then

$$y_0 = \lim_{z \to \infty} Y(z).$$

(b) If $Y(z)$ exists for $|z| > 1$ and $(z-1)Y(z)$ is analytic at $z = 1$, then

$$\lim_{k \to \infty} y_k = \lim_{z \to 1} (z-1)Y(z).$$

Proof. Part (a) follows immediately from the definition of the z-transform. To prove part (b), consider

$$\begin{aligned} Z(y_{k+1} - y_k) &= \sum_{k=0}^{\infty} y_{k+1} z^{-k} - \sum_{k=0}^{\infty} y_k z^{-k} \\ &= \lim_{n \to \infty} \left[\sum_{k=0}^{n} y_{k+1} z^{-k} - \sum_{k=0}^{n} y_k z^{-k} \right] \\ &= \lim_{n \to \infty} \left[-y_0 + y_1(1 - z^{-1}) + y_2(z^{-1} - z^{-2}) \right. \\ &\quad \left. + \cdots + y_n \left(z^{-n+1} - z^{-n} \right) + y_{n+1} z^{-n} \right]. \end{aligned}$$

Thus,

$$\lim_{z \to 1} \left(Z(y_{k+1}) - Z(y_k) \right) = \lim_{n \to \infty} (y_{n+1} - y_0).$$

From the shifting theorem

$$\lim_{z \to 1} \left[zY(z) - zy_0 - Y(z) \right] = \lim_{k \to \infty} y_k - y_0.$$

3.7 The z-Transform

Hence

$$\lim_{k\to\infty} y_k = \lim_{z\to 1}(z-1)Y(z).$$ ∎

Example 3.39. Verify directly the last theorem for the sequence $y_k = 1$.

$$1 = y_0 = \lim_{z\to\infty} Z(1) = \lim_{z\to\infty} \frac{z}{z-1} = 1,$$
$$1 = \lim_{z\to 1} y_k = \lim_{z\to 1}(z-1)Z(1)$$
$$= \lim_{z\to 1} \frac{(z-1)z}{z-1} = 1.$$

Theorem 3.15. If $Z(y_k) = Y(z)$ for $|z| > r$, then for constants $a \neq 0$,

$$Z(a^k y_k) = Y\left(\frac{z}{a}\right)$$

for $|z| > r|a|$.

Proof. Observe that

$$Z(a^k y_k) = \sum_{k=1}^{\infty} \frac{a^k y_k}{z^k} = \sum_{k=1}^{\infty} y_k \left(\frac{z}{a}\right)^{-k}$$
$$= Y\left(\frac{z}{a}\right).$$ ∎

Example 3.40. Find $Z(3^k \sin 4k)$.

$$Z(3^k \sin 4k) = Z(\sin 4k)]_{\frac{z}{3}}$$
$$= \frac{\frac{z}{3}\sin 4}{\frac{z^2}{9} - 2(\cos 4)\frac{z}{3} + 1}$$
$$= \frac{3z \sin 4}{z^2 - 6z \cos 4 + 9}.$$

Chapter 3. Linear Difference Equations

Example 3.41. Solve the following initial value problem using z-transforms:

$$y_{k+1} - 3y_k = 4,$$
$$y_0 = 1.$$

Taking the z-transform of both sides of the difference equation, we have

$$zY(z) - zy_0 - 3Y(z) = 4\frac{z}{z-1}$$

$$(z-3)Y(z) = z + \frac{4z}{z-1} = \frac{z^2 + 3z}{z-1}$$

$$Y(z) = z \cdot \frac{z+3}{(z-1)(z-3)}$$

$$= z\left[\frac{-2}{z-1} + \frac{3}{z-3}\right]$$

$$= -2\frac{z}{z-1} + 3\frac{z}{z-3}.$$

From Table 1 we find the solution

$$y_k = -2 + 3^{k+1}.$$

Example 3.42. Solve the initial value problem

$$y_{k+1} - 3y_k = 3^k,$$
$$y_0 = 2.$$

Since 3^k is a solution of the homogeneous equation, we expect the

3.7 The z-Transform

solution of this problem to involve the function $k3^k$.

$$zY(z) - 2z - 3Y(z) = \frac{z}{z-3}$$

$$(z-3)Y(z) = 2z + \frac{z}{z-3} = \frac{2z^2 - 5z}{z-3}$$

$$Y(z) = z\frac{2z-5}{(z-3)^2}$$

$$= z\left[\frac{2}{z-3} + \frac{1}{(z-3)^2}\right]$$

$$= 2\frac{z}{z-3} + \frac{1}{3}\frac{3z}{(z-3)^2}.$$

Then

$$y_k = 2 \cdot 3^k + \frac{1}{3}k3^k.$$

We can use the z-transform to solve some difference equations with variable coefficients.

Example 3.43. Solve the initial value problem

$$(k+1)y_{k+1} - (50-k)y_k = 0, \quad y_0 = 1.$$

Taking the z-transform of both sides,

$$zZ(ky_k) - 50Y(z) - zY'(z) = 0$$
$$-z^2Y'(z) - zY'(z) = 50Y(z)$$
$$\frac{Y'(z)}{Y(z)} = \frac{-50}{z(z+1)}$$
$$= -\frac{50}{z} + \frac{50}{z+1}$$
$$\log Y(z) = -50\log z + 50\log(z+1) + C$$
$$Y(z) = \left(\frac{z+1}{z}\right)^{50}.$$

By Exercise 3.107(b),

$$y_k = \binom{50}{k}.$$

Example 3.44. Solve the second order initial value problem

$$y_{k+2} + y_k = 10 \cdot 3^k$$
$$y_0 = 0, \quad y_1 = 0.$$

By the shifting theorem

$$z^2 Z(y_k) - y_0 z^2 - y_1 z + Z(y_k) = \frac{10z}{z-3}$$
$$(z^2 + 1) Z(y_k) = \frac{10z}{z-3},$$

so we have

$$\begin{aligned}
Z(y_k) &= \frac{10z}{(z-3)(z^2+1)} \\
&= z \left[\frac{A}{z-3} + \frac{Bz+C}{z^2+1} \right] \\
&= z \left[\frac{1}{z-3} - \frac{z+3}{z^2+1} \right] \\
&= \frac{z}{z-3} - \frac{z^2}{z^2+1} - 3\frac{z}{z^2+1} \\
&= \frac{z}{z-3} - \frac{z^2 - z\cos\frac{\pi}{2}}{z^2 - 2z\cos\frac{\pi}{2} + 1} - 3\frac{z\sin\frac{\pi}{2}}{z^2 - 2z\cos\frac{\pi}{2} + 1}.
\end{aligned}$$

Hence

$$y_k = 3^k - \cos(\frac{\pi}{2}k) - 3\sin(\frac{\pi}{2}k).$$

3.7 The z-Transform

Example 3.45. Solve the system

$$u_{k+1} - v_k = 3k3^k,$$
$$u_k + v_{k+1} - 3v_k = k3^k,$$
$$u_0 = 0, \quad v_0 = 3.$$

By Theorem 3.12

$$zU(z) - zu_0 - V(z) = \frac{9z}{(z-3)^2},$$
$$U(z) + zV(z) - zv_0 - 3V(z) = \frac{3z}{(z-3)^2},$$

or

$$zU(z) - V(z) = \frac{9z}{(z-3)^2},$$
$$U(z) + (z-3)V(z) = 3z + \frac{3z}{(z-3)^2}.$$

Multiplying both sides of the first equation by $z - 3$ and adding, we get

$$(z^2 - 3z + 1)U(z) = 3z + \frac{3z}{(z-3)^2} + \frac{9z^2 - 27z}{(z-3)^2}$$
$$= \frac{3z(z^2 - 3z + 1)}{(z-3)^2}$$
$$U(z) = \frac{3z}{(z-3)^2},$$

and one of the unknowns is given by

$$u_k = k3^k.$$

120 Chapter 3. Linear Difference Equations

Furthermore, the other unknown is

$$\begin{aligned} v_k &= u_{k+1} - 3k3^k \\ &= (k+1)3^{k+1} - 3k3^k \\ &= 3^{k+1}. \end{aligned}$$

Example 3.46. Find the currents i_k, $0 \leq k \leq n$, in the ladder network shown in Fig. 3.7.

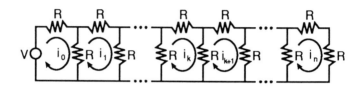

Fig. 3.7 A ladder network

We begin by applying Kirchhoff's law to the initial loop in Fig. 3.7:

$$V = Ri_0 + R(i_0 - i_1).$$

Solving for i_1 we obtain

$$i_1 = 2i_0 - \frac{V}{R}.$$

Now in general we apply Kirchhoff's law to the loop corresponding to i_{k+1} and obtain

$$R(i_{k+1} - i_{k+2}) + R(i_{k+1} - i_k) + Ri_{k+2} = 0.$$

Simplifying, we have

$$i_{k+2} - 3i_{k+1} + i_k = 0,$$

3.7 The z-Transform

for $0 \leq k \leq n-2$. If we apply the z-transform to both sides of the preceding equation, we get

$$(z^2 I(z) - z^2 i_0 - z i_1) - 3(z I(z) - z i_0) + I(z) = 0$$

or

$$(z^2 - 3z + 1)I(z) = i_0 z^2 + (i_1 - 3i_0)z.$$

Using the equation for i_1, we have

$$I(z) = i_0 \frac{z^2 - \left(1 + \frac{V}{i_0 R}\right)z}{z^2 - 3z + 1}.$$

Let a be the positive solution of $\cosh a = \frac{3}{2}$; then $\sinh a = \frac{\sqrt{5}}{2}$. Note that

$$I(z) = i_0 \frac{z^2 - z\cosh(a)}{z^2 - 2z\cosh(a) + 1} + \left(\frac{i_0}{2} + \frac{V}{R}\right) \frac{2}{\sqrt{5}} \frac{z\sinh(a)}{z^2 - 2z\cosh(a) + 1}.$$

It follows that

$$i_k = i_0 \cosh(ak) + \left(\frac{i_0}{2} + \frac{V}{R}\right) \frac{2}{\sqrt{5}} \sinh(ak),$$

$0 \leq k \leq n$. Using Kirchhoff's law for the last loop in Fig. 3.6, we get that $i_{n-1} = 3i_n$. This additional equation uniquely determines i_0 and hence all the i_k's for $0 \leq k \leq n$.

We now define the unit impulse sequence $\delta(n)$, $n \geq 1$, by

$$\delta_k(n) = \begin{cases} 1 & k = n \\ 0 & k \neq n. \end{cases}$$

It follows immediately from the definition of the z-transform that

$$Z(\delta_k(n)) = \frac{1}{z^n}.$$

Example 3.47. Solve the initial value problem

$$y_{k+1} - 2y_k = 3\delta_k(4), \quad y_0 = 1.$$

Taking the z-transform of both sides we have

$$zY(z) - z - 2Y(z) = \frac{3}{z^4}$$

$$(z-2)Y(z) = z + \frac{3}{z^4}$$

$$Y(z) = \frac{z}{z-2} + \frac{3}{z^4(z-2)}$$

$$= \frac{z}{z-2} + 3z^{-5}\frac{z}{z-2}.$$

An application of the inverse z-transform results in

$$y_k = 2^k + 3 \cdot 2^{k-5} u_k(5).$$

We could also write this in the form

$$y_k = \begin{cases} 2^k, & 0 \leq k \leq 4 \\ 2^k + 3 \cdot 2^{k-5}, & k \geq 5. \end{cases}$$

We define the convolution of two sequences $\{u_k\}$ and $\{v_k\}$ by

$$\{u_k\} * \{v_k\} = \left\{ \sum_{m=0}^{k} u_{k-m} v_m \right\}.$$

Briefly we write

$$u_k * v_k = \sum_{m=0}^{k} u_{k-m} v_m.$$

3.7 The z-Transform

Theorem 3.16. (Convolution Theorem)

If $U(z)$ exits for $|z| > a$ and $V(z)$ exists for $|z| > b$, then

$$Z(u_k * v_k) = U(z)V(z)$$

for $|z| > \max\{a,b\}$.

Proof. For $|z| > \max\{a,b\}$,

$$\begin{aligned}
U(z)V(z) &= \sum_{k=0}^{\infty} \frac{u_k}{z^k} \sum_{k=0}^{\infty} \frac{v_k}{z^k} \\
&= \sum_{k=0}^{\infty} \sum_{m=0}^{k} \frac{u_{k-m} v_m}{z^k} \\
&= Z(u_k * v_k).
\end{aligned}$$ ∎

Since $\sum_{m=0}^{k} y_m = 1 * y_k$, Theorem 3.14 gives us

$$\begin{aligned}
Z\left(\sum_{m=0}^{k} y_m\right) &= Z(1) Z(y_k) \\
&= \frac{z}{z-1} Z(y_k).
\end{aligned}$$

Corollary 3.3. If $Z(y_k)$ exists for $|z| > r$, then

$$Z\left(\sum_{m=0}^{k} y_m\right) = \frac{z}{z-1} Z(y_k)$$

for $|z| > \max\{1, r\}$.

Example 3.48. Find

$$Z\left(\sum_{m=0}^{k} 3^m\right).$$

By Corollary 3.3,

$$Z\left(\sum_{m=0}^{k} 3^m\right) = \frac{z}{z-1} Z(y_k),$$

$$= \frac{z^2}{(z-1)(z-3)}, \quad (|z| > 3).$$

Now consider the Volterra summation equation of convolution type

$$y_k = f_k + \sum_{m=0}^{k-1} u_{k-m-1} y_m \quad (k \geq 0), \tag{3.24}$$

where f_k and u_{k-m-1} are given. The term u_{k-m-1} is called the kernel of the summation equation. The equation is said to be homogeneous if $f_k \equiv 0$ and nonhomogeneous otherwise. Such an equation can often be solved by use of the z-transform.

To see this, replace k by $k+1$ in Eq. (3.24) to get

$$y_{k+1} = f_{k+1} + \sum_{m=0}^{k} u_{k-m} y_m$$

or

$$y_{k+1} = f_{k+1} + u_k * y_k.$$

Taking the z-transform of both sides and using the fact that $y_0 = f_0$, we have

$$zY(z) = zF(z) + V(z)Y(z).$$

Hence

$$Y(z) = \frac{zF(z)}{z - V(z)}.$$

The desired solution y_k is then obtained if we can compute the inverse

3.7 The z-Transform

transform. The next example is of this type.

Example 3.49. Solve the Volterra summation equation

$$y_k = 1 + 16 \sum_{m=0}^{k-1}(k - m - 1)y_m, \quad k \geq 0.$$

Replacing k by $k+1$ we have

$$y_{k+1} = 1 + 16 \sum_{m=0}^{k}(k - m)y_m$$
$$= 1 + 16k * y_k.$$

Taking the z-transform of both sides, we obtain $Y(z)$ as follows:

$$zY(z) - z = \frac{z}{z-1} + 16\frac{z}{(z-1)^2}Y(z)$$

$$\left[1 - \frac{16}{(z-1)^2}\right]Y(z) = 1 + \frac{1}{z-1}$$

$$\frac{z^2 - 2z - 15}{(z-1)^2}Y(z) = \frac{z}{z-1}$$

$$Y(z) = \frac{z(z-1)}{(z-5)(z+3)}$$

$$= z\left[\frac{\frac{1}{2}}{z-5} + \frac{\frac{1}{2}}{z+3}\right]$$

$$= \frac{1}{2}\frac{z}{z-5} + \frac{1}{2}\frac{z}{z+3}.$$

Then

$$y_k = \frac{1}{2}5^k + \frac{1}{2}(-3)^k.$$

A related equation is the Fredholm summation equation

$$y_k = f_k + \sum_{m=a}^{b} K_{k,m} y_m \quad (a \leq k \leq b). \tag{3.25}$$

Here a and b are integers, and the kernel $K_{k,m}$ and the sequence f_k are given. Since this equation is actually a linear system of $b - a + 1$ equations in $b - a + 1$ unknowns $y_a, \ldots, y_b$, it can be solved by matrix methods. If $b - a$ is large, this might not be the best way to solve this equation. If $K_{k,m}$ is separable, then the following procedure may yield a more efficient method of solution.

We say $K_{k,m}$ is separable provided

$$K_{k,m} = \sum_{i=1}^{p} \alpha_i(k)\beta_i(m), \quad (a \leq k, m \leq b).$$

Substituting this expression into Eq. (3.25) we obtain

$$y_k = f_k + \sum_{i=1}^{p} \alpha_i(k) \left(\sum_{m=a}^{b} \beta_i(m) y_m \right).$$

Hence

$$y_k = f_k + \sum_{i=1}^{p} c_i \alpha_i(k), \quad a \leq k \leq b, \tag{3.26}$$

where

$$c_i = \sum_{m=a}^{b} \beta_i(m) y_m.$$

By multiplying both sides of Eq. (3.26) by $\beta_i(k)$ and summing

3.7 The z-Transform

from a to b, we obtain

$$\sum_{k=a}^{b} \beta_i(k) y_k = \sum_{k=a}^{b} \beta_i(k) f_k + \sum_{i=1}^{p} c_i \left(\sum_{k=a}^{b} \alpha_i(k) \beta_j(k) \right).$$

Hence

$$c_j = u_j + \sum_{i=1}^{p} a_{ji} c_i, \quad 1 \leq j \leq p, \tag{3.27}$$

where

$$u_j = \sum_{k=a}^{b} f_k \beta_j(k)$$

and

$$a_{ij} = \sum_{k=a}^{b} \alpha_j(k) \beta_i(k).$$

Let A be the p by p matrix $A = (a_{ij})$, let $\vec{c} = [c_1, \cdots, c_p]^T$, and $\vec{u} = [u_1, \cdots, u_p]^T$. Then Eq. (3.27) becomes

$$\vec{c} = \vec{u} + A\vec{c}.$$

But this equation is equivalent to

$$(I - A)\vec{c} = \vec{u}, \tag{3.28}$$

where I is the p by p identity matrix. We have essentially proved the following theorem.

Theorem 3.17. The Fredholm equation (3.25) with a separable kernel has a solution y_k if and only if Eq. (3.28) has a solution $\vec{c}$. If $\vec{c} = (c_1, \cdots, c_p)^T$ is a solution of Eq. (3.28) then a corresponding solution y_k of Eq. (3.25) is given by Eq. (3.26).

Example 3.50. Solve the Fredholm summation equation

$$y_k = 1 + \sum_{m=0}^{19}(1+km)y_m, \quad 0 \le k \le 19.$$

Here we have the separable kernel

$$K_{k,m} = 1 + km.$$

Take

$$\alpha_1(k) = 1, \ \beta_1(m) = 1,$$
$$\alpha_2(k) = k, \ \beta_2(m) = m.$$

Then

$$a_{11} = \sum_{k=0}^{19} 1 = 20$$

$$a_{12} = a_{21} = \sum_{k=0}^{19} k = 190$$

$$a_{22} = \sum_{k=0}^{19} k^2 = 2,470.$$

Furthermore,

$$u_1 = \sum_{k=0}^{19} 1 = 20$$

$$u_2 = \sum_{k=0}^{19} k = 190.$$

3.7 The z-Transform

Equation (3.28) in this case is

$$\begin{bmatrix} -19 & -190 \\ -190 & -2,469 \end{bmatrix} \begin{bmatrix} c_1 \\ c_2 \end{bmatrix} = \begin{bmatrix} 20 \\ 190 \end{bmatrix}.$$

Solving for c_1 and c_2 we obtain

$$c_1 = \frac{-13,280}{10,811}, \quad c_2 = \frac{10}{569}.$$

From Theorem 3.17 we obtain the unique solution

$$\begin{aligned} y_k &= 1 + \frac{-13,280}{10,811} \cdot 1 + \frac{10}{589} \cdot k \\ &= \frac{2,469}{10,811} + \frac{10}{569} k, \quad 0 \le k \le 19. \end{aligned}$$

Example 3.51. Solve the Fredholm summation equation

$$y_k = 2 + \lambda \sum_{m=0}^{29} \frac{m}{29} y_m, \quad 0 \le k \le 29,$$

for all values of λ.

Take

$$\alpha_1(k) = \lambda, \quad \beta_1(m) = \frac{m}{29};$$

then

$$a_{11} = \sum_{m=0}^{29} \lambda \frac{m}{29} = 15\lambda,$$

$$u_1 = \sum_{m=0}^{29} \frac{2m}{29} = 60.$$

Hence Eq. (3.28) is

$$(1 - 15\lambda)c = 60.$$

For $\lambda = \frac{1}{15}$ there is no solution of this summation equation. For $\lambda \neq \frac{1}{15}$, $c = \frac{60}{1-15\lambda}$. The corresponding solution is

$$y_k = 2 + \frac{60\lambda}{1 - 15\lambda}, \quad 0 \leq k \leq 29.$$

Now consider the homogeneous Fredholm equation

$$y_k = \lambda \sum_{m=a}^{b} K_{k,m} y_m, \quad a \leq k \leq b, \qquad (3.29)$$

where λ is a parameter. We say that λ_0 is an eigenvalue of this equation provided for this value of λ there is a nontrivial solution y_k, called an eigensequence. We say (λ_0, y_k) is an eigenpair for Eq. (3.29). Note that $\lambda = 0$ is not an eigenvalue. We say that $K_{k,m}$ is symmetric provided

$$K_{k,m} = K_{m,k}$$

for $a \leq k, m \leq b$. Several properties of eigenpairs for Eq. (3.29) with a symmetric kernel are given in the following theorem.

Theorem 3.18. If $K_{k,m}$ is real and symmetric, then all the eigenvalues of Eq. (3.29) are real. If (λ_i, u_k) (λ_j, v_k) are eigenpairs with $\lambda_i \neq \lambda_j$, then u_k and v_k are orthogonal, i.e.,

$$\sum_{k=a}^{b} u_k v_k = 0.$$

Corresponding to each eigenvalue we can always pick a real eigensequence.

Proof. Let (μ, u_k), (ν, v_k) be eigenpairs of Eq. (3.29). Then $\mu, \nu \neq 0$.

3.7 The z-Transform

Since (μ, u_k) is an eigenpair for Eq. (3.29),

$$u_k = \mu \sum_{m=a}^{b} K_{k,m} u_m.$$

Multiplying by v_k and summing from a to b, we obtain

$$\begin{aligned}\sum_{k=a}^{b} u_k v_k &= \mu \sum_{k=a}^{b} \sum_{m=a}^{b} K_{k,m} u_m v_k \\ &= \mu \sum_{m=a}^{b} \left(\sum_{k=a}^{b} K_{m,k} v_k \right) u_m \\ &= \frac{\mu}{\nu} \sum_{m=a}^{b} v_m u_m,\end{aligned}$$

since (ν, v_k) is an eigenpair for Eq. (3.29). It follows that

$$(\nu - \mu) \sum_{k=a}^{b} u_k v_k = 0. \tag{3.30}$$

If $\mu \neq \nu$, then we get the orthogonality result

$$\sum_{k=a}^{b} u_k v_k = 0.$$

If (λ_i, y_k) is an eigenpair of Eq. (3.29), then $(\overline{\lambda}_i, \overline{y}_k)$ is an eigenpair of Eq. (3.29). With $(\mu, u_k) = (\lambda_i, y_k)$ and $(\nu, v_k) = (\overline{\lambda}_i, \overline{y}_k)$, Eq. (3.30) becomes

$$(\overline{\lambda} - \lambda) \sum_{k=a}^{b} y_k \overline{y}_k = 0.$$

It follows that $\lambda = \overline{\lambda}$, and hence every eigenvalue of Eq. (3.29) is real. The last statement of the Theorem is left as an exercise. ∎

Table 1. z-Transforms

Sequence	z – transform
1	$\frac{z}{z-1}$
a^k	$\frac{z}{z-a}$
k	$\frac{z}{(z-1)^2}$
k^2	$\frac{z(z+1)}{(z-1)^3}$
$k^{(n)}$	$\frac{n!z}{(z-1)^{n+1}}$
$\sin ak$	$\frac{z \sin a}{z^2 - 2z \cos a + 1}$
$\cos ak$	$\frac{z^2 - z \cos a}{z^2 - 2z \cos a + 1}$
$\sinh ak$	$\frac{z \sinh a}{z^2 - 2z \cosh a + 1}$
$\cosh ak$	$\frac{z^2 - z \cosh a}{z^2 - 2z \cosh a + 1}$
$\delta_k(n)$	$\frac{1}{z^n}$
$u_k(n)$	$\frac{z^{1-n}}{z-1}$
ky_k	$-zY'(z)$
$u_k * v_k$	$U(z)V(z)$
$\sum_{m=0}^{k} y_i$	$\frac{z}{z-1}Y(z)$
$a^k y_k$	$Y\left(\frac{z}{a}\right)$
y_{k+n}	$z^n Y(z) - \sum_{m=0}^{n-1} y_m z^{n-m}$
$y_{k-n} u_k(n)$	$z^{-n} Y(z)$

Exercises

Section 3.1.

3.1 Show that the equation $\Delta y(t) + y(t) = e^t$ can not be put in the form of Eq. (3.1) and so is not a first order linear difference equation.

3.2 Solve by iteration for $t = 1, 2, 3, \cdots$:
(a) $u(t+1) = \frac{t}{t+1} u(t)$,
(b) $u(t+1) = \frac{3t+1}{3t+7} u(t)$.

3.3 Find all solutions:
(a) $u(t+1) - e^{3t} u(t) = 0$,
(b) $u(t+1) - e^{\cos 2t} u(t) = 0$.

3.4 Show that a general solution of the constant coefficient first order difference equation $u(t+1) - cu(t) = 0$ is $u(t) = Ac^t$. Use this result to solve the nonhomogeneous equations
(a) $y(t+1) - 2y(t) = 5$,
(b) $y(t+1) - 4y(t) = 3 \cdot 2^t$,
(c) $y(t+1) - 5y(t) = 5^t$.

3.5 Let $y(t)$ represent the total number of squares of all dimensions on an t by t checkerboard.
(a) Show that $y(t)$ satisfies

$$y(t+1) = y(t) + t^2 + 2t + 1.$$

(b) Solve for $y(t)$.

3.6 Suppose $y(1) = 2$ and find the solution of

$$y(t+1) - 3y(t) = e^t \quad (t = 1, 2, 3, \cdots).$$

3.7 Solve for $t = 1, 2, \cdots$:

$$y(t+1) - \frac{3t+1}{3t+7} y(t) = \frac{t}{(3t+4)(3t+7)}.$$

3.8 Consider for $t = 1, 2, \cdots$ the equation $y(t+1) - ty(t) = 1$.

(a) Show that the solution is

$$y(t) = (t-1)! \left[\sum_{k=1}^{t-1} \frac{1}{k!} + y(1) \right].$$

(b) Using the fact that $\sum_{k=1}^{\infty} \frac{1}{k!} = e - 1$, derive another expression for $y(t)$.

3.9 (a) Show that the solutions of the equation $y_{n+1}(x) + \frac{x}{n} y_n(x) = \frac{e^{-x}}{n}$ ($n = 1, 2, \cdots$) are

$$y_n(x) = \frac{(-x)^{n-1}}{(n-1)!} \left[C + e^x \sum_{k=1}^{n-1} (-1)^k (k-1)! \left(\frac{1}{x} \right)^k \right].$$

(b) For what value of C is $y_n(x) = E_n(x)$ the exponential integral? (See Exercise 1.15.)

3.10 If we invest $\$1,000$ at an annual interest rate of 10% for 10 years, how much money will we have if the interest is compounded
  (a) annually, (b) semiannually, (c) quarterly,
  (d) monthly, (e) daily?

3.11 Assume we invest a certain amount of money at 8% a year compounded annually. How long does it take for our money to double? triple?

3.12 What is the present value of an annuity where we deposited $\$900$ at the beginning of each year for nine years if the annual interest rate is 9%?

3.13 A man aged 40 wishes to accumulate a fund for retirement by depositing $\$1200$ at the beginning of each year for 25 years until he retires at age 65. If the annual interest rate is 7%, how much will he accumulate for his retirement?

3.14 In Example 3.2, suppose we are allowed to increase our deposit by 5% each year. How much will we have in the IRA at the end of the tth year? How much will we have after 20 years?

3.15 Let y_n denote the number of multiplications needed to compute the determinant of an n by n matrix by cofactor expansion.
  (a) Show that $y_{n+1} = (n+1)(y_n + 1)$.
  (b) Compute y_n.

Exercises

3.16 In an elementary economics model of the market place, the price p_n of a product after n years is related to the supply s_n after n years by $p_n = a - bs_n$, where a and b are positive constants, since a large supply causes the price to be low in a given year. Assume that price and supply in alternate years are proportional: $kp_n = s_{n+1}$ ($k > 0$).

(a) Show p_n satisfies $p_{n+1} + bkp_n = a$.

(b) Solve for p_n.

(c) If $bk < 1$, show that the price stabilizes, i.e., show that p_n converges to a limit as $n \to \infty$. What happens if $bk > 1$?

3.17 Let $y(x) = \sum_{n=0}^{\infty} a_n x^n$ in the differential equation $y'(x) = y(x) + e^x$.

(a) Show that $\{a_n\}$ satisfies the difference equation

$$a_{n+1} = \frac{a_n}{n+1} + \frac{1}{(n+1)!}.$$

(b) Use the solution of the equation in (a) to compute $y(x)$.

3.18 Show by substitution that the function

$$u(t) = C(t) a^t \frac{\Gamma(t-r_1) \cdots \Gamma(t-r_n)}{\Gamma(t-s_1) \cdots \Gamma(t-s_m)}$$

with $\Delta C(t) = 0$ satisfies the equation

$$u(t+1) = a \frac{(t-r_1) \cdots (t-r_n)}{(t-s_1) \cdots (t-s_m)} u(t).$$

3.19 Solve $u(t+1) = \frac{2t^3}{3(t+1)^2} u(t)$ in terms of the gamma function. Simplify your answer.

3.20 An example of a "full history" difference equation is

$$y_n = n + \sum_{k=1}^{n-1} y_k \quad (n = 2, 3, \cdots).$$

Solve for y_n, assuming $y_1 = 1$. (Hint: compute $y_{n+1} - y_n$.)

3.21 Find a solution of

$$y(t-1) - ty(t) = -t$$

that has the form of a factorial series. Show the series converges for all $t \neq 0, -1, -2, \cdots$.

Section 3.2.

3.22 What is the order of the equation

$$\Delta^3 y(t) + \Delta^2 y(t) - \Delta y(t) - y(t) = 0?$$

3.23 Give proofs of Theorem 3.3 and Corollary 3.1.

3.24 Show that $u_1(t) = 2^t$ and $u_2(t) = 3^t$ are linearly independent solutions of

$$u(t+2) - 5u(t+1) + 6u(t) = 0.$$

3.25 Use the result of Exercise 3.24 to find the unique solution of the initial value problem

$$u(t+2) - 5u(t+1) + 6u(t) = 0,$$

$$u(3) = 0, \quad u(4) = 12,$$

where $t = 3, 4, 5, \cdots$.

3.26 Verify that the Casoratian satisfies Eq. (3.5).

3.27 (a) Show that $u_1(t) = t^2 + 2$, $u_2(t) = t^2 - 3t$ and $u_3(t) = 2t - 1$ are solutions of $\Delta^3 u(t) = 0$.

(b) Compute the Casoratian of the functions in (a) and determine whether they are linearly independent.

3.28 Are $u_1(t) = 2^t \cos \frac{2\pi t}{3}$ and $u_2(t) = 2^t \sin \frac{2\pi t}{3}$ linearly independent solutions of $u(t+2) + 2u(t+1) + 4u(t) = 0$?

Exercises

3.29 Use Exercise 3.24 and Theorem 3.6 to solve

$$y(t+2) - 5y(t+1) + 6y(t) = 2^t.$$

3.30 Use Theorem 3.6 to solve the first order equation

$$y(t+1) - 2y(t) = 2^t \binom{t}{5}.$$

3.31 Find all solutions of

$$y(t+2) - 7y(t+1) + 6y(t) = 2t - 1.$$

(See Example 3.8.)

3.32 Given that $1, (-1)^t$, and 2^t solve the homogeneous equation, use variation of parameters to solve

$$y(t+3) - 2y(t+2) - y(t+1) + 2y(t) = 8 \cdot 3^t.$$

3.33 (a) Show that

$$a_1(t) = -\sum_{k=a}^{t-1} \frac{r(k)}{p_2(k)} \frac{u_2(k+1)}{w(k+1)}$$

and

$$a_2(t) = \sum_{k=a}^{t-1} \frac{r(k)}{p_2(k)} \frac{u_1(k+1)}{w(k+1)}$$

are solutions of Eqs. (3.6) and (3.7).

(b) Use part (a) to prove Corollary 3.2.

3.34 For $n = 2$, define the Cauchy function $K(t, k)$ for Eq. (3.4) to be the function defined for $t, k = a, a+1, \cdots$, such that for each fixed k, $K(t, k)$ is the solution of Eq. (3.4)' satisfying $K(k+1, k) = 0$, $K(k+2, k) = (p_2(k))^{-1}$.

(a) Show that

$$K(t,k) = \frac{-1}{p_2(k)w(k+1)} \det \begin{bmatrix} u_1(t) & u_2(t) \\ u_1(k+1) & u_2(k+1) \end{bmatrix},$$

where $u_1(t), u_2(t)$ are linearly independent solutions of Eq. (3.4)' with Casoratian $w(t)$.

(b) Use Corollary 3.2 to show that the solution of the initial value problem Eq. (3.4), $y(a) = y(a+1) = 0$, is given by

$$y(t) = \sum_{k=a}^{t-1} K(t,k)r(k).$$

(A related formulation is given in Chapter 6.)

3.35 (a) Use Corollary 3.2 to solve

$$y(t+2) - 5y(t+1) + 6y(t) = 2^t, \quad y(1) = y(2) = 0.$$

(b) Check your answer using the result of Exercise 3.29.

3.36 Which of the following sets are linearly independent on R^1?
(a) $u_1(t) = \cos 2\pi t$, $u_2(t) = \sin 2\pi t$,
(b) $u_1(t) = 1$, $u_2(t) = t$, $u_3(t) = t^2$,
(c) $u_1(t) = 2^t$, $u_2(t) = 3^t$.

Section 3.3.

3.37 In the case that the characteristic roots $\lambda_1, \cdots, \lambda_n$ are distinct, show that the solutions $\lambda_1^t, \cdots, \lambda_n^t$ of Eq. (3.8) are linearly independent. (Hint: use the value of the Vandermonde determinant:

$$\det \begin{bmatrix} 1 & 1 & \cdots & 1 \\ c_1 & c_2 & \cdots & c_n \\ c_1^2 & c_2^2 & \cdots & c_n^2 \\ \vdots & \vdots & \ddots & \vdots \\ c_1^{n-1} & c_2^{n-1} & \cdots & c_n^{n-1} \end{bmatrix} = \prod_{j>i}(c_j - c_i).)$$

Exercises

3.38 Solve the following equations:
  (a) $(E - 6)^5 u(t) = 0$,
  (b) $u(t + 2) + 6u(t + 1) + 3u(t) = 0$,
  (c) $u(t + 3) - 4u(t + 2) + 5u(t + 1) - 2u(t) = 0$,
  (d) $u(t + 4) - 8u(t + 2) + 16u(t) = 0$.

3.39 Find all real solutions:
  (a) $u(t + 2) + u(t) = 0$,
  (b) $u(t + 2) - 8u(t + 1) + 32u(t) = 0$,
  (c) $u(t + 4) + 2u(t + 2) + u(t) = 0$,
  (d) $u(t + 6) + 2u(t + 3) + u(t) = 0$.

3.40 Compute the sequence of coefficients $\{a_n\}_{n=0}^{\infty}$ so that

$$\frac{2 - 3t}{1 - 3t + 2t^2} = \sum_{n=0}^{\infty} a_n t^n$$

on some open interval about $t = 0$. Find the radius of convergence of the infinite series.

3.41 Find a homogeneous equation with constant coefficients for which one solution is
  (a) $(t + \sqrt{2})^2$,
  (b) t^5,
  (c) $t(-3)^t$,
  (d) $\frac{\sin \frac{2\pi}{3} t}{2^t}$.

3.42 Solve by the annihilator method
  (a) $8y(t + 2) - 6y(t + 1) + y(t) = 2^t$,
  (b) $y(t + 2) - 2y(t + 1) + y(t) = 3t + 5$,
  (c) $y(t + 2) + y(t + 1) - 12y(t) = t3^t$.

3.43 Solve by the annihilator method

$$y(t + 2) + 4y(t) = \cos t.$$

3.44 Solve the initial value problem

$$y_{n+2} - 4y_{n+1} + 3y_n = n4^n, \quad y_1 = \frac{2}{9}, \quad y_2 = \frac{1}{9}.$$

3.45 Use the annihilator method to solve
$$(E^2 - E + 2)y(t) = 3^t + t3^t.$$

3.46 Rework Exercise 3.32 by the annihilator method.

3.47 Solve the homogeneous system
$$\begin{aligned} u(t+1) - 3u(t) + v(t) &= 0, \\ -u(t) + v(t+1) - v(t) &= 0. \end{aligned}$$

3.48 Find all $u(t)$ and $v(t)$ that satisfy
$$\begin{aligned} u(t+2) - 3u(t) + 2v(t) &= 0, \\ u(t) + v(t+2) - 2v(t) &= 0. \end{aligned}$$

3.49 Use the annihilator method to solve
$$\begin{aligned} u(t+1) - 4u(t) - v(t) &= 3^t, \\ u(t+1) - 2u(t) + v(t+1) - 2v(t) &= 2. \end{aligned}$$

Section 3.4.

3.50 Find a formula for the sum of the first n Fibonacci numbers.

3.51 Show that the generating function for the Fibonacci sequence is $\frac{1}{1-t-t^2}$.

3.52 If $u(t)$ is the tth Fibonacci number, then show that
$$u(t+1)u(t-1) - u(t)^2 = (-1)^t \quad (t \geq 1).$$

3.53 A strip is one unit wide by n units long. We want to paint this strip with one by one squares that are red or blue. In how many ways can we paint the strip if we do not allow consecutive red squares?

3.54 In how many ways can a one by n hallway be tiled if we use one by one blue tiles and one by two red tiles?

3.55 (a) Solve the difference equation in Example 3.15 for the case $\frac{w^2 m}{4k} > 1$.

(b) Show that most of the solutions in part (a) are unbounded as $n \to \infty$.

3.56 In Example 3.16, how many years elapse before the prey population becomes extinct? When does the predator population die out?

3.57 Solve the problem

$$\Delta x(t) = -.5x(t) - .3y(t)$$
$$\Delta y(t) = -.2x(t) - .6y(t)$$

if $x(0) = 2$, $y(0) = 5$.

3.58 (a) Use the method of Section 3.3 to solve

$$u_{n+2} - 2x u_{n+1} + u_n = 0, \quad u_0 = 1, \quad u_1 = x.$$

(b) Show that the u_n obtained in part (a) is the same as $T_n(x)$.

3.59 Show that $T_n(x)$ is a polynomial of degree n.

3.60 Show that the generating function for $\{T_n(x)\}$ is $\frac{1-xt}{1-2xt+t^2}$.

3.61 The Chebyshev polynomials of the second kind are defined by

$$U_n(x) = \frac{\sin\left((n+1)\cos^{-1} x\right)}{\sqrt{1-x^2}} \quad (n \geq 0).$$

Show that $U_n(x)$ satisfies the same difference equation as $T_n(x)$.

3.62 Show that the Chebyshev polynomials of the second kind are orthogonal on $[-1,1]$ with respect to the weight function $\sqrt{1-x^2}$.

3.63 Suppose that in Example 3.18 we want to compute only the quantity of water in the topsoil at 9PM each day. Find the solution by solving a first order equation.

3.64 Solve Example 3.18 with the assumption that only a fourth of the total amount of water in the topsoil is lost between 9AM and 9PM.

3.65 Compute the determinant in Example 3.19 for the case $a^2 - 4bc > 0$.

3.66 Compute the determinant in Example 3.19 for the case $a^2 = 4bc$.

3.67 Solve the equation $y_{n+1} = \sum_{k=0}^{n}(\varepsilon + y_{n-k})A_k$ in Example 3.20 if $A_0 = A_1 = c > 0$ and $A_k = 0$ for $k \geq 2$.

3.68 Use the method of generating functions to solve the equation

$$u_{n+1} = \sum_{k=0}^{n} \frac{u_{n-k}}{2^k}$$

if $u_0 = 1$.

3.69 Let $x(n)$ be the number of ways a four by n hallway can be tiled using two by one tiles.

(a) Find a system of three equations in three unknowns (one of which is $x(n)$) that model the problem.

(b) Use your equations iteratively to find the number of ways to tile a four by ten hallway.

3.70 Three products A, B, and C compete for the same (fixed) market. Let $x(t)$, $y(t)$, and $z(t)$ be the respective percentages of the market for these products after t months. If the changes in the percentages are given by

$$\Delta x(t) = x(t) + \frac{1}{3}y(t) + \frac{2}{3}z(t),$$
$$\Delta y(t) = \frac{1}{3}x(t) + \frac{5}{3}y(t) + \frac{1}{3}z(t),$$
$$\Delta z(t) = \frac{5}{3}z(t),$$

and if initially product A has 50% of the market, product B has 30% of the market, and product C has 20% of the market, find the percentages for each product after t months.

3.71 Consider a game with two players A and B, where player A has probability p of winning a chip from B and player B has probability $1 - p$ of winning a chip from A on each turn. The game ends when one player has all the chips.

(a) Let $u(t)$ be the probability that A will win the game given that

A has t chips. Show that $u(t)$ satisfies

$$u(t) = pu(t+1) + (1-p)u(t-1).$$

(b) Suppose that at the beginning of the game A has a chips and B has b chips. Find the probability that A wins the game.

3.72 Let

$$I_n = \int_0^\pi \frac{\cos n\theta - \cos n\varphi}{\cos \theta - \cos \varphi} d\theta, \quad (n = 0, 1, \cdots).$$

(a) Show that I_n satisfies the equation

$$I_{n+2} - 2(\cos \varphi)I_{n+1} + I_n = 0, \quad (n = 0, 1, \cdots).$$

(b) Compute I_n for $n = 0, 1, 2, \cdots$.

Section 3.5.

3.73 Find the general solution of

$$(E - (t+1))(E+1)u(t) = 0.$$

3.74 Solve by the method of factoring:
  (a) $u_{n+2} - (2n+1)u_{n+1} + n^2 u_n = 0$,
  (b) $u_{n+2} - (e^n + 1)u_{n+1} + e^n u_n = 0$.

3.75 Factor and solve:

$$u_{n+2} - \frac{3n-2}{n-1}u_{n+1} + \frac{2n}{n-1}u_n = n2^n.$$

3.76 In the nth-order equation $\sum_{k=0}^n p_k(t)u(t+k) = 0$, suppose a solution $u_1(t)$ is known. Make the substitution $u = u_1 v$ and use Theorem 2.8 with $a_k = p_k(t)u_1(t+k)$, $b_k = v(t+k)$ to obtain an $(n-1)^{\text{st}}$-order equation with unknown Δv.

3.77 Find general solutions of
  (a) $2t(t+1)\Delta^2 u(t) + 8t\Delta u(t) + 4u(t) = 0$,

(b) $t(t+1)\Delta^2 u(t) - 3t\Delta u(t) + 4u(t) = 0$.

3.78 Solve the equation

$$t(t+1)\Delta^2 u - 2t\Delta u + 2u = t.$$

3.79 Verify Eqs. (3.15) and (3.16).

3.80 Use the method of generating functions to solve

$$3(n+2)u_{n+2} - (3n+4)u_{n+1} + u_n = 0$$

if $u_0 = 3u_1$.

3.81 One solution of $(n+1)u_{n+2} + (2n-1)u_{n+1} - 3nu_n = 0$ is easy to find. What is the general solution?

3.82 Check that $u_n = 2^n$ solves

$$nu_{n+2} - (1+2n)u_{n+1} + 2u_n = 0,$$

and find a second independent solution.

3.83 Use generating functions to solve $(n+2)(n+1)u_{n+2} - 3(n+1)u_{n+1} + 2u_n = 0$.

3.84 Find a factorial series solution of the form $\sum_{k=0}^{\infty} a_k t^{(-k)}$ for

$$u(t+2) - 3(t+2)(t+1)u(t+1) + 3(t+2)(t+1)u(t) = 0.$$

3.85 (a) Compute a formal series solution $u(t) = \sum_{k=0}^{\infty} a_k t^{(-k+\frac{1}{2})}$ for $t\Delta u(t) - \frac{1}{2}u(t) = 0$.
(Hint: Use the identity $tt^{(r)} = t^{(r+1)} + rt^{(r)}$.)

(b) Show that the trial solution $u(t) = \sum_{k=0}^{\infty} a_k t^{-k}$ leads to the zero solution.

Section 3.6.

3.86 Solve the Riccati equations
(a) $y(t+1)y(t) + 2y(t+1) + 7y(t) + 20 = 0$,
(b) $y(t+1)y(t) - 2y(t) + 2 = 0$.

Exercises

3.87 Use the change of variable $v(t) = \frac{1}{y(t)}$ to solve the Riccati equation
$$ty(t+1)y(t) + y(t+1) - y(t) = 0.$$

3.88 Use a logarithm to solve
  (a) $\frac{y_{n+1}}{y_n} = 2y_n^{\frac{1}{n}}$,
  (b) $y_{n+2} = y_{n+1} y_n^2$.

3.89 Solve: $(t+1)y^2(t+1) - ty^2(t) = 1$.

3.90 Use the change of variable $y_n = \sin z_n$ to solve $y_{n+1} = 2y_n\sqrt{1 - y_n^2}$.

3.91 Solve the equation
$$y(t+1) = y(t)\left(y^2(t) + 3y(t) + 3\right)$$
by trying $\xi(y) = cy + d$ in Eq. (3.22).

3.92 Find the most general equation $y(t+1) = f(y(t))$ that can be solved using $\xi(y) = cy + d$ in Eq. (3.22).

3.93 Let a be a positive constant. Then Newton's method for computing $\sqrt{a^2} = a$ is
$$y_{n+1} = \frac{1}{2}\left(y_n + \frac{a^2}{y_n}\right).$$

  (a) Find D so that $\xi(y) = a - \frac{y^2}{a}$ solves Eq. (3.22) for this difference equation.
  (b) Use the change of variable Eq. (3.23) to solve the difference equation.
  (c) Show that the solution $y_n \to a$ as $n \to \infty$.

3.94 Solve the difference equation
$$y_{n+1} = \frac{1}{2}\left(y_n - \frac{a^2}{y_n}\right).$$

(Hint: try $\xi(y) = -a - \frac{y^2}{a}$.)

3.95 Solve $y(t+1) = (1 - 2y(t))^2$. (Hint: see Example 3.30.)

3.96 Consider the equation $y(t+1) = f(t, y(t))$. Suppose that $\xi(t, y)$ and $D(t)$ satisfy

$$D(t)\xi(t+1, f(t,y)) = \xi(t,y)\frac{\partial f}{\partial y}(t,y).$$

Show that a change of variable transforms the difference equation into a first order linear equation.

3.97 Solve the equation $y(t+1) = (y(t) + t - 1)^t - t$ by choosing $\xi = y+t-1$ in the last exercise.

3.98 Use $\xi = \sqrt{y(t-y)}$ to solve $y(t) = \frac{4(1+t)}{t^2}y(t)(t-y(t))$.

Section 3.7.

3.99 Find the z-transform of each of the following:
   (a) $y_k = 2 + 3k$,
   (b) $u_k = 3^k \cos 2k$,
   (c) $v_k = \sin(2k - 3)$,
   (d) $y_k = k^3$,
   (e) $u_k = 3y_{k+3}$,
   (f) $v_k = k \cos \frac{k\pi}{2}$,
   (g) $y_k = \frac{1}{k!}$,
   (h) $u_k = \begin{cases} \frac{(-1)^{\frac{k}{2}}}{(k+1)!}, & k \text{ even} \\ 0, & k \text{ odd.} \end{cases}$

3.100 Find $Z(\cosh at)$ using Theorem 3.10.

3.101 Find $Z(\cos at)$ using Theorem 3.10.

3.102 Find the sequences whose z-transforms are
   (a) $Y(z) = \frac{2z^2 - 3z}{z^2 - 3z - 4}$,
   (b) $U(z) = \frac{3z^2 - 4z}{z^2 - 3z + 2}$,
   (c) $V(z) = \frac{2z^2 + z}{(z-1)^2}$,
   (d) $Y(z) = \frac{z}{2z^2 - 2\sqrt{2}z + 2}$,
   (e) $U(z) = \frac{2z^2 - z}{2z^2 - 2z + 2}$,
   (f) $V(z) = \frac{z^2 + 3z}{(z-3)^2}$,

(g) $W(z) = \frac{3z^2+5}{z^4}$,

(h) $Y(z) = e^{\frac{1}{z^2}}$.

3.103 Use Theorem 3.11 to show that

$$Z(k^n) = (-1)^n \left(z\frac{d}{dz}\right)^n \frac{z}{z-1}.$$

Use this formula to find $Z(k^3)$.

3.104 Use Theorem 3.13 to find $Z(k^2)$ and $Z(k^3)$.

3.105 Derive the formula for $Z(\delta_k(n))$ by expressing $\delta_k(n)$ in terms of step functions.

3.106 Find the z-transform of each of the following sequences:
(a) $y_1 = 1$, $y_3 = 4$, $y_5 = 2$, $y_k = 0$ otherwise,
(b) $y_{2k+1} = 0$, $y_{2k} = 1$, $k = 0, 1, 2, \cdots$,
(c) $y_{2k} = 0$, $y_{2k+1} = 1$, $k = 0, 1, 2, \cdots$.

3.107 (a) Use Theorem 3.13 to show that for n a positive integer

$$Z\left(\binom{k}{n}\right) = \frac{z}{(z-1)^{n+1}}, \quad |z| > 1.$$

(b) Use the Binomial Theorem to show that

$$Z\left(\binom{r}{k}\right) = \frac{(z+1)^r}{z^r}, \quad |z| > 1.$$

3.108 Solve the following first order initial value problems using z-transforms.
(a) $y_{k+1} - 3y_k = 4^k$, $y_0 = 0$,
(b) $y_{k+1} + 4y_k = 10$, $y_0 = 3$,
(c) $y_{k+1} - 5y_k = 5^{k+1}$, $y_0 = 0$,
(d) $y_{k+1} - 2y_k = 3 \cdot 2^k$, $y_0 = 3$,
(e) $y_{k+1} + 3y_k = 4\delta_k(2)$, $y_0 = 2$.

3.109 Solve the following second order initial value problems using z-transforms:
(a) $y_{k+2} - 5y_{k+1} + 6y_k = 0$, $y_0 = 1$, $y_1 = 0$,

(b) $y_{k+2} - y_{k+1} - 6y_k = 0$, $y_0 = 5$, $y_1 = -5$,
(c) $y_{k+2} - 8y_{k+1} + 16y_k = 0$, $y_0 = 0$, $y_1 = 4$,
(d) $y_{k+2} - y_k = 16 \cdot 3^k$, $y_0 = 2$, $y_1 = 6$,
(e) $y_{k+2} - 3y_{k+1} + 2y_k = u_k(4)$.

3.110 Solve the following systems using z-transforms:

(a)
$$u_{k+1} - 2v_k = 2 \cdot 4^k$$
$$-4u_k + v_{k+1} = 4^{k+1}$$
$$u_0 = 2, \quad v_0 = 3,$$

(b)
$$u_{k+1} - v_k = 0$$
$$u_k + v_{k+1} = 0$$
$$u_0 = 0, \quad v_0 = 1,$$

(c)
$$u_{k+1} - v_k = 2k$$
$$-u_k + v_{k+1} = 2k + 2$$
$$u_0 = 0, \quad v_0 = 1,$$

(d)
$$u_{k+1} - v_k = -1$$
$$-u_k + v_{k+1} = 3$$
$$u_0 = 0, \quad v_0 = 2.$$

3.111 Use Theorem 3.12 to prove Corollary 3.3.

3.112 Prove that the convolution product is commutative ($u_k * v_k = v_k * u_k$) and associative (($u_k * v_k) * w_k = u_k * (v_k * w_k)$).

3.113 Calculate the following convolutions:
(a) $1 * 1$,
(b) $1 * k$,
(c) $k * k$.

3.114 Solve the following summation equations for $k \geq 0$:

(a) $y_k = 3 \cdot 5^k - 4\sum_{m=0}^{k-1} 5^{k-m-1} y_m,$

(b) $y_k = k + 4\sum_{m=0}^{k-1}(k-m-1)y_m,$

(c) $y_k = 3 + 12\sum_{m=0}^{k+1}\left(2^{k-m-1} - 1\right) y_m.$

3.115 Solve the following equations for $k \geq 0$:

(a) $y_k = 2^k + \sum_{m=0}^{k-1} 2^{k-m-1} y_m,$

(b) $y_k = 3 + 9\sum_{m=0}^{k-1}(k-m-1)y_m,$

(c) $y_k = 2^k + 12\sum_{m=0}^{k-1}\left(3^{k-m-1} - 2^{k-m-1}\right) y_m.$

3.116 Solve

$$y_k = 2 + \lambda \sum_{m=0}^{24} \frac{k}{50} y_m$$

for all values of λ for which the equation has a solution.

3.117 Solve the following Fredholm summation equations:

(a) $y_k = 10 + \sum_{m=0}^{20} km y_m,$

(b) $y_k = k + \sum_{m=1}^{15} m y_m,$

(c) $y_k = k + \sum_{m=1}^{15} k y_m,$

(d) $y_k = \lambda \sum_{m=1}^{19} km y_m.$

3.118 Solve the following Fredholm summation equations:

(a) $y_k = 1 + \sum_{m=1}^{15}(1 - km)y_m,$

(b) $y_k = k + \sum_{m=1}^{10}(m^2 + km)y_m.$

3.119 Prove the last statement in Theorem 3.18.

3.120 Find the currents in the ladder network obtained from Fig. 3.7 by replacing the resistor at the top of each loop by a resistor having resistance $R_0 \neq R$.

Chapter 4
Stability Theory

4.1 Initial Value Problems for Linear Systems

Most of our discussion up to this point has been restricted to a single difference equation with one unknown function. However, mathematical models frequently involve several unknown quantities with (usually) an equal number of equations. We consider systems of the form

$$u_1(t+1) = a_{11}(t)u_1(t) + \cdots + a_{1n}(t)u_n(t) + f_1(t)$$
$$u_2(t+1) = a_{21}(t)u_1(t) + \cdots + a_{2n}(t)u_n(t) + f_2(t)$$
$$\vdots$$
$$u_n(t+1) = a_{n1}(t)u_1(t) + \cdots + a_{nn}(t)u_n(t) + f_n(t),$$

for $t = a, a+1, a+2, \cdots$. This system can be written as an equivalent vector equation

$$u(t+1) = A(t)u(t) + f(t), \qquad (4.1)$$

where

$$u(t) = \begin{bmatrix} u_1(t) \\ \vdots \\ u_n(t) \end{bmatrix}, \quad A(t) = \begin{bmatrix} a_{11}(t) & \cdots & a_{1n}(t) \\ \vdots & \ddots & \vdots \\ a_{n1}(t) & \cdots & a_{nn}(t) \end{bmatrix}, \quad f(t) = \begin{bmatrix} f_1(t) \\ \vdots \\ f_n(t) \end{bmatrix}.$$

The study of Eq. (4.1) includes the nth order scalar equation

$$p_n(t)y(t+n) + \cdots + p_0(t)y(t) = r(t) \qquad (4.2)$$

as a special case. To see this, let $y(t)$ solve Eq. (4.2) and define

$$u_i(t) = y(t+i-1)$$

for $1 \leq i \leq n$, $t = a, a+1, \cdots$. Then the vector function $u(t)$ with components $u_i(t)$ satisfies Eq. (4.1) if

$$A(t) = \begin{bmatrix} 0 & 1 & 0 & \cdots & 0 \\ 0 & 0 & 1 & \cdots & 0 \\ \vdots & \vdots & \ddots & \ddots & \vdots \\ 0 & 0 & \cdots & 0 & 1 \\ -\frac{p_0(t)}{p_n(t)} & -\frac{p_1(t)}{p_n(t)} & -\frac{p_2(t)}{p_n(t)} & \cdots & -\frac{p_{n-1}(t)}{p_n(t)} \end{bmatrix}, \quad f(t) = \begin{bmatrix} 0 \\ 0 \\ \vdots \\ r(t) \end{bmatrix}.$$
(4.3)

The matrix $A(t)$ in Eq. (4.3) is called the "companion matrix" of Eq. (4.2). Conversely, if $u(t)$ solves Eq. (4.1) with $A(t)$ and $f(t)$ given in Eq. (4.3), then $y(t) = u_1(t)$ is a solution of Eq. (4.2).

Given an initial vector $u(t_0) = u_0$ for some t_0 in $\{a, a+1, \cdots\}$, Eq. (4.1) can be solved iteratively for $u(t_0+1)$, $u(t_0+2)$, $\cdots$.

Theorem 4.1. For each t_0 in $\{a, a+1, \cdots\}$ and each n-vector u_0, Eq. (4.1) has a unique solution $u(t)$ defined for $t = t_0, t_0+1, \cdots$, so that $u(t_0) = u_0$.

Now assume that A is independent of t (i.e., all coefficients in the system are constants) and $f(t) = 0$. Then the solution $u(t)$ of

$$u(t+1) = Au(t), \qquad (4.4)$$

satisfying the initial condition $u(0) = u_0$, is $u(t) = A^t u_0$, ($t = 0, 1, 2, \cdots$). Hence the solutions of Eq. (4.4) can be found by calculating powers of A. We have chosen to take the initial condition at $t = 0$ for simplicity since an arbitrary initial condition can be shifted to 0 by translation along the t axis.

There are a number of concepts from linear algebra that will be needed in our calculations. The equation

$$Au = \lambda u, \qquad (4.5)$$

where λ is a parameter, always has the trivial solution $u = 0$. If Eq. (4.5) has a nontrivial solution u for some λ then λ is called an eigenvalue of A

4.1 Initial Value Problems for Linear Systems

and u is called a corresponding eigenvector of A. The eigenvalues of A satisfy the characteristic equation

$$\det(\lambda I - A) = 0,$$

where I is the n by n identity matrix. An eigenvalue is said to be simple if its multiplicity as a root of the characteristic equation is one. The spectrum of A, denoted $\sigma(A)$, is the set of eigenvalues of A, and the spectral radius of A is

$$r(A) = \max\{|\lambda| : \lambda \text{ is in } \sigma(A)\}.$$

Example 4.1. Find the eigenvalues, eigenvectors, and spectral radius for

$$A = \begin{bmatrix} 0 & 1 \\ -2 & -3 \end{bmatrix}.$$

The characteristic equation of A is

$$\det \begin{bmatrix} \lambda & -1 \\ 2 & \lambda + 3 \end{bmatrix} = 0,$$

or

$$\lambda^2 + 3\lambda + 2 = 0,$$

so $\sigma(A) = \{-2, -1\}$. To find the eigenvectors corresponding to $\lambda = -2$, we solve

$$(-2I - A)u = 0$$

or

$$\begin{bmatrix} -2 & -1 \\ 2 & 1 \end{bmatrix} \begin{bmatrix} u_1 \\ u_2 \end{bmatrix} = \begin{bmatrix} 0 \\ 0 \end{bmatrix}.$$

The eigenvectors are all nonzero multiples of the vector with $u_1 = 1$, $u_2 = -2$. Similarly, the eigenvectors corresponding to $\lambda = -1$ are all nonzero multiples of the vector with $u_1 = 1$, $u_2 = -1$. Finally, the spectral radius of A is

$$r(A) = \max\{|-2|, |-1|\} = 2.$$

Now let λ be an eigenvalue of A and let u be a corresponding eigenvector. For $t = 0, 1, 2, \cdots$, we have

$$A^t u = \lambda^t u,$$

so $u(t) = \lambda^t u$ satisfies Eq. (4.4) with initial vector u. More generally, if u_0 can be written as a linear combination of the eigenvectors of A, say

$$u_0 = b_1 u^1 + \cdots + b_k u^k,$$

where each u^i is an eigenvector corresponding to λ_i, then the solution of Eq. (4.4) is

$$u(t) = b_1 \lambda_1^t u^1 + \cdots + b_k \lambda_k^t u^k. \tag{4.6}$$

As a result, if A has n linearly independent eigenvectors (this is necessarily the case if A has n distinct eigenvalues or if A is symmetric), then every solution of the system can be calculated in this way.

Example 4.1. (continued) Solve Eq. (4.4) if $A = \begin{bmatrix} 0 & 1 \\ -2 & -3 \end{bmatrix}$.

Let $u_0 = \begin{bmatrix} u_1 \\ u_2 \end{bmatrix}$ be an initial vector and recall that $\begin{bmatrix} 1 \\ -2 \end{bmatrix}$ is an eigenvector for $\lambda = -2$ and $\begin{bmatrix} 1 \\ -1 \end{bmatrix}$ is an eigenvector for $\lambda = -1$. Now set

$$\begin{bmatrix} u_1 \\ u_2 \end{bmatrix} = b_1 \begin{bmatrix} 1 \\ -2 \end{bmatrix} + b_2 \begin{bmatrix} 1 \\ -1 \end{bmatrix} = \begin{bmatrix} 1 & 1 \\ -2 & -1 \end{bmatrix} \begin{bmatrix} b_1 \\ b_2 \end{bmatrix}.$$

4.1 Initial Value Problems for Linear Systems

The solution of this linear system is

$$\begin{bmatrix} b_1 \\ b_2 \end{bmatrix} = \begin{bmatrix} 1 & 1 \\ -2 & -1 \end{bmatrix}^{-1} \begin{bmatrix} u_1 \\ u_2 \end{bmatrix}$$
$$= \begin{bmatrix} -1 & -1 \\ 2 & 1 \end{bmatrix} \begin{bmatrix} u_1 \\ u_2 \end{bmatrix}$$
$$= \begin{bmatrix} -u_1 - u_2 \\ 2u_1 + u_2 \end{bmatrix}.$$

By Eq. (4.6), the solution of Eq. (4.4) with initial vector u_0 is

$$u(t) = -(u_1 + u_2)(-2)^t \begin{bmatrix} 1 \\ -2 \end{bmatrix} + (2u_1 + u_2)(-1)^t \begin{bmatrix} 1 \\ -1 \end{bmatrix}.$$

Before solving Eq. (4.4) in general, we recall an important and beautiful result from linear algebra (see Grossman [93]).

The Cayley-Hamilton Theorem. Every square matrix satisfies its characteristic equation.

Example 4.2. Verify the Cayley-Hamilton Theorem for

$$A = \begin{bmatrix} 1 & 2 \\ 3 & 4 \end{bmatrix}.$$

The characteristic equation for A is

$$\det \begin{bmatrix} \lambda - 1 & -2 \\ -3 & \lambda - 4 \end{bmatrix} = \lambda^2 - 5\lambda - 2 = 0.$$

Now

$$A^2 - 5A - 2I$$
$$= \begin{bmatrix} 7 & 10 \\ 15 & 22 \end{bmatrix} - \begin{bmatrix} 5 & 10 \\ 15 & 20 \end{bmatrix} - \begin{bmatrix} 2 & 0 \\ 0 & 2 \end{bmatrix}$$
$$= \begin{bmatrix} 0 & 0 \\ 0 & 0 \end{bmatrix},$$

156   Chapter 4. Stability Theory

and A does satisfy its characteristic equation.

Remark. The Cayley-Hamilton Theorem implies that A^n can be written as a linear combination of I, A, A^2, $\cdots$, A^{n-1}, if A is an n by n matrix. It follows that every power of A also can be written as a linear combination of I, A, A^2, $\cdots$, A^{n-1}.

Let $\lambda_1, \cdots, \lambda_n$ be the (not necessarily distinct) eigenvalues of A, with each eigenvalue repeated as many times as its multiplicity. Define

$$\begin{aligned} M_0 &= I, \\ M_i &= (A - \lambda_i I)M_{i-1}, \quad (1 \leq i \leq n). \end{aligned} \quad (4.7)$$

It follows from the Cayley-Hamilton Theorem that $M_n = 0$.

Definition (4.7) implies that each A^i is a linear combination of M_0, $\cdots$, M_i for $i = 0, \cdots, n-1$, and by the remark above the same is true for every power of A. Then we can write

$$A^t = \sum_{i=0}^{n-1} c_{i+1}(t) M_i$$

for $t \geq 0$, where the $c_{i+1}(t)$ are to be determined. Since $A^{t+1} = A \cdot A^t$,

$$\begin{aligned} \sum_{i=0}^{n-1} c_{i+1}(t+1) M_i &= A \sum_{i=0}^{n-1} c_{i+1}(t) M_i \\ &= \sum_{i=0}^{n-1} c_{i+1}(t) [M_{i+1} + \lambda_{i+1} M_i] \quad \text{(from Eq. (4.7))} \\ &= \sum_{i=1}^{n-1} c_i(t) M_i + \sum_{i=0}^{n-1} c_{i+1}(t) \lambda_{i+1} M_i \end{aligned}$$

where we have replaced i by $i - 1$ in the first sum and used the fact that $M_n = 0$. The preceding equation is satisfied if the $c_i(t)$, $(i = 1, \cdots, n)$, are chosen to satisfy the system:

4.1 Initial Value Problems for Linear Systems

$$\begin{bmatrix} c_1(t+1) \\ \vdots \\ c_n(t+1) \end{bmatrix} = \begin{bmatrix} \lambda_1 & 0 & 0 & \cdots & 0 \\ 1 & \lambda_2 & 0 & \cdots & 0 \\ 0 & 1 & \lambda_3 & \cdots & 0 \\ \vdots & \ddots & \ddots & \ddots & \vdots \\ 0 & \cdots & 0 & 1 & \lambda_n \end{bmatrix} \begin{bmatrix} c_1(t) \\ \vdots \\ c_n(t) \end{bmatrix}. \tag{4.8}$$

Since $A^0 = I = c_1(0)I + \cdots + c_n(0)M_{n-1}$, we must have

$$\begin{bmatrix} c_1(0) \\ c_2(0) \\ \vdots \\ c_n(0) \end{bmatrix} = \begin{bmatrix} 1 \\ 0 \\ \vdots \\ 0 \end{bmatrix}. \tag{4.9}$$

By Theorem 4.1, the initial value problem (4.8), (4.9) has a unique solution. We have proved the following theorem:

Theorem 4.2. The solution of Eq. (4.4) with initial vector u_0 is

$$u(t) = \sum_{i=0}^{n-1} c_{i+1}(t) M_i u_0,$$

where the M_i are given by Eq. (4.7) and the $c_i(t)$, $(i = 1, \cdots, n)$, are uniquely determined by Eqs. (4.8) and (4.9).

Example 4.3. Use Theorem 4.2 to solve

$$u(t+1) = \begin{bmatrix} 1 & 1 \\ -1 & 3 \end{bmatrix} u(t), \quad u(0) = \begin{bmatrix} \alpha \\ \beta \end{bmatrix}.$$

The characteristic equation is

$$\det \begin{bmatrix} \lambda - 1 & -1 \\ 1 & \lambda - 3 \end{bmatrix} = 0, \text{ or}$$

$$\lambda^2 - 4\lambda + 4 = 0.$$

The matrix has an eigenvalue $\lambda = 2$ of multiplicity two. By Eq. (4.7),

$$M_0 = I,$$
$$M_1 = A - 2I = \begin{bmatrix} -1 & 1 \\ -1 & 1 \end{bmatrix}.$$

From Eqs. (4.8) and (4.9),

$$c_1(t+1) = 2c_1(t), \quad c_1(0) = 1,$$

so $c_1(t) = 2^t$. Using Eqs. (4.8) and (4.9) again,

$$c_2(t+1) = 2c_2(t) + 2^t, \quad c_2(0) = 0,$$

which has the solution $c_2(t) = t2^{t-1}$.

By Theorem 4.2,

$$u(t) = (c_1(t)I + c_2(t)M_1) \begin{bmatrix} \alpha \\ \beta \end{bmatrix}$$

$$= \left(2^t \begin{bmatrix} 1 & 0 \\ 0 & 1 \end{bmatrix} + t2^{t-1} \begin{bmatrix} -1 & 1 \\ -1 & 1 \end{bmatrix} \right) \begin{bmatrix} \alpha \\ \beta \end{bmatrix}$$

$$= 2^t \begin{bmatrix} 1 - \frac{t}{2} & \frac{t}{2} \\ -\frac{t}{2} & 1 + \frac{t}{2} \end{bmatrix} \begin{bmatrix} \alpha \\ \beta \end{bmatrix}.$$

This problem cannot be solved using Eq. (4.6) since the matrix $\begin{bmatrix} 1 & 1 \\ -1 & 3 \end{bmatrix}$ has only one independent eigenvector corresponding to $\lambda = 2$.

Occasionally, it is possible to compute the powers of a matrix very quickly by writing it as the sum of two commuting matrices, one of which is easy to raise to its powers (e.g., a diagonal matrix) and the other of which is nilpotent, i.e., all of its powers beyond some point are the zero matrix. Our next example is of this type.

4.1 Initial Value Problems for Linear Systems

Example 4.4. Compute all powers of

$$A = \begin{bmatrix} 2 & 0 \\ 1 & 2 \end{bmatrix}.$$

Write

$$A^t = \left(2I + \begin{bmatrix} 0 & 0 \\ 1 & 0 \end{bmatrix}\right)^t.$$

Now since $\begin{bmatrix} 0 & 0 \\ 1 & 0 \end{bmatrix}$ is nilpotent and commutes with I, the Binomial Theorem yields

$$\begin{aligned} A^t &= 2^t I + t 2^{t-1} \begin{bmatrix} 0 & 0 \\ 1 & 0 \end{bmatrix} \\ &= \begin{bmatrix} 2^t & 0 \\ t 2^{t-1} & 2^t \end{bmatrix}. \end{aligned}$$

Finally, we return to the nonhomogeneous system

$$u(t+1) = Au(t) + f(t). \tag{4.10}$$

The next theorem contains a "variation of parameters" formula for solving Eq. (4.10).

Theorem 4.3. The solution of Eq. (4.10) satisfying the initial condition $u(0) = u_0$ is

$$u(t) = A^t u_0 + \sum_{s=0}^{t-1} A^{t-s-1} f(s). \tag{4.11}$$

Proof. By Theorem 4.1, it is enough to show that Eq. (4.11) satisfies the initial value problem. First we have

$$\sum_{s=0}^{-1} A^{-s-1}f(s) = 0$$

by the usual convention, so $u(0) = u_0$.

For $t \geq 1$,

$$\begin{aligned}
u(t+1) &= A^{t+1}u_0 + \sum_{s=0}^{t} A^{t-s}f(s) \\
&= A^{t+1}u_0 + \sum_{s=0}^{t-1} A^{t-s}f(s) + f(t) \\
&= A\left[A^t u_0 + \sum_{s=0}^{t-1} A^{t-s-1}f(s)\right] + f(t) \\
&= Au(t) + f(t).
\end{aligned}$$ ∎

4.2 Stability of Linear Systems

The solution of an initial value problem for a system of equations with n unknowns is represented geometrically by a sequence of points $\{(u_1(t), \cdots, u_n(t))\}_{t=0}^{\infty}$ in $\mathcal{R}^n$. In many of the applications of this subject, it is useful to know the general location of those points for large values of t. Of course, there are numerous possibilities: the sequence could converge to a point or at least remain near a point, the sequence could oscillate among values near several points, the sequence might become unbounded, or the sequence might remain in a bounded set but jump around in seemingly unpredictable fashion.

The study of these matters is called *stability theory*. We will present some of the elements of this theory for homogeneous linear systems in the present section and generalize to nonlinear systems in the next section.

The first result is fundamental.

Theorem 4.4. Let A be an n by n matrix with $r(A) < 1$. Then every solution $u(t)$ of Eq. (4.4) satisfies $\lim_{t\to\infty} u(t) = 0$.

4.2 Stability of Linear Systems

Furthermore, if $r(A) < \delta < 1$, then there is a constant $C > 0$ so that

$$|u(t)| \leq C\delta^t |u_0| \tag{4.12}$$

for $t \geq 0$ and every solution of u of Eq. (4.4).

Proof. Fix δ so that $r(A) < \delta < 1$. From Theorem 4.2, the solution of Eq. (4.4), $u(0) = u_0$ is

$$u(t) = \sum_{i=0}^{n-1} c_{i+1}(t) M_i u_0.$$

By Eq. (4.8),

$$|c_1(t+1)| \leq r(A)|c_1(t)|.$$

Iterating this inequality and using $c_1(0) = 1$, we have

$$|c_1(t)| \leq (r(A))^t \leq \delta^t, \quad (t \geq 0).$$

Again by Eq. (4.8),

$$|c_2(t+1)| \leq r(A)|c_2(t)| + |c_1(t)|$$
$$\leq r(A)|c_2(t)| + (r(A))^t.$$

It follows from iteration and $c_2(0) = 0$ that

$$|c_2(t)| \leq t \cdot (r(A))^{t-1}, \quad (t \geq 0)$$
$$\leq t \left(\frac{r(A)}{\delta}\right)^{t-1} \delta^{t-1}.$$

L'Hospital's rule implies that

$$t \left(\frac{r(A)}{\delta}\right)^{t-1} \to 0, \quad (t \to \infty),$$

so there is a constant $B_1 > 0$ so that

$$|c_2(t)| \leq B_1 \delta^t, \quad (t \geq 0).$$

Similarly we can show that for $t \geq 0$

$$|c_3(t)| \leq \frac{t(t-1)}{2} r(A)^{t-2},$$

from which it follows that there is a B_2 so that

$$|c_3(t)| \leq B_2 \delta^t.$$

Continuing in this way (by induction), we obtain a constant $B^* > 0$ so that

$$|c_i(t)| \leq B^* \delta^t, \quad (t \geq 0),$$

for $i = 1, 2, \cdots, n$.

Now for any matrix M, there is a constant $D > 0$ so that

$$|Mv| \leq D|v|$$

for all v in $\mathcal{R}^n$. Finally, the solution $u(t)$ of Eq. (4.4), $u(0) = u_0$ satisfies

$$|u(t)| \leq \sum_{i=0}^{n-1} |c_{i+1}(t)||M_i u(0)|$$
$$\leq B^* \delta^t |u_0| \sum_{i=0}^{n-1} D_i$$
$$\leq C \delta^t |u_0|$$

for $C = B^* \sum_{i=0}^{n-1} D_i$. Consequently, Eq. (4.12) holds. Since $0 < \delta < 1$, $\lim_{t \to \infty} u(t) = 0$. ∎

When all solutions of the system go to the origin as t goes to infinity, the origin is said to be "asymptotically stable." A more precise definition of asymptotic stability is given in the next section.

4.2 Stability of Linear Systems

The next theorem shows that $r(A) < 1$ is a necessary condition for this type of stability.

Theorem 4.5. If $r(A) \geq 1$, then some solution $u(t)$ of Eq. (4.4) does not go to the origin as t goes to infinity.

Proof. Since $r(A) \geq 1$, there is an eigenvalue λ of A so that $|\lambda| \geq 1$. Let v be a corresponding eigenvector. Then $u(t) = \lambda^t v$ is a solution of Eq. (4.4), and $|u(t)| = |\lambda|^t |v| \not\to 0$ as $t \to \infty$. ∎

Example 4.5. $u(t+1) = \begin{bmatrix} 1 & -5 \\ .25 & -1 \end{bmatrix} u(t).$

The characteristic equation for $A = \begin{bmatrix} 1 & -5 \\ .25 & -1 \end{bmatrix}$ is $\lambda^2 + \frac{1}{4} = 0$. Then $\sigma(A) = \{\frac{i}{2}, -\frac{i}{2}\}$ and $r(A) = \frac{1}{2}$, so all solutions of this system converge to the origin as $t \to \infty$. Fig. 4.1 illustrates how the solution starting at $\begin{bmatrix} 10 \\ 1 \end{bmatrix}$ spirals in towards the origin.

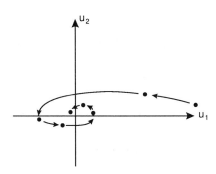

Fig. 4.1 A spiral solution portrait

If the matrix A has spectral radius $r(A) \leq 1$, then under certain conditions the system exhibits a weaker form of stability.

Theorem 4.6. Assume
 a) $r(A) \leq 1$,
 b) each eigenvalue λ of A with $|\lambda| = 1$ is simple. Then there is a constant $C > 0$ so that

$$|u(t)| \leq C|u_0| \qquad (4.13)$$

for $t \geq 0$ and every solution u of Eq. (4.4).

Proof. Label the eigenvalues of A so that $|\lambda_i| = 1$ for $i = 1, \cdots, k-1$, and $|\lambda_i| < 1$ for $i = k, \cdots, n$. From Eqs. (4.8) and (4.9),

$$c_1(t) = \lambda_1^t.$$

Next, c_2 satisfies

$$c_2(t+1) = \lambda_2 c_2(t) + \lambda_1^t,$$

so (as in the annihilator method)

$$(E - \lambda_1)(E - \lambda_2)c_2(t) = 0.$$

Since $\lambda_1 \neq \lambda_2$,

$$c_2(t) = B_{12}\lambda_1^t + B_{22}\lambda_2^t,$$

for some constants B_{12}, B_{22}. Continuing in this way, we have

$$c_i(t) = B_{1i}\lambda_1^t + \cdots + B_{ii}\lambda_i^t$$

for $i = 1, \cdots, k-1$. Consequently, there is a constant $D > 0$ so that

$$|c_i(t)| \leq D$$

for $i = 1, \cdots, k-1$ and $t \geq 0$.
From Eq. (4.8),

$$c_k(t+1) = \lambda_k c_k(t) + c_{k-1}(t),$$
$$|c_k(t+1)| \leq |\lambda_k||c_k(t)| + D.$$

4.2 Stability of Linear Systems

Choose $\delta = \max\{|\lambda_k|, \cdots, |\lambda_n|\} < 1$. Then

$$|c_k(t+1)| \leq \delta|c_k(t)| + D.$$

By iteration and the initial condition $c_k(0) = 0$,

$$|c_k(t)| \leq D \sum_{j=0}^{t-1} \delta^j$$
$$\leq \frac{D}{1-\delta}$$

for $t \geq 0$. In a similar manner, we find that there is a constant D^* so that

$$|c_i(t)| \leq D^*$$

for $i = 1, \cdots, n$ and $t \geq 0$.

From Theorem 4.2, the solution of Eq. (4.4), $u(0) = u_0$, is given by

$$u(t) = \sum_{i=0}^{n-1} c_{i+1}(t) M_i u_0$$

and

$$|u(t)| \leq D^* \sum_{i=0}^{n-1} |M_i u_0|$$
$$\leq C|u_0|$$

for $t \geq 0$ and some $C > 0$. ∎

The preceding theorem is useful in the analysis of multistep methods for the numerical approximation of solutions of initial value problems for differential equations (see [31]). A partial converse is given in Exercise 4.17.

Example 4.6. Consider the system

$$u(t+1) = \begin{bmatrix} \cos\theta & \sin\theta \\ -\sin\theta & \cos\theta \end{bmatrix} u(t),$$

where θ is a fixed angle. The matrix A in this example is a rotation matrix. When it is multiplied by a vector u, the resulting vector has the same length as u, but its direction is θ radians clockwise from u. Consequently, every solution u of the system has all of its values on a circle centered at the origin of radius $|u(0)|$. (See Fig. 4.2 for the case that $\theta = \frac{\pi}{2}$.)

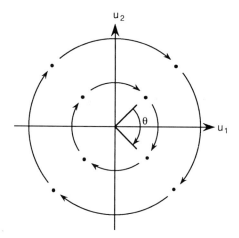

Fig. 4.2 Clockwise rotation through $\theta = \frac{\pi}{2}$

The eigenvalues of the rotation matrix are $\lambda = \cos\theta \pm i\sin\theta$. Then the hypotheses of Theorem 4.6 are satisfied, and in fact Eq. (4.13) holds with $C = 1$.

Next, we want to investigate the behavior of the solutions of a system in which some, but not all, of the eigenvalues have absolute value less than one. Some additional concepts and results from linear algebra will be needed. (See, for example, Hirsch and Smale [127].)

4.2 Stability of Linear Systems

Let λ be an eigenvalue of A of multiplicity m. Then the generalized eigenvectors of A corresponding to λ are the nontrivial solutions v of

$$(A - \lambda I)^m v = 0.$$

Of course, every eigenvector of A is also a generalized eigenvector. The set of all generalized eigenvectors corresponding to λ, together with the zero vector, is a generalized eigenspace and is a vector space having dimension m. The intersection of any two generalized eigenspaces is the zero vector. Finally, A times a generalized eigenvector is a vector in the same generalized eigenspace.

Example 4.7. What are the generalized eigenvectors for

$$A = \begin{bmatrix} 3 & 1 & 0 \\ 0 & 3 & 0 \\ 0 & 0 & 2 \end{bmatrix}?$$

A has eigenvalues $\lambda_1 = 3$ (multiplicity two) and $\lambda_2 = 2$. The generalized eigenvectors corresponding to $\lambda_1 = 3$ are solutions of

$$(A - 3I)^2 v = 0$$

or

$$\begin{bmatrix} 0 & 0 & 0 \\ 0 & 0 & 0 \\ 0 & 0 & 1 \end{bmatrix} \begin{bmatrix} v_1 \\ v_2 \\ v_3 \end{bmatrix} = 0.$$

The generalized eigenspace consists of all vectors with $v_3 = 0$. This is a two dimensional space, and

$$\begin{bmatrix} 1 \\ 0 \\ 0 \end{bmatrix}, \begin{bmatrix} 0 \\ 1 \\ 0 \end{bmatrix},$$

are basis vectors.

Corresponding to $\lambda_2 = 2$, there is a one-dimensional (generalized) eigenspace spanned by the eigenvector

$$\begin{bmatrix} 0 \\ 0 \\ 1 \end{bmatrix}.$$

Theorem 4.7. (The Stable Subspace Theorem) Let $\lambda_1, \cdots, \lambda_n$ be the (not necessarily distinct) eigenvalues of A arranged so that $\lambda_1, \cdots, \lambda_k$ are the eigenvalues with $|\lambda_i| < 1$. Let S be the k-dimensional space spanned by the generalized eigenvectors corresponding to $\lambda_1, \cdots, \lambda_k$. If u is a solution of Eq. (4.4) with $u(0)$ in S, then $u(t)$ is in S for $t \geq 0$ and

$$\lim_{t \to \infty} u(t) = 0.$$

Proof. Let u be a solution of Eq. (4.4) with $u(0)$ in S. Since A takes every generalized eigenspace into itself, it also takes S into itself. Then $u(t)$ is in S for $t \geq 0$.

Choose δ so that

$$\max\{|\lambda_1|, \cdots, |\lambda_k|\} < \delta < 1.$$

As in the proof of Theorem 4.4, there is a constant $B > 0$ such that

$$|c_i(t)| \leq B\delta^t$$

for $t \geq 0$, $1 \leq i \leq k$. By Theorem 4.2

$$u(t) = \sum_{i=0}^{n-1} c_{i+1}(t) M_i u(0).$$

Recalling the definition of M_i, Eq. (4.7), and the fact that $u(0)$ is a linear combination of generalized eigenvectors corresponding to $\lambda_1, \cdots, \lambda_k$, we have, for $i \geq k$,

$$M_i u(0) = 0.$$

4.2 Stability of Linear Systems

Then

$$|u(t)| \leq \sum_{i=0}^{k-1} |c_{i+1}(t)| |M_i u(0)|$$
$$\leq B\delta^t \sum_{i=0}^{k-1} |M_i u(0)|$$
$$\leq C\delta^t |u(0)|, \quad (t \geq 0),$$

for some constant C, so $\lim_{t \to \infty} u(t) = 0$. ∎

The set S in Theorem 4.7 is called the "stable subspace" for Eq. (4.4). It can be shown that every solution of the system that goes to the origin as t tends to infinity must have its initial point in S. Thus S can be described as the union of all sequences $\{u(t)\}_{t=0}^{\infty}$ that solve the system and satisfy $\lim_{t \to \infty} u(t) = 0$.

Example 4.8. What is the stable subspace for the system

$$u(t+1) = \begin{bmatrix} .5 & 0 & 0 \\ 1 & .5 & 0 \\ 0 & 1 & 2 \end{bmatrix} u(t) \ ?$$

The characteristic equation is

$$\det \begin{bmatrix} \lambda - .5 & 0 & 0 \\ -1 & \lambda - .5 & 0 \\ 0 & -1 & \lambda - 2 \end{bmatrix} = (\lambda - .5)^2 (\lambda - 2) = 0.$$

The stable subspace has dimension two and consists of the solutions of

$$(A - .5I)^2 v = 0$$

or

$$\begin{bmatrix} 0 & 0 & 0 \\ 0 & 0 & 0 \\ 1 & \frac{3}{2} & \frac{9}{4} \end{bmatrix} \begin{bmatrix} v_1 \\ v_2 \\ v_3 \end{bmatrix} = \begin{bmatrix} 0 \\ 0 \\ 0 \end{bmatrix}.$$

Thus S is the plane

$$4v_1 + 6v_2 + 9v_3 = 0.$$

(See Fig. 4.3). From Theorem 4.7, every solution that originates in this plane remains in the plane for all values of t and converges to the origin as $t \to \infty$. Since $\begin{bmatrix} 0 \\ 0 \\ 1 \end{bmatrix}$ is an eigenvector corresponding to $\lambda = 2$, the solutions originating on the v_3 axis are given by

$$u(t) = 2^t \begin{bmatrix} 0 \\ 0 \\ v_3 \end{bmatrix}, \quad (t \geq 0).$$

These remain on the v_3 axis and approach infinity in the positive or negative direction, depending on whether v_3 is positive or negative.

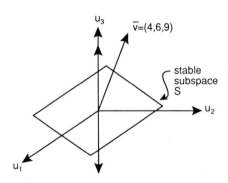

Fig. 4.3 A stable subspace

If some of the eigenvalues λ of A with $|\lambda| < 1$ are complex numbers, then the corresponding generalized eigenvectors will also be complex, and the stable subspace is a complex vector space. However, those generalized eigenvectors occur in conjugate pairs, and it is not difficult to verify that the real and imaginary parts of these vectors are real vectors that generate a real stable subspace of the same dimension.

4.3 Stability of Nonlinear Systems

In keeping with our concept of a difference equation articulated in the first chapter, we assume that we can write such an equation so that the value of the unknown at the largest value of the independent variable is isolated on one side of the equation. For example,

$$y(t+3) = y^2(t+2)y(t+1) - 5\sin y(t).$$

It is always possible to rewrite such an equation as an equivalent first order system. In the present example, set $u_1(t) = y(t)$, $u_2(t) = y(t+1)$, and $u_3(t) = y(t+2)$ to obtain

$$\begin{bmatrix} u_1 \\ u_2 \\ u_3 \end{bmatrix}(t+1) = \begin{bmatrix} u_2(t) \\ u_3(t) \\ u_3^2(t)u_2(t) - 5\sin u_1(t) \end{bmatrix}.$$

Note that t does not appear explicitly on the right-hand side of the equation. In this case, the system is said to be "autonomous." For an autonomous system, $u(t)$ is a solution for $t \geq 0$ if and only if $u(t - t_0)$ is a solution for $t \geq t_0$. (Verify this!)

We will study stability theory for the general autonomous system

$$u(t+1) = f(u(t)) \quad (t = 0, 1, 2 \cdots), \tag{4.14}$$

where u is an n-vector and f is a function from $\mathcal{R}^n$ into $\mathcal{R}^n$. The domain of f need not be all of $\mathcal{R}^n$, but f should map its domain into itself so that Eq. (4.14) is meaningful for all initial points in the domain and all t. Of course, Eq. (4.4) is a special case of Eq. (4.14).

Definition 4.1. A vector u in $\mathcal{R}^n$ is a "fixed point" of f if $f(u) = u$. A vector $v \in \mathcal{R}^n$ is a "periodic point" of f if there is a positive integer k so that $f^k(v) = v$, and k is a "period" for v. Here $f^k(v) = \underbrace{f(\cdots(f(f(v)))\cdots)}_{k \text{ times}}$.

Note that fixed points of f represent constant solutions of Eq. (4.14), and periodic points yield periodic solutions. The period k is not unique since any integral multiple of k is also a period, but every periodic point has a least period.

Example 4.9.

(a) $f(u) = 2u(1-u)$ has fixed points $u = 0$ and $u = \frac{1}{2}$.

(b) $f\left(\begin{bmatrix} u_1 \\ u_2 \end{bmatrix}\right) = \begin{bmatrix} \frac{u_2}{u_1} \\ \frac{u_2}{u_1} \end{bmatrix}$ with domain $\left\{ \begin{bmatrix} u_1 \\ u_2 \end{bmatrix} : u_1 \neq 0, u_2 \neq 0 \right\}$.

Every vector in the domain of f is a periodic point with period six (see Example 3.28). The only fixed point of f is $\begin{bmatrix} 1 \\ 1 \end{bmatrix}$. Three points have least period three (find them), but no points have least period two.

(c) $f\left(\begin{bmatrix} u_1 \\ u_2 \end{bmatrix}\right) = \begin{bmatrix} u_2 \\ -u_1 \end{bmatrix}$. This function rotates every vector $90°$ clockwise (see Fig. 4.2 and Example 4.6). Consequently, $\begin{bmatrix} 0 \\ 0 \end{bmatrix}$ is the only fixed point and every other point is periodic with least period four.

(d) The following system has been proposed as a discrete predator-prey model:

$$u_1(t+1) = (1+r)u_1(t) - \frac{\alpha u_1(t) u_2(t)}{1 + \beta u_1(t)},$$

$$u_2(t+1) = (1-d)u_2(t) + \frac{c\alpha u_1(t) u_2(t)}{1 + \beta u_1(t)}.$$

(See Freedman[80].) Here u_1 and u_2 denote the numbers of prey and predators, respectively, r is the birth rate minus the natural death rate of the prey, d is the death rate of the predators and α, β and c are positive constants. The nonlinear term in the first equation represents the number of prey devoured, while the nonlinear term in the second equation is the number of predators born.

The fixed points are found (with a little algebra) to be $\begin{bmatrix} 0 \\ 0 \end{bmatrix}$ and $\frac{1}{c\alpha - d\beta} \begin{bmatrix} d \\ cr \end{bmatrix}$. The second of these is ecologically interesting if $c\alpha - d\beta > 0$ since it is then located in a region where u_1 and u_2 are positive.

Finding periodic points is in general not a simple matter since it involves solving nonlinear equations of the form $f^k(v) = v$.

The following definition introduces the fundamental concepts in stability theory.

4.3 Stability of Nonlinear Systems

Definition 4.2.
  (a) Let u belong to $\mathcal{R}^n$ and $r > 0$. The "open ball centered at u with radius r" is the set

$$B(u,r) = \{v \text{ in } \mathcal{R}^n : |v - u| < r\}.$$

  (b) Let v be a fixed point of f. Then v is "stable" provided that, given any ball $B(v,\epsilon)$, there is a ball $B(v,\delta)$ so that if u is in $B(v,\delta)$, then $f^t(u)$ is in $B(v,\epsilon)$ for $t \geq 0$. If v is not stable, then it is "unstable."
  (c) If, in addition to the conditions in part (b), there is a ball $B(v,r)$ so that $f^t(u) \to v$ as $t \to \infty$ for all u in $B(v,r)$, then v is "asymptotically stable."
  (d) Let w be a periodic point of f with period k. Then w is "stable" ("asymptotically stable") if $w, f(w), \cdots, f^{k-1}(w)$ are stable (asymptotically stable) as fixed points of f^k.

Intuitively, a fixed point v is stable if points close to v do not wander far from v under all iterations of f. Asymptotic stability of v requires the additional condition that all solutions of Eq. (4.14) that start near v converge to v.

In Theorems 4.4 and 4.5, we showed that $r(A) < 1$ is a necessary and sufficient condition for the origin to be an asymptotically stable fixed point for the homogeneous linear system of Eq. (4.4). Actually, this is a strong type of asymptotic stability (called global asymptotic stability) since all solutions of Eq. (4.4) converge to the origin. Under the weaker conditions of Theorem 4.6, the origin is stable since the inequality (4.13) implies that if $u(0)$ is in $B(0,\delta)$, then $u(t)$ is in $B(0,C\delta)$ for $t \geq 0$.

There is an elementary graphical technique known as the "staircase method" that is useful for stability analysis in the case of a single nonlinear equation of first order.

Example 4.10. $u(t+1) = 2u(t)(1 - u(t))$.

First, we graph $y = 2u(1-u) = f(u)$ and $y = u$ on the same coordinate axes (see Fig. 4.4). The fixed points of f are the intersection points of the two graphs. Choose any initial value $u(0)$ in $(0, \frac{1}{2})$ and proceed vertically to the point $(u(0), f(u(0))) = (u(0), u(1))$ on the graph of f. Now move horizontally to $(u(1), u(1))$ on the line $y = u$. Since $u(2) = f(u(1))$, another vertical movement brings us to $(u(1), u(2))$ on the graph of f. By alternating vertical motion to the graph of f with horizontal motion to the line in this way, we generate the solution sequence $\{u(t)\}$.

In this case, Fig. 4.4 shows that the sequence rapidly converges to the fixed point $\frac{1}{2}$. Similarly, if we begin with an initial value in $(\frac{1}{2}, 1)$, then the solution converges to $\frac{1}{2}$. It follows that $\frac{1}{2}$ is asymptotically stable.

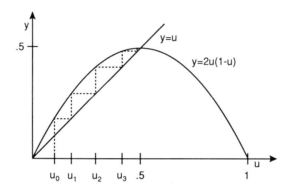

Fig. 4.4 Staircase method for $f(u) = 2u(1-u)$

Recall from Example 3.27 that a general solution of $u(t+1) = 2u(t)(1-u(t))$ is $u(t) = \frac{1}{2}(1 - A^{2^t})$. Then $A = 1 - 2u(0)$, so

$$u(t) = \frac{1}{2}\left(1 - (1 - 2u(0))^{2^t}\right).$$

If $0 < u(0) < 1$, then $-1 < 1 - 2u(0) < 1$, and $\lim_{t \to \infty} u(t) = \frac{1}{2}$ in agreement with the result of the graphical analysis. The fixed point $u = 0$ is clearly unstable.

Example 4.11. Use the staircase method to analyze solutions of $u(t+1) = \cos u(t)$.

Figure 4.5 illustrates that the solution $u(t)$ with initial value $u(0) = 1.3$ satisfies $\lim_{t \to \infty} u(t) = y_0$, where $y_0 \simeq 0.739$ is the unique fixed point of $\cos u$. This behavior is also easily demonstrated using a calculator by entering 1.3 (radians) and pushing the cosine button repeatedly. Try some other initial conditions. What happens if you use degrees instead of radians?

4.3 Stability of Nonlinear Systems

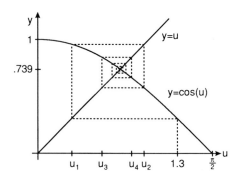

Fig. 4.5 Staircase method for $f(u) = \cos u$

The next result is an analytic method for checking asymptotic stability in the scalar case.

Theorem 4.8. Suppose f has a continuous first derivative in some open interval containing a fixed point v.

(a) If $|f'(v)| < 1$, then v is asymptotically stable.
(b) If $|f'(v)| > 1$, then v is unstable.

Proof.

(a) Assume $|f'(v)| < 1$. By continuity of f', there is an α so that $|f'(u)| \leq \alpha < 1$ on some interval $I = (v - \delta, v + \delta)$, $\delta > 0$. The Mean Value Theorem gives

$$|f(u) - f(w)| = |f'(c)| \, |u - w|$$
$$\leq \alpha |u - w|$$

if u, w are in I. For each u in I,

$$|f(u) - v| \leq \alpha |u - v| < \delta,$$

so $f(u)$ is in I, and we can conclude that v is stable. Furthermore,

$$|f^{t+1}(u) - v| \leq \alpha |f^t(u) - v|$$

for each $t \geq 0$ and u in I, so by induction

$$|f^t(u) - v| \leq \alpha^t |u - v|$$

for $t \geq 0$ and u in I. Since $\lim_{t \to \infty} \alpha^t = 0$, each solution of Eq. (4.14) that originates in I converges to v as $t \to \infty$, and v is asymptotically stable.

(b) Assume $|f'(v)| > 1$. Choose $\lambda > 1$ and $I = (v - \epsilon, v + \epsilon)$ for some $\epsilon > 0$ so that

$$\begin{aligned} |f(u) - f(w)| &= |f'(u)||u - w| \\ &\geq \lambda |u - w| \end{aligned}$$

for all u, w in I. By induction,

$$|f^t(u) - v| \geq \lambda^t |u - v|$$

as long as $f^t(u)$ is in I. Since $\lambda > 1$, it follows that all solutions of (4.14) which originate in I, except for the constant solution $u(t) = v$, must leave I for sufficiently large t. Then v is unstable. ∎

Example 4.12.

(a) Let $f(u) = 2u(1 - u)$. Then $f'(u) = 2 - 4u$, so $f'(0) = 2$, $f'(\frac{1}{2}) = 0$. and as in Example 4.10 we find that 0 is unstable while $\frac{1}{2}$ is asymptotically stable.

(b) If $f(u) = \cos u$, then $f'(u) = -\sin u$, and $|f'(u)| < 1$ if $u \neq \frac{\pi}{2} + 2n\pi$. The fixed point $v \simeq 0.739$ is asymptotically stable.

If $|f'(v)| = 1$, then v might be asymptotically stable, merely stable or unstable (see Exercise 4.25).

Theorem 4.8 can also be used to test the stability of periodic points. For example, suppose that v is a point of period two for the function f. By the Chain Rule and Theorem 4.8, v is asymptotically stable if

$$|(f^2)'(v)| = |f'(f(v))\, f'(v)| < 1.$$

Note that this calculation also implies that the periodic point $f(v)$ is asymptotically stable. Similar remarks apply to points of higher periods.

The most powerful method for establishing stability and asymptotic stability for nonlinear systems is due to the Soviet mathematician A.M. Liapunov.

4.3 Stability of Nonlinear Systems

Definition 4.3. Let v be a fixed point of f. A real-valued continuous function V on some ball B about v is called a "Liapunov function" for f at v provided $V(v) = 0$, $V(u) > 0$ for $u \neq v$ in B, and

$$\Delta_t V(u) \equiv V(f(u)) - V(u) \leq 0 \qquad (4.15)$$

for all u in B. If the inequality (4.15) is strict for $u \neq v$, then V is a "strict Liapunov function."

Let u be a solution of Eq. (4.14) with $u(0)$ in B. Then Eq. (4.15) requires that $V(u(t))$ is nonincreasing as a function of t as long as $u(t)$ is in B. When f is continuous, the existence of a Liapunov function at v implies that v is stable.

Theorem 4.9. Let v be a fixed point of f, and assume f is continuous on some ball about v. If there is a Liapunov function for f at v, then v is stable. If there is a strict Liapunov function for f at v, then v is asymptotically stable.

Proof. (Note: the proof requires some topological methods which can be found in undergraduate textbooks on analysis or topology; see Bartle [21].)

Let V be a Liapunov function for v on a ball B about v. Suppose $B(v, \epsilon)$ is properly contained in B and f is continuous on $B(v, \epsilon)$. Then

$$m \equiv \min\{V(u) : |u - v| = \epsilon\}$$

is positive since V is positive and continuous on $\{u : |u - v| = \epsilon\}$, which is a closed and bounded set.

Choose $\delta > 0$ small enough so that $B(v, \delta)$ is contained in the open set $U \equiv \{u \in B : V(u) < \frac{m}{2}\}$. For each u in $B(v, \delta)$,

$$V(f(u)) \leq V(u) < \frac{m}{2}, \quad \text{(by Eq.(4.15))}.$$

Let W be the maximal open connected subset of U that contains v. Note that $W \subseteq B(v, \epsilon)$.

Let u belong to $B(v, \delta)$. Since f is continuous and $B(v, \delta)$ is connected, $f(B(v, \delta))$ is connected. Then $f(B(v, \delta)) \subseteq W$, so $f(u)$ is in $B(v, \epsilon)$. In a similar way, we can show $f^t(u)$ belongs to $B(v, \epsilon)$ for $t \geq 2$. Therefore v is stable.

Suppose further that V is a strict Liapunov function. We assume v is not asymptotically stable and seek to arrive at a contradiction. There is a w in $B(v,\delta)$ so that $f^t(w)$ does not converge to v as $t \to \infty$. However, each $f^t(w)$ is in $B(v,\epsilon)$, so there is a subsequence $f^{t_n}(w)$ that converges to some $\overline{v} \neq v$ in $B(v,\epsilon)$ as $n \to \infty$. For each t, some $t_n > t$, so

$$V(f^t(w)) > V(f^{t_n}(w)) > V(\overline{v}) \qquad (4.16)$$

by Eq. (4.15) (with strict inequality) and the continuity of V.

Again by Eq. (4.15), $V(f(\overline{v})) < V(\overline{v})$. Since $V \circ f$ is continuous, there is a $\gamma > 0$ so that

$$V(f(u)) < V(\overline{v})$$

for u in $B(\overline{v}, \gamma)$. Choose n large enough that $f^{t_n}(w)$ in $B(\overline{v}, \gamma)$. Then

$$V\left(f^{t_n+1}(w)\right) < V(\overline{v}),$$

which contradicts Eq. (4.16). We then conclude that v is asymptotically stable. ∎

Exercise 4.34 contains a generalization of the second portion of this theorem.

In practice, the difficulty in making use of Theorem 4.9 is finding a Liapunov function. However, if one can be found, then we may gain, in addition to stability, further information about solutions of Eq. (4.14). For example, the following corollary may allow us to locate solutions that converge to the fixed point. The proof is similar to the last part of the proof of Theorem 4.9.

Corollary 4.1. Suppose there is a strict Liapunov function for f at v on B and f is continuous on B. Then every solution of Eq. (4.14) that remains in B for $t \geq t_0$ must converge to v.

Example 4.13. Use Theorem 4.9 to show that the origin is stable for

$$u(t+1) = \begin{bmatrix} \cos\theta & \sin\theta \\ -\sin\theta & \cos\theta \end{bmatrix} u(t).$$

Recall that stability was established earlier (Example 4.6) by computing the eigenvalues.

4.3 Stability of Nonlinear Systems

Here we define V on $\mathcal{R}^2$ by

$$V\left(\begin{bmatrix} u_1 \\ u_2 \end{bmatrix}\right) = u_1^2 + u_2^2.$$

Then $V\left(\begin{bmatrix} 0 \\ 0 \end{bmatrix}\right) = 0$, $V(u) > 0$ otherwise, and

$$\begin{aligned}
\Delta_t V(u) &= V\left(\begin{bmatrix} u_1 \cos\theta + u_2 \sin\theta \\ -u_1 \sin\theta + u_2 \cos\theta \end{bmatrix}\right) - V\left(\begin{bmatrix} u_1 \\ u_2 \end{bmatrix}\right) \\
&= (u_1 \cos\theta + u_2 \sin\theta)^2 + (-u_1 \sin\theta + u_2 \cos\theta)^2 \\
&\quad - u_1^2 - u_2^2 \\
&= 0.
\end{aligned}$$

Consequently, V is a Liapunov function, and the origin is stable.

Note that since $V(u)$ is the square of the length of u, the equation $\Delta_t V(u) = 0$ tells us that each solution of the system remains on a fixed circle centered at the origin.

Example 4.14. What can be said about the stability of the origin for

$$u(t+1) = \begin{bmatrix} u_2(t) - u_2(t)(u_1^2(t) + u_2^2(t)) \\ u_1(t) - u_1(t)(u_1^2(t) + u_2^2(t)) \end{bmatrix}?$$

Again, we try $V(u) = u_1^2 + u_2^2$. Now

$$\begin{aligned}
\Delta_t V(u) &= \left[u_2\left(1 - (u_1^2 + u_2^2)\right)\right]^2 + \left[u_1\left(1 - (u_1^2 + u_2^2)\right)\right]^2 - u_1^2 - u_2^2 \\
&= (u_2^2 + u_1^2)\left(1 - 2(u_1^2 + u_2^2) + (u_1^2 + u_2^2)^2\right) - u_1^2 - u_2^2 \\
&= (u_1^2 + u_2^2)^2\left(-2 + (u_1^2 + u_2^2)\right) \\
&< 0,
\end{aligned}$$

when u is in $B(0, \sqrt{2})$ and $u \neq 0$. It follows that V is a strict Liapunov function and the origin is asymptotically stable.

Furthermore, since $\Delta_t V(u) < 0$ is equivalent to $|f(u)| < |u|$, every solution originating in $B(0, \sqrt{2})$ must remain there. By Corollary 4.1, each such solution must converge to the origin.

In some cases, the asymptotic stability of a fixed point can be established by showing that the "linear" part of the function satisfies the condition for asymptotic stability of a linear system. Assume f has the form

$$f(u) = Au + g(u), \qquad (4.17)$$

where A is an n by n constant matrix and g satisfies

$$\lim_{u \to 0} \frac{|g(u)|}{|u|} = 0. \qquad (4.18)$$

Equation (4.18) implies that $g(0) = 0$ and that g contains no nontrivial linear terms.

The conditions (4.17) and (4.18) mean that f is differentiable at $u = 0$. This is necessarily the case if f has continuous first order partial derivatives at $u = 0$ (see Bartle [21]). The matrix A is the Jacobian matrix of f at 0 given by

$$A = \begin{bmatrix} \frac{\partial f_1}{\partial u_1} & \frac{\partial f_1}{\partial u_2} & \cdots & \frac{\partial f_1}{\partial u_n} \\ \frac{\partial f_2}{\partial u_1} & \frac{\partial f_2}{\partial u_2} & \cdots & \frac{\partial f_2}{\partial u_n} \\ \vdots & \vdots & \ddots & \vdots \\ \frac{\partial f_n}{\partial u_1} & \frac{\partial f_n}{\partial u_2} & \cdots & \frac{\partial f_n}{\partial u_n} \end{bmatrix},$$

where $f_1, \cdots, f_n$ are the components of f and all the partial derivatives are evaluated at the origin.

Theorem 4.10. If f is defined by Eq. (4.17) with $r(A) < 1$ and g satisfying Eq. (4.18), then the origin is asymptotically stable.

Proof. From Theorem 4.4, there is a constant $C > 0$ so that

$$|A^t u| \leq C\delta^t |u|,$$

for $t \geq 0$, u in $\mathcal{R}^n$, if $r(A) < \delta < 1$. Since $\det(A - \lambda I) = \det(A - \lambda I)^T = \det(A^T - \lambda I)$, A and A^T have the same eigenvalues.

4.3 Stability of Nonlinear Systems

Then there is a constant $D > 0$ so that

$$|(A^T)|^t|u| \leq D\delta^t|u|,$$

for $t \geq 0$, u in $\mathcal{R}^n$. Now

$$|(A^T)^t A^t u| \leq D\delta^t |A^t u|$$
$$\leq DC\delta^{2t}|u|,$$

for all t and u. The geometric series $\sum_{t=0}^{\infty}(\delta^2)^t$ converges, so by the comparison test

$$Bu \equiv \sum_{t=0}^{\infty}(A^T)^t A^t u$$

converges for each u. Since B is linear, we can represent it by an n by n matrix.

Define V on $\mathcal{R}^n$ by

$$V(u) = u^T B u = \sum_{t=0}^{\infty}|A^t u|^2.$$

It is easily checked that $V(0) = 0$ and $V(u) > 0$ for $u \neq 0$. Finally, consider

$$\Delta_t V(u) = V(Au + g(u)) - V(u)$$
$$= (Au + g(u))^T B(Au + g(u)) - u^T B u.$$

Since B is symmetric,

$$(x + y)^T B(x + y) = x^T B x + 2x^T B y + y^T B y.$$

Using this formula, we have

$$\Delta_t V(u) = u^T (A^T B A) u + 2 u^T A^T B g(u)$$
$$+ g(u)^T B g(u) - u^T B u.$$

182   Chapter 4.   Stability Theory

Note that

$$\begin{aligned}A^T B A u &= \sum_{t=0}^{\infty} A^T (A^T)^t A^t A u \\ &= \sum_{t=0}^{\infty} (A^T)^{t+1} A^{t+1} u \\ &= \sum_{t=1}^{\infty} (A^T)^t A^t u \\ &= Bu - u,\end{aligned}$$

where we have used a change of index. Thus

$$\Delta_t V(u) = u^T B u - u^T u + 2 u^T A^T B g(u)$$

$$+ g(u)^T B g(u) - u^T B u$$

$$= |u|^2 \left[-1 + 2 \frac{u^T}{|u|} A^T B \frac{g(u)}{|u|} + \frac{g(u)^T}{|u|} B \frac{g(u)}{|u|} \right].$$

Using Eq. (4.18), $\Delta_t V(u) < 0$ for $|u|$ sufficiently small and $u \neq 0$. By Theorem 4.9, the origin is asymptotically stable. ∎

In case $r(A) > 1$ and g satisfies Eq. (4.18), it can be shown that the origin is unstable for f given by Eq. (4.17) (see LaSalle [160]). If $r(A) = 1$, then linearization gives no information, and some other method must be used to investigate stability. For instance, the function f in Example 4.14 is

$$F\left(\begin{bmatrix} u_1 \\ u_2 \end{bmatrix} \right) = \begin{bmatrix} 0 & 1 \\ 1 & 0 \end{bmatrix} \begin{bmatrix} u_1 \\ u_2 \end{bmatrix} - (u_1^2 + u_2^2) \begin{bmatrix} u_2 \\ u_1 \end{bmatrix}.$$

Here the matrix A has spectral radius one and g satisfies Eq. (4.18) since

$$\frac{|g(u)|}{|u|} = \frac{(u_1^2 + u_2^2)|u|}{|u|} = u_1^2 + u_2^2 \to 0$$

4.4 Chaotic Behavior

as $\begin{bmatrix} u_1 \\ u_2 \end{bmatrix} \to 0$. Even though linearization fails, we were able to establish asymptotic stability by use of a Liapunov function.

Example 4.15. For arbitrary constants α and β consider the system

$$u_1(t+1) = 0.5u_1(t) + \alpha u_1(t)u_2(t),$$
$$u_2(t+1) = -0.7u_2(t) + \beta u_1(t)u_2(t).$$

The Jacobian matrix evaluated at the origin for this system is

$$A = \begin{bmatrix} 0.5 & 0 \\ 0 & -0.7 \end{bmatrix}.$$

Note that $r(A) = 0.7 < 1$. Since the function $f(u)$ for this system has continuous first partial derivatives with respect to u_1 and u_2, Theorem 4.10 implies that the origin is asymptotically stable.

Now suppose that f has fixed point $v \neq 0$. Let $w = u - v$. The equation $u(t+1) = f(u(t))$ is transformed into

$$w(t+1) = f(w(t) + v) - v \equiv h(w(t)).$$

Then the origin is a fixed point for h, and the Jacobian matrix of h at 0 is the same as the Jacobian matrix of f at v. Consequently, we can test v for asymptotic stability by computing the eigenvalues of the Jacobian matrix of f at v.

Although we have concentrated on stability questions for fixed points in this section, the same methods can be used to test periodic points since, by Definition 4.2, periodic points have the stability properties of their iterates as fixed points of f^k. We will examine periodic points more closely in the next section.

Additional information on stability theory for nonlinear systems can be found in LaSalle [160]. See Lakshmikantham and Trigiante [157] for the case of nonautonomous systems.

4.4 Chaotic Behavior

In Examples 3.29 and 3.30, we showed that the equation

$$u(t+1) = au(t)((1 - u(t))), \quad (t = 0, 1, \cdots), \tag{4.19}$$

can be solved explicitly for $u(t)$ when $a = 2$ and $a = 4$. The solutions are

$$u(t) = \frac{1}{2}\left[1 - (1 - 2u(0))^{2^t}\right], \quad (a = 2),$$

and

$$u(t) = \sin^2\left[2^{t-1} \cos^{-1}(1 - 2u(0))\right], \quad (a = 4), \qquad (4.20)$$

where $0 \leq u(0) \leq 1$. The behavior of these solutions is remarkably different. For $a = 2$, each solution $u(t)$ of Eq. (4.19) with $0 < u(0) < 1$ converges rapidly to the asymptotically stable fixed point $u = \frac{1}{2}$. In the case $a = 4$, most solutions seem to jump randomly about the interval $(0, 1)$ (see Fig. 4.6). However, there are also many periodic points such as $u = \frac{3}{4}$ (period 1) and $u = \frac{1}{2}(1 - \cos\frac{2\pi}{5})$ (period 2).

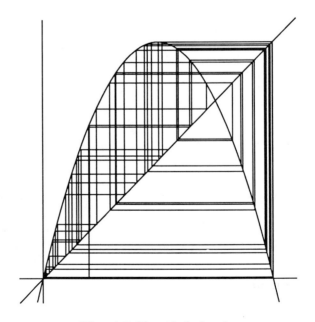

Fig. 4.6 Chaotic behavior

4.4 Chaotic Behavior

Let us investigate the nature of the solutions for intermediate values of a. Define $f(u) = au(1-u)$. Setting $f(u) = u$, we obtain the fixed points $u = \frac{a-1}{a}$ and $u = 0$. Now $f'(0) = a > 2$ and $f'\left(\frac{a-1}{a}\right) = 2 - a$, so 0 is unstable while $\frac{a-1}{a}$ is asymptotically stable for $2 \leq a < 3$. Thus there is little change in the behavior of solutions as a increases from 2 to 3.

Consider the composite function

$$f(f(u)) = a^2 u(1-u)(1 - au + au^2).$$

If we set $f(f(u)) = u$, the roots are

$$0, \quad \frac{a-1}{a}, \quad \frac{a+1 \pm \sqrt{(a+1)(a-3)}}{2a}.$$

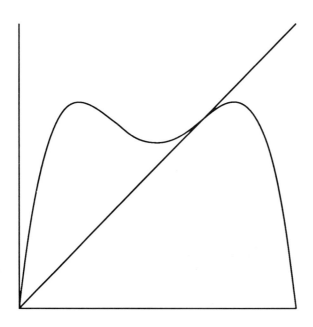

Figure 4.7a $f(f(u))$ for $a < 3$

The first two roots are the fixed points of f, and the remaining two are points of period two for $a > 3$. Figure 4.7 illustrates how these new periodic points occur as the slope of the tangent line to $f(f(u))$ at $u = \frac{a-1}{a}$ increases beyond one when a passes through 3.

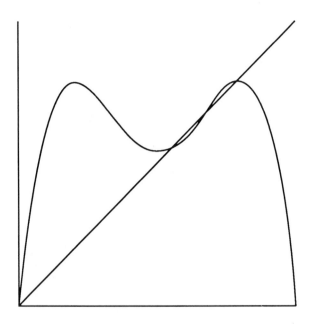

Figure 4.7b $f(f(u))$ for $a > 3$

These points of period 2 are asymptotically stable (see Exercise 4.38) as a ranges from 3 to about 3.45. A staircase diagram with a stable pair of periodic points is shown in Fig. 4.8. At the same value of a where these become unstable, stable points of period 4 appear. These remain stable over a short interval of a values before giving way to stable points of period 8. This phenomenon of period doubling continues with the ratio of consecutive lengths of intervals of stability approaching "Feigenbaum's number" $4.6692\cdots$. At approximately $a = 3.57$ all points whose periods are powers of two are unstable, and the situation becomes complicated. For some values of $a > 3.57$, the motion of solutions appears random, while there are small a intervals on which asymptotically stable periodic solutions (having periods different from 2^n) control the behavior of solutions. Near $a = 4$ most solutions bounce erratically around the interval $(0, 1)$.

4.4 Chaotic Behavior

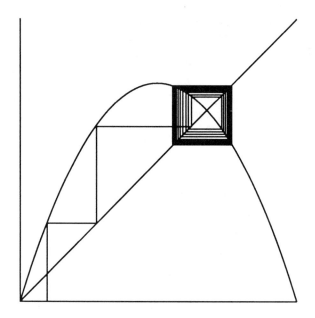

Fig. 4.8 Stable points of period 2

Many of these properties associated with the special equation (4.19) have been observed for a variety of equations that exhibit a transition from stability to "chaotic" behavior. Even Feigenbaum's number and the order of occurrence of periods for periodic points are quite generic phenomena (see Feigenbaum [74] and Devaney [57]).

The study of chaos is a fairly new and rapidly evolving branch of mathematics. There is no general agreement on a definition of chaos. In this section we will not be concerned with general results but will focus on some of the characteristics of chaos and on some of the methods for recognizing chaotic behavior.

We begin by asking whether the motion of solutions of Eq. (4.19) with $a = 4$ is random (or more accurately "pseudo-random") in some sense. Let

$$\begin{aligned}\theta(t) &= 2^{t-1}\cos^{-1}(1 - 2u(0)) \bmod \pi \\ &= 2^{t-1}\theta(1) \bmod \pi.\end{aligned}$$

This equation says that $\theta(t)$ is in the interval $[0, \pi)$ and that $2^{t-1}\theta(1) - \theta(t)$ is an integral multiple of π. From Eq. (4.20), $u(t) = \sin^2 \theta(t)$ since the square of the sine function has period π.

A graph of $\theta(t)$ against $\theta(1)$ is shown in Fig. 4.9. We see that for large values of t, small $\theta(1)$ intervals are linearly expanded into much larger intervals. It is reasonable to conjecture that as t grows large, $\theta(t)$ is approximately uniformly distributed in the following sense: given a subinterval of $[0, \pi)$ and a $\theta(1)$ chosen at random from that subinterval, the corresponding value of $\theta(t)$ has approximately the same chance of occurring in each subinterval of $[0, \pi)$ of a prescribed length.

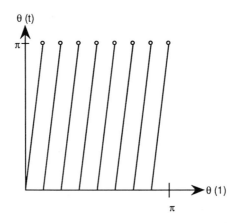

Fig. 4.9 $\theta(t)$ **as a function of** $\theta(1)$

Since $u(t) = \sin^2 \theta(t)$, the probability density function p for $u(t)$ is given by

$$\int p(u(t))\, du(t) \;=\; \frac{2}{\pi} \int d\theta(t),$$

where the factor of two appears because $\sin^2 \theta$ maps two points to one point in $[0, 1]$.

4.4 Chaotic Behavior

Thus

$$p(u(t)) = \frac{2}{\pi \frac{du(t)}{d\theta(t)}}$$

$$= \frac{2}{\pi 2 \sin\theta(t) \cos\theta(t)}$$

$$= \frac{1}{\pi \sqrt{u(t)(1-u(t))}}.$$

It is easily checked that $\int_0^1 p(u)du = 1$, so p is a probability density function on $[0, 1]$.

If we use p to subdivide $[0, 1]$ into subintervals of equal probability, then we expect to find after many iterations about the same number of values of a solution in each subinterval. For example to obtain four subintervals of equal probability, set

$$\int_0^b \frac{1}{\pi\sqrt{u(1-u)}} = 0.25,$$

from which it follows that

$$b = \sin^2\left(\frac{0.25\pi}{2}\right) \simeq 0.14645.$$

The subintervals are $[0, 0.14645]$, $[0.14645, 0.5]$, $[0.5, 0.85355]$, and $[0.85355, 1]$. As a rough check, we computed 600 iterations of Eq. (4.19) with $a = 4$ and $u(0) = 0.2$ and obtained 158, 143, 155, and 144 points in the first, second, third, and fourth subintervals respectively.

Perhaps the most fundamental characteristic of chaotic motion is the property of sensitive dependence on initial conditions. From a practical point of view, this property implies that any small error in the initial condition can lead to much larger errors in the values of a solution as t increases. We give a precise definition for the equation

$$u(t+1) = f(u(t)), \qquad (4.21)$$

where f maps an interval I of real numbers onto itself.

Definition 4.4. The solutions of Eq. (4.21) have "sensitive dependence on initial conditions" if there is a $d > 0$ so that for each u_0 in I and every open interval J containing u_0, there is a v_0 in J so that the solutions u and v of Eq. (4.21) with $u(0) = u_0$ and $v(0) = v_0$ satisfy $|u(t) - v(t)| > d$ for some t.

The property of sensitive dependence on initial conditions is sometimes called the "butterfly effect" because if the laws of meteorology have this property, then the motion of a butterfly's wings can have large scale effects on the weather.

The solutions of Eq. (4.19) with $a = 4$ have sensitive dependence on initial conditions because of the angle doubling that occurs in Eq. (4.20). To be specific, consider initial values u_0 and v_0, which lie near each other in $(0, 1)$. The corresponding angles $\theta(1) = \cos^{-1}(1 - 2u_0)$ and $\varphi(1) = \cos^{-1}(1 - 2v_0)$ are also close together (mod π). For $t \geq 1$,

$$\theta(t) - \varphi(t) = 2^{t-1}(\theta(1) - \varphi(1)) \bmod \pi,$$

so the difference doubles (mod π) with each iteration. It follows that the solutions $u(t) = \sin^2 \theta(t)$ and $v(t) = \sin^2 \varphi(t)$ will not be near each other for most values of t.

Note that this result casts some doubt on the validity of the experiment mentioned earlier where 600 iterations were computed and sorted into four intervals. Since the computations necessarily contain round-off errors after the first few steps, the computed solution is actually quite different from the exact solution! This example illustrates a difficulty in studying chaotic behavior computationally.

Example 4.16. Let f be the "tent map"

$$f(u) = \begin{cases} 2u, & 0 \leq u < \frac{1}{2} \\ 2(1-u), & \frac{1}{2} \leq u \leq 1. \end{cases}$$

(See Fig. 4.10).

4.4 Chaotic Behavior

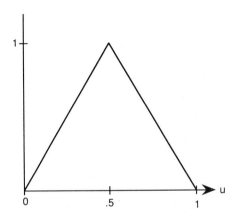

Fig. 4.10 The tent map

Consider intervals of the form $\left(\frac{k-1}{2^n}, \frac{k}{2^n}\right)$, where n is any nonnegative integer and k is an integer between 1 and 2^n. It is easy to see that f maps each of these intervals ($n > 0$) onto another interval of the same type but with twice the original length. Thus solutions of Eq. (4.21) have sensitive dependence on initial conditions.

Another characteristic of chaotic motion is the existence of many unstable periodic points. Consider again Eq. (4.20) written in the form $u(t) = \sin^2[2^{t-1}\theta(1)]$. If there is an integer m so that

$$2^{t-1}\theta(1) = \theta(1) + m\pi$$

for some t, then $\theta(1)$ yields a periodic point. We have

$$\theta(1) = \frac{m\pi}{2^{t-1} - 1}$$

for all integers $t \geq 2$ and integers m so that $0 \leq m \leq 2^{t-1} - 1$. We conclude that, not only are there infinitely many periodic points, but they are dense in the interval $[0, 1]$. All of them are unstable because of the butterfly effect.

In order to study the periodic points of the tent map f in Example 4.16, we introduce the method of "symbolic dynamics." For each initial point u_0 in $[0,1]$, define a sequence $\{b(t)\}_{t=0}^{\infty}$ of zeros and ones as follows:

$$b(t) = \begin{cases} 0 & \text{if } u(t) < \frac{1}{2}, \\ 1 & \text{if } u(t) \geq \frac{1}{2}, \end{cases}$$

where $u(t)$ is the solution of Eq. (4.21) with $u(0) = u_0$. Thus $b(t)$ contains information about which half-interval holds the tth iterate of u_0 under f. Because of the interval-doubling property of f, two distinct initial points must be associated with two different binary sequences.

Definition 4.5. The "shift-operator" σ acts on binary sequences as follows:

$$\sigma(b(t)) = a(t),$$

where $a(t) = b(t+1)$.

The shift operator deletes the first number $b(0)$ in the sequence and shifts every other number in the sequence one position to the left. If we denote by h the one-to-one function that maps each u_0 in $[0,1]$ to the corresponding sequence $b(t)$, then we have

$$h \circ f = \sigma \circ h.$$

This relation indicates that the action of σ on the sequence space is equivalent to the action of f on $[0,1]$.

A sequence $b(t)$ is a periodic "point" for σ if and only if it is repeating, i.e., $b(t+m) = b(t)$ for all t. Consequently, σ has 2^m periodic points of period m. Furthermore, each of these periodic sequences is associated with a unique point u in $[0,1]$ that is a periodic point for f of period m.

For example, consider the periodic binary sequence $0,1,0,1,0,1,\cdots$. Since $b(0) = 0$, u is in $[0, \frac{1}{2}]$. Next, $b(1) = 1$ implies u belongs to $[\frac{1}{4}, \frac{1}{2}]$. Since $b(2) = 0$, u must be in $[\frac{3}{8}, \frac{1}{2}]$, and so forth. In fact, u is the unique point in the intersection of the resulting nested sequence of closed intervals. To see that u has period 2, note that

$$\begin{aligned} f(f(u)) &= (h^1 \circ \sigma \circ h) \circ (h^{-1} \circ \sigma \circ h)(u) \\ &= h^{-1}(\sigma^2(h(u))) \\ &= h^{-1}(h(u)) \\ &= u. \end{aligned}$$

4.4 Chaotic Behavior

In this manner, one can show that f has 2^m points of period m. Thus f has infinitely many periodic points. These points are dense in $[0, 1]$ since there is exactly one point of period m in each of the intervals $[0, \frac{1}{2^m})$, $(\frac{1}{2^m}, \frac{2}{2^m})$, $\ldots$, $(\frac{2^m-1}{2^m}, 1)$.

Systems of difference equations can exhibit the phenomena that we have described for single equations as well as a variety of other intricate behavior. Consider first the system

$$\begin{aligned} u_1(t+1) &= 1 + u_2(t) - au_1^2(t), \\ u_2(t+1) &= bu_1(t), \end{aligned} \qquad (4.22)$$

which was discovered by the astronomer Hénon [121]. If $b = 0$, then Eq. (4.22) reduces to the scalar equation $u_1(t+1) = 1 - au_1^2(t)$, which can be transformed into Eq. (4.19) by a linear change of variable. Thus Hénon's example contains the quadratic equation as a special case! For $b \neq 0$ the function $f(u_1, u_2) = (1 + u_2 - au_1^2, bu_1)$ is truly two-dimensional, mapping the plane one-to-one onto itself.

Fix $b = 0.3$ and let a increase from zero. The fixed points of f are

$$u_1 = \frac{-.7 \pm \sqrt{.49 + 4a}}{2a} = \frac{u_2}{.3}. \qquad (4.23)$$

The Jacobian matrix for f is

$$\begin{bmatrix} -2au_1 & 1 \\ .3 & 0 \end{bmatrix},$$

with eigenvalues $\lambda = -au_1 \pm \sqrt{a^2 u_1^2 + 0.3}$. From Theorem 4.10, a fixed point is asymptotically stable if $|\lambda| < 1$ for both eigenvalues. It follows that the fixed point in Eq. (4.23) obtained by selecting the plus sign is asymptotically stable for $a < .3675$. As a increases beyond $.3675$, a pair of asymptotically stable points of period two appear, and they continue to attract nearby solutions until $a = .9125$, where we begin to observe asymptotically stable points of period four. This doubling of period continues, as in the case of Eq. (4.19), up to a certain value of a where behavior of solutions becomes more complex.

Around $a = 1.4$, an interesting phenomenon occurs. There is a set of points of parabolic shape (see Fig. 4.11) that attracts all nearby solutions (a strange attractor?)

Furthermore, solutions seem to jump around the attractor randomly and have sensitive dependence on initial conditions.

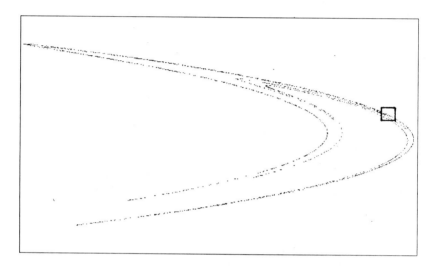

Fig. 4.11 The Hénon attractor

This Hénon attractor appears to display the characteristics of a "fractal." Closer examination of the boxed region in Fig. 4.11, which seems to contain three dotted lines, reveals that the lower line is single, the middle line is actually double, and the upper line a triple of lines (see Fig. 4.12). Further magnification of the triple lines yields a similar structure. This similarity appears to extend to all scalings, provided that enough points are plotted.

Hénon's work was motivated by a famous system of three differential equations studied by Lorenz [172]:

$$\frac{dx}{dt} = s(y - x),$$
$$\frac{dy}{dt} = rx - y - xz,$$
$$\frac{dz}{dt} = -bz + xy.$$

4.4 Chaotic Behavior   195

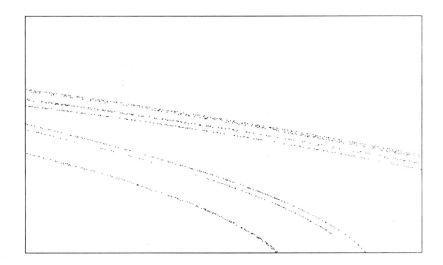

Fig. 4.12 Enlargement of a portion of the attractor

These equations approximately model the motion of fluid in a horizontal layer that is heated from below. For the parameter values $s = 10$, $b = \frac{8}{3}$, and $r = 30$, there is a complicated, double-lobed set which attracts nearby solutions (see Fig. 4.13). The motion of solutions near this attractor can be studied by computing the successive intersections (in a fixed direction) of a solution with the plane $z = 29$. The resulting sequence of points, called the "Poincaré map" after Henri Poincaré, is plotted in Fig. 4.14. The intersection points form a pattern of numerous closely packed line segments much like those in the Hénon attractor. Other characteristics of chaotic motion are also present.

Another interesting example is the predator-prey model (Smith [232])

$$\begin{aligned} u_1(t+1) &= au_1(t)(1-u_1(t)) - u_1(t)u_2(t), \\ u_2(t+1) &= \frac{1}{b}u_1(t)u_2(t), \end{aligned} \quad (4.24)$$

which reduces to Eq. (4.19) when $u_2 = 0$. Let $b = .31$ and let a increase from a value of one. Up to about $a = 2.6$, the system has an asymptotically

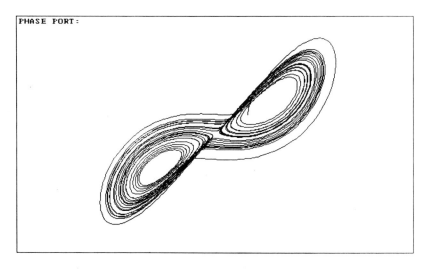

Fig. 4.13 The Lorenz attractor

Fig. 4.14 A Poincaré section of the attractor

4.4 Chaotic Behavior

stable fixed point. As a increases beyond 2.6, the fixed point expands into an "invariant circle," a set topologically like a circle which captures the points of a solution sequence (see Fig. 4.15). When $a > 3.44$, the invariant circle breaks up into a complicated attracting set (see Fig. 4.16).

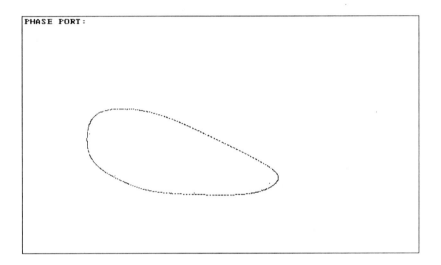

Fig. 4.15 An invariant circle

The study of chaotic behavior, strange attractors and other geometrically complicated phenomena is an emerging branch of mathematics. Many of the terms associated with these studies were coined in the 1970s and 1980s. The modern availability of small, powerful computers has made experimentation with systems of nonlinear equations possible for anyone who wishes to investigate these phenomena. In addition, the importance of chaos in mathematical modeling is being recognized in fields ranging from business cycle theory to the physiology of the heart. As new methods for describing and analyzing chaotic behavior continue to be developed, we are beginning to come to a better understanding of the global behavior of nonlinear systems.

Fig. 4.16 A predator-prey attractor

Exercises

Section 4.1

4.1 Convert the second order system

$$v(t+2) - 6v(t+1) + 4w(t+1) - 3v(t) + w(t) = 0,$$
$$w(t+2) + w(t+1) + 3v(t+1) - 2w(t) = t3^t$$

into a first order system of Eq. (4.1).

4.2 Show that the characteristic equation for

$$y(t+2) + ay(t+1) + by(t) = 0$$

is the same as the characteristic equation of the companion matrix.

4.3 Find the spectrum and the spectral radius:

(a) $\begin{bmatrix} 0 & 1 \\ -2 & -3 \end{bmatrix}$, (b) $\begin{bmatrix} 3 & 2 & 4 \\ 2 & 0 & 2 \\ 4 & 2 & 3 \end{bmatrix}$,

(c) $\begin{bmatrix} 3 & 2 \\ -5 & 1 \end{bmatrix}$, (d) $\begin{bmatrix} 2 & -6 & -6 \\ -1 & 1 & 2 \\ 3 & -6 & -7 \end{bmatrix}$.

Also, find the eigenvectors in parts (a) and (b).

4.4 Use Eq. (4.6) to solve $u(t+1) = Au(t)$ in case
(a) A is the matrix in Exercise 4.3(a),
(b) A is the matrix in 4.3(b).

4.5 Verify the Cayley-Hamilton Theorem for $A = \begin{bmatrix} a & b \\ c & d \end{bmatrix}$.

4.6 Use Theorem 4.2 to solve

$$y(t+2) + 3y(t+1) + 2y(t) = 0,$$
$$y(0) = -1, \quad y(1) = 7.$$

4.7 Find A^t, $(t \geq 0)$, for each of the following matrices A:

(a) $\begin{bmatrix} 2 & 2 \\ 2 & -1 \end{bmatrix}$,

(b) $\begin{bmatrix} 1 & 0 & 0 \\ 1 & 1 & 0 \\ 0 & 1 & 1 \end{bmatrix}$,

(c) $\begin{bmatrix} 0 & 1 & 0 \\ 0 & 1 & 0 \\ 0 & 1 & 1 \end{bmatrix}$,

(d) $\begin{bmatrix} 1 & 1 & 0 \\ 0 & 1 & 0 \\ 0 & 1 & 1 \end{bmatrix}$.

4.8 Use Theorem 4.2 in solving the initial value problems

(a) $u(t+1) = \begin{bmatrix} 6 & -2 \\ -2 & 9 \end{bmatrix} u(t)$, $u(0) = \begin{bmatrix} 2 \\ 1 \end{bmatrix}$;

(b) $u(t+1) = \begin{bmatrix} 2 & 1 \\ -1 & 4 \end{bmatrix} u(t)$, $u(0) = [-1 \ \ 1]$.

4.9 If A is the block diagonal matrix $A = \begin{bmatrix} A_1 & & & \\ & A_2 & & \\ & & \ddots & \\ & & & A_k \end{bmatrix}$, where each A_i is a square matrix and all other entries in A are zero, then $A^t = \begin{bmatrix} A_1^t & & & \\ & A_2^t & & \\ & & \ddots & \\ & & & A_k^t \end{bmatrix}$. Using this fact and Theorem 4.2, find A^t if $A = \begin{bmatrix} 2 & 0 & 0 \\ 0 & -3 & 4 \\ 0 & 4 & 3 \end{bmatrix}$.

4.10 Compute A^t, $(t \geq 0)$, if

$$A = \begin{bmatrix} 3 & 1 \\ -13 & -3 \end{bmatrix}.$$

(Note: A has complex eigenvalues.)

Exercises

4.11 Solve, using Theorem 4.3,

(a) $u(t+1) = \begin{bmatrix} 2 & 2 \\ 2 & -1 \end{bmatrix} u(t) + \begin{bmatrix} 0 \\ 1 \end{bmatrix}$, $u(0) = \begin{bmatrix} 0 \\ 0 \end{bmatrix}$;

(b) $u(t+1) = \begin{bmatrix} 0 & 1 & 0 \\ 0 & 1 & 0 \\ 0 & 1 & 1 \end{bmatrix} u(t) + \begin{bmatrix} t^2 \\ 0 \\ t \end{bmatrix}$, $u(0) = \begin{bmatrix} 1 \\ 2 \\ 3 \end{bmatrix}$.

4.12 Use Theorem 4.3 to solve the system in Example 3.13.

Section 4.2

4.13 For which of the following systems do all solutions converge to the origin as $t \to \infty$?

(a) $u(t+1) = \begin{bmatrix} .9 & .2 \\ -.1 & .6 \end{bmatrix} u(t)$.

(b) $u(t+1) = \begin{bmatrix} 0 & 1 & 0 \\ 0 & 0 & 1 \\ 1 & -3 & 3 \end{bmatrix} u(t)$.

(c) $u(t+1) = \begin{bmatrix} 0 & 1 & 0 \\ 0 & 0 & 1 \\ -\frac{3}{8} & -1 & -1 \end{bmatrix} u(t)$.

4.14 If $r(A) > 1$, then show that some real solution $u(t)$ of Eq. (4.4) satisfies $\lim_{t \to \infty} |u(t)| = \infty$.

4.15 For which of the following systems does Eq. (4.13) hold for every solution?

(a) $u(t+1) = \begin{bmatrix} 2 & 1 \\ -3 & -2 \end{bmatrix} u(t)$.

(b) $u(t+1) = \begin{bmatrix} 2 & 1 \\ 4 & 2 \end{bmatrix} u(t)$.

(c) $u(t+1) = \begin{bmatrix} 1 & 0 & 0 \\ 0 & -1 & 0 \\ 0 & 0 & 1 \end{bmatrix} u(t)$.

4.16 Show that the converse of Theorem 4.6 is not true. (Hint: consider Exercise 4.15(c).)

4.17 Prove the following: if the characteristic equation for

$$y(t+n) + p_{n-1}y(t+n-1) + \cdots + p_0 y(t) = 0$$

has a multiple characteristic root λ with $|\lambda| = 1$, then the difference equation has an unbounded solution.

4.18 Find the stable subspace S for each of the following:

(a) $u(t+1) = \begin{bmatrix} 0 & 1 \\ \frac{2}{3} & -\frac{5}{3} \end{bmatrix} u(t),$

(b) $u(t+1) = \begin{bmatrix} .1 & 1 & 0 \\ 0 & .1 & 1 \\ 0 & 0 & 2 \end{bmatrix} u(t),$

(c) $u(t+1) = \begin{bmatrix} 0 & 1 & 0 \\ 0 & 0 & 1 \\ \frac{1}{4} & \frac{3}{4} & 0 \end{bmatrix} u(t).$

4.19 For the system in Exercise 4.18(b) show that if $u(0)$ is not in the stable subspace, then $|u(t)| \to \infty$ as $t \to \infty$.

4.20 Find the real two-dimensional stable subspace for

$$u(t+1) = \begin{bmatrix} 0 & \frac{1}{2} & -1 \\ \frac{3}{2} & 1 & 0 \\ 0 & -\frac{5}{6} & 1 \end{bmatrix} u(t).$$

Section 4.3

4.21 Show that the solutions of
   (a) $u(t+1) = 2u(t) + \sqrt{100 - u(t)}$ with $u(0) = 10$,
   (b) $u(t+1) = \frac{2}{1-u(t)}$ with $u(0) = 3$, are not defined for all $t \geq 0$.

4.22 Find all fixed points for the following functions:
   (a) $f(u) = 4u(1-u)$,
   (b) $f(u) = u + \sin 8u$.

4.23 Find all periodic points and their periods.
   (a) $f(u) = -u^3$,
   (b) $f(u) = u - u^2$,
   (c) $f\left(\begin{bmatrix} u_1 \\ u_2 \end{bmatrix}\right) = \frac{1}{2} \begin{bmatrix} \sqrt{2} & \sqrt{2} \\ -\sqrt{2} & \sqrt{2} \end{bmatrix} \begin{bmatrix} u_1 \\ u_2 \end{bmatrix}.$

Exercises

4.24 Use the staircase method to investigate the asymptotic behavior of solutions of $u(t+1) = f(u(t))$ for
(a) $f = u^2$, (b) $f = u^3$, (c) $f = 3u + 2$, (d) $f = -u + 1$.

4.25 Use Theorem 4.8 to test stability of the fixed points for the functions in Exercise 4.24.

4.26 Show $f = \frac{1}{2}(u^3 + u)$ has one asymptotically stable fixed point and two unstable fixed points using
(a) the staircase method,
(b) Theorem 4.8.

4.27 Despite the fact that each of the functions u, $u - u^3$, $u + u^3$ has derivative one at $u = 0$, show that their stability properties at $u = 0$ are different.

4.28 Show that the periodic points of

$$f\left(\begin{bmatrix} u_1 \\ u_2 \end{bmatrix}\right) = \frac{1}{2} \begin{bmatrix} \sqrt{2} & \sqrt{2} \\ -\sqrt{2} & \sqrt{2} \end{bmatrix} \begin{bmatrix} u_1 \\ u_2 \end{bmatrix}$$

are stable but not asymptotically stable.

4.29 Define

$$f(u) = \begin{cases} \frac{1}{u}, & u \neq 0 \\ 0, & u = 0 \end{cases}, \text{ and } V(u) = \frac{u^2}{1 + u^4}.$$

Show that V is a Liapunov function for f at 0 but 0 is not stable. Does this example contradict Theorem 4.9?

4.30 Let α and β be any numbers and let $n > 1$. Use a Liapunov function to show the origin is asymptotically stable for

$$f\begin{bmatrix} u_1 \\ u_2 \end{bmatrix} = \begin{bmatrix} \alpha u_1^n \\ \beta u_2^n \end{bmatrix}.$$

For what initial points do the solutions converge to the origin?

4.31 Carry out the instructions in Exercise 4.30 for the system in Example 4.15.

4.32 Show $\begin{bmatrix} 0 \\ 0 \end{bmatrix}$ is asymptotically stable for

$$\begin{bmatrix} u_1 \\ u_2 \end{bmatrix}(t+1) = \begin{bmatrix} \frac{\alpha u_2(t)}{1+u_1^2(t)} \\ \frac{\beta u_1(t)}{1+u_2^2(t)} \end{bmatrix}$$

if $\alpha^2, \beta^2 < 1$ using
  (a) Theorem 4.9,
  (b) Theorem 4.10.

4.33 Use the remark following the proof of Theorem 4.10 to show that both of the fixed points of

$$\begin{bmatrix} u_1 \\ u_2 \end{bmatrix}(t+1) = \begin{bmatrix} 2u_1(t) - \frac{2u_1(t)u_2(t)}{1+u_1(t)} \\ \frac{2u_1(t)u_2(t)}{1+u_1(t)} \end{bmatrix}$$

are unstable. (This is a special predator-prey model; see Example 4.9(d).)

4.34 Suppose that $f(v) = v$, f is continuous in a ball B about v and V is a Liapunov function for f in B such that for each $u \neq v$ in B there is a positive integer T so that $V(f^T(u)) - V(u) < 0$. Show that v is asymptotically stable.

4.35 Use Exercise 4.34 to show the origin is asymptotically stable for the system in Exercise 4.32 if $\alpha^2 = 1, \beta^2 < 1$ or $\alpha^2 < 1, \beta^2 = 1$.

Section 4.4

4.36 Find a periodic point of least period four for $u(t+1) = 4u(t)(1-u(t))$.

4.37 Show that $u = \frac{a-1}{a}$ is unstable as a fixed point of $f \circ f$ for $a > 3$, if $f = au(1-u)$.

4.38 Show that the periodic points of least period two are asymptotically stable for Eq. (4.19) if $3 < a < 1 + \sqrt{6}$.

4.39
  (a) Show that $u(t) = \cot(2^t \theta_0)$ solves the equation $u(t+1) = \frac{1}{2}(u(t) - \frac{1}{u(t)})$.
  (b) Find a probability density function for $u(t)$ and use it to break the real line into four intervals of equal probability.

Exercises

(c) Test your answer in part (b) by computing the first 500 iterations of $u(t)$ for some initial value $u(0)$ and sorting the values into the four subintervals.

4.40 Show that the solutions in Exercise 4.39 have sensitive dependence on initial conditions.

4.41 Consider $u(t+1) = g(u(t))$, where g is the "baker map"

$$g(u) = \begin{cases} 2u, & 0 \le u \le 0.5 \\ 2u - 1, & 0.5 < u \le 1. \end{cases}$$

Show that solutions exhibit sensitive dependence on initial conditions.

4.42 Show that the four points of period 2 for the tent map in Example 4.16 have exactly one representative in each of the intervals $[0, .25)$, $(.25, .5)$, $(.5, .75)$ and $(.75, 1)$.

4.43 Prove that the function h defined for the tent map is not onto, i.e., that there is a binary sequence that is not associated with any initial point in $[0, 1]$.

4.44 Show that the baker map in Exercise 4.41 has 2^m points of period m.

4.45 Another indication of chaotic behavior of a function f is the presence of a "snap-back repellor." This is an unstable fixed point p such that the repelling domain for p (the set of u so that $|f(u) - p| > |u - p|$) contains a point q so that $f^t(q) = p$ for some positive integer t. Show that Eq. (4.19) with $a = 4$ has two snap-back repellors.

4.46 Verify that Eq. (4.22) with $b = .3$ has an asymptotically stable fixed point for $0 < a < .3675$.

4.47 Verify that Eq. (4.22) with $b = .3$ has asymptotically stable points of period two if $.3675 < a < .9125$.

4.48 Use a computer to check the claims made about the predator-prey system Eq. (4.24).

Chapter 5
Asymptotic Methods

5.1 Introduction

In the last chapter we saw that it is often possible to predict the behavior of solutions of difference equations for large values of the independent variable, even for systems of nonlinear equations. We will go a step further in this chapter and try to find approximations of solutions that are accurate for large t. For simplicity, the discussion will be limited to single equations. This introductory section does not deal with difference equations but does present a number of the basic concepts and tools of asymptotic analysis.

Suppose we wish to describe the asymptotic behavior of the function $(4t^2+t)^{\frac{3}{2}}$ as $t \to \infty$. Since $4t^2$ is much larger than t when t is large, a good approximation should be given by $(4t^2)^{\frac{3}{2}} = 8t^3$. The relative error, which determines the number of significant digits in the approximation, is given by

$$\frac{(4t^2+t)^{\frac{3}{2}} - 8t^3}{8t^3}.$$

By the Mean Value Theorem,

$$(1+x)^{\frac{3}{2}} = 1 + \frac{3}{2}(1+c)^{\frac{1}{2}}x,$$

for some $0 < c < x$, so

$$\frac{(4t^2+t)^{\frac{3}{2}} - 8t^3}{8t^3} = \frac{8t^3(1+\frac{1}{4t})^{\frac{3}{2}} - 8t^3}{8t^3}$$

$$= (1+\frac{1}{4t})^{\frac{3}{2}} - 1 = \frac{3}{2}(1+c)^{\frac{1}{2}}\left(\frac{1}{4t}\right),$$

207

where $0 < c < \frac{1}{4t}$, and the relative error goes to zero as $t \to \infty$. This fact can also be expressed in the form

$$\lim_{t \to \infty} \frac{(4t^2 + t)^{\frac{3}{2}}}{8t^3} = 1.$$

Definition 5.1. If $\lim_{t \to \infty} \frac{y(t)}{z(t)} = 1$, then we say that "$y(t)$ is asymptotic to $z(t)$ as t tends to infinity" and write

$$y(t) \sim z(t), \quad (t \to \infty).$$

We have $(4t^2 + t)^{\frac{3}{2}} \sim 8t^3$, $(t \to \infty)$. Some other elementary examples are $\frac{1}{3t^2+2t} \sim \frac{1}{3t^2}$, $(t \to \infty)$, and $\sinh t \sim \frac{e^t}{2}$, $(t \to \infty)$. A famous nontrivial example is the Prime Number Theorem. Let

$$\Pi(t) = \text{the number of primes less than } t.$$

Then

$$\Pi(t) \sim \frac{t}{\log t}, \quad (t \to \infty).$$

(See [228].)

Definition 5.2. If $\lim_{t \to \infty} \frac{u(t)}{v(t)} = 0$, then we say that "$u(t)$ is much smaller than $v(t)$ as t tends to infinity" and write

$$u(t) \ll v(t), \quad (t \to \infty).$$

The following expressions are equivalent:

$$y(t) \sim z(t), \quad (t \to \infty),$$

and

$$y(t) - z(t) \ll z(t), \quad (t \to \infty).$$

5.1 Introduction

Also, if $y(t) \sim z(t)$, $(t \to \infty)$, and if $u(t) \ll z(t)$, $(t \to \infty)$, then $y(t)+u(t) \sim z(t)$, $(t \to \infty)$. For example, we have

$$(4t^2 + t)^{\frac{3}{2}} + t^2 \log t \sim 8t^3, \quad (t \to \infty)$$

since $t^2 \log t \ll 8t^3$, $(t \to \infty)$.

Now let's return to our initial example and ask how we might display more precise information about the asymptotic behavior of $(4t^2 + t)^{\frac{3}{2}}$. Since there is a constant $M > 0$ so that

$$\left| \frac{(4t^2 + t)^{\frac{3}{2}} - 8t^3}{8t^3} \right| \leq \frac{M}{t}$$

for $t \geq 1$ ($M = \frac{15}{32}$ will do), we say that the relative error goes to zero as $t \to \infty$ at a rate proportional to $\frac{1}{t}$.

Definition 5.3. If there are constants M and t_0 so that $|u(t)| \leq M|v(t)|$ for $t \geq t_0$, then we say that "$u(t)$ is big *oh* of $v(t)$ as t tends to infinity" and write

$$u(t) = \mathcal{O}(v(t)), \quad (t \to \infty).$$

We have

$$\frac{(4t^2 + t)^{\frac{3}{2}} - 8t^3}{8t^3} = \mathcal{O}\left(\frac{1}{t}\right), \quad (t \to \infty),$$

which is frequently written

$$(4t^2 + t)^{\frac{3}{2}} = 8t^3 \left(1 + \mathcal{O}\left(\frac{1}{t}\right)\right), \quad (t \to \infty).$$

If a better approximation is desired, then we first use Taylor's formula to find

$$(1 + x)^{\frac{3}{2}} = 1 + \frac{3}{2}x + \frac{3}{4}(1 + d)^{-\frac{1}{2}} \frac{x^2}{2}$$

for some d between 0 and x. Then

$$(4t^2+t)^{\frac{3}{2}} = 8t^3\left(1+\frac{1}{4t}\right)^{\frac{3}{2}}$$

$$= 8t^3\left[1+\frac{3}{8t}+\frac{3}{128}(1+d)^{-\frac{1}{2}}\frac{1}{t^2}\right],$$

so

$$(4t^2+t)^{\frac{3}{2}} = 8t^3\left[1+\frac{3}{8t}+\mathcal{O}\left(\frac{1}{t^2}\right)\right], \quad (t\to\infty).$$

Since the "big oh" notation suppresses the constant M in Definition 5.3, we can in general deduce only that an $\mathcal{O}(\frac{1}{t^2})$ approximation is better than an $\mathcal{O}(\frac{1}{t})$ approximation if t is "sufficiently large." Of course in specific examples we can often be more precise. For instance, we have

$$(4t^2+t)^{\frac{3}{2}} = 8t^3+3t^2+\frac{3}{16}(1+d)^{-\frac{1}{2}}t$$
$$> 8t^3+3t^2$$
$$> 8t^3$$

for all $t>0$, so the $\mathcal{O}(\frac{1}{t^2})$ estimate $8t^3+3t^2$ is closer to $(4t^2+t)^{\frac{3}{2}}$ than the $\mathcal{O}(\frac{1}{t})$ estimate $8t^3$ for all $t>0$.

Example 5.1. Recall from Exercise 1.15 that the exponential integral is

$$E_n(x) = \int_1^\infty \frac{e^{-xt}}{t^n}dt, \quad (x>0).$$

We can use integration by parts to analyze the asymptotic behavior of $E_n(x)$ as $x\to\infty$:

$$\int_1^\infty \frac{e^{-xt}}{t^n}dt = \frac{e^{-x}}{x} - \frac{n}{x}\int_1^\infty \frac{e^{-xt}}{t^{n+1}}dt.$$

5.1 Introduction

Now

$$\int_1^\infty \frac{e^{-xt}}{t^{n+1}} dt \le \int_1^\infty e^{-xt} dt = \frac{e^{-x}}{x},$$

so for each fixed n,

$$E_n(x) = \frac{e^{-x}}{x}\left(1 + \mathcal{O}\left(\frac{1}{x}\right)\right), \quad (x \to \infty).$$

By using integration by parts repeatedly, we obtain for each positive integer k

$$E_n(x) = \frac{e^{-x}}{x}\left[1 - \frac{n}{x} + \frac{n(n+1)}{x^2} - \cdots + (-1)^k \frac{n(n+1)\cdots(n+k-1)}{x^k} + \mathcal{O}\left(\frac{1}{x^{k+1}}\right)\right], \quad (x \to \infty).$$

The series in brackets is called an "asymptotic series" since it gives improved information about asymptotic behavior with the addition of each new term. However, note that the infinite series

$$\sum_{k=1}^\infty (-1)^k \frac{n(n+1)\cdots(n+k-1)}{x^k}$$

diverges for each x by the ratio test. It follows that for each fixed x, some finite number of terms yields an optimal estimate.

Next, let's investigate the behavior of $E_n(x)$ for large n. Using integration by parts with the roles of e^{-xt} and $\frac{1}{t^n}$ interchanged, we have

$$\int_1^\infty \frac{e^{-xt}}{t^n} dt = \frac{e^{-x}}{n-1} - \int_1^\infty \frac{x}{n-1} \frac{e^{-xt}}{t^{n-1}} dt.$$

Since

$$\int_1^\infty \frac{e^{-xt}}{t^{n-1}} dt \le \int_1^\infty \frac{e^{-x}}{t^{n-1}} dt = \frac{e^{-x}}{n-2},$$

$$E_n(x) = \frac{e^{-x}}{n-1}\left[1 + \mathcal{O}\left(\frac{1}{n-2}\right)\right], \quad (n \to \infty),$$

where x is any fixed positive number. A second integration by parts gives

$$E_n(x) = \frac{e^{-x}}{n-1}\left[1 - \frac{x}{n-2} + \mathcal{O}\left(\frac{1}{(n-2)(n-3)}\right)\right], \quad (n \to \infty),$$

and the calculation can be continued to any number of terms. Note that we could write $\mathcal{O}(\frac{1}{n})$ instead of $\mathcal{O}(\frac{1}{n-2})$ and $\mathcal{O}(\frac{1}{n^2})$ instead of $\mathcal{O}\left(\frac{1}{(n-2)(n-3)}\right)$ without any loss of information.

5.2 Asymptotic Analysis of Sums

Chapter 3 presented a number of methods for solving linear difference equations in which the solutions involved sums such as $\sum_{k=1}^{n} f(k)$. Since these sums do not usually have an explicit closed form representation, it is of interest to find asymptotic approximations of sums for large n. In the next example, as in Section 5.1, Taylor's formula is an essential ingredient in the approximation.

Example 5.2.

The solution of the equation

$$y_{n+1} - ny_n = 1, \quad (n = 1, 2, 3, \cdots),$$

is

$$y_n = (n-1)!\left[y_1 + \sum_{k=1}^{n-1} \frac{1}{k!}\right], \quad (n \geq 2).$$

The sum in brackets is a partial sum of the Taylor series for e^θ with $\theta = 1$.

5.2 Asymptotic Analysis of Sums

From Taylor's formula,

$$e = \sum_{k=0}^{n-1} \frac{1}{k!} + \frac{e^c}{n!}, \quad (0 < c < 1),$$

so

$$\sum_{k=1}^{n-1} \frac{1}{k!} = e - 1 - \frac{e^c}{n!}.$$

Then

$$y_n = (n-1)! \left[y_1 + e - 1 - \frac{e^c}{n!} \right]$$
$$= (n-1)! \left[y_1 + e - 1 + \mathcal{O}\left(\frac{1}{n!}\right) \right], \quad (n \to \infty).$$

We conclude that for large n, y_n is approximately $(y_1 + e - 1)(n-1)!$, and the relative error goes to zero like $\frac{1}{n!}$.

The procedure used to obtain an asymptotic approximation in this example was successful because the infinite series converged to a known quantity and an error estimate for the difference between the finite sum and the infinite sum was available. In contrast, the next example is a highly divergent series in which the largest term determines the asymptotic behavior.

Example 5.3. What is the asymptotic behavior of

$$\sum_{k=1}^{n} k^k ?$$

We begin by factoring out the largest term:

$$\sum_{k=1}^{n} k^k = n^n \left[1 + \left\{ \frac{(n-1)^{n-1}}{n^n} + \frac{(n-2)^{n-2}}{n^n} + \cdots + \frac{1}{n^n} \right\} \right].$$

The sum in braces is less than

$$\frac{1}{n} + \frac{1}{n^2} + \cdots + \frac{1}{n^{n-1}} = \frac{1}{n}\sum_{k=0}^{n-2}\left(\frac{1}{n}\right)^k$$

$$= \frac{1}{n}\frac{1-\left(\frac{1}{n}\right)^{n-1}}{1-\frac{1}{n}}.$$

Since the expression $\frac{1-\left(\frac{1}{n}\right)^{n-1}}{1-\frac{1}{n}}$ is bounded, we have

$$\sum_{k=1}^{n} k^k = n^n \left[1 + \mathcal{O}\left(\frac{1}{n}\right)\right], \quad (n \to \infty).$$

Then the asymptotic value of $\sum_{k=1}^{n} k^k$ is given by the largest term with a relative error that approaches zero like $\frac{1}{n}$ as $n \to \infty$.

In a similar way, it can be shown that

$$\sum_{k=1}^{n} k^k = n^n \left[1 + \left(\frac{n-1}{n}\right)^{n-1}\frac{1}{n} + \mathcal{O}\left(\frac{1}{n^2}\right)\right], \quad (n \to \infty).$$

Consequently, the two largest terms of the series yield an $\mathcal{O}(\frac{1}{n^2})$ asymptotic estimate.

For sums that are not one of the extreme cases considered in the last two examples, asymptotic information is usually more difficult to obtain. The summation by parts method can be useful since it may separate out the dominant portion of the sum.

Example 5.4.

$$\sum_{k=2}^{n-1} 2^k \log k.$$

5.2 Asymptotic Analysis of Sums

We begin by applying Theorem 2.8(a):

$$\sum_{k=2}^{n-1} 2^k \log k = (\log n) \sum_{k=2}^{n-1} 2^k - \sum_{k=2}^{n-1} \left(\sum_{i=2}^{k} 2^i \right) \Delta \log k \quad (5.1)$$

$$= (\log n)(2^n - 4) - \sum_{k=2}^{n-1} (2^{k+1} - 4) \Delta \log k$$

$$= 2^n \log n - \sum_{k=2}^{n-1} 2^{k+1} \Delta \log k$$

$$- 4 \log n + \sum_{k=2}^{n-1} 4 \Delta \log k.$$

Note that the first two terms are asymptotically much larger than the last two terms, which we will neglect.

By the Mean Value Theorem,

$$\Delta \log k < \frac{1}{k},$$

so

$$\sum_{k=2}^{n-1} 2^{k+1} \Delta \log k < \sum_{k=2}^{n-1} 2^{k+1} \left(\frac{1}{k} \right)$$

$$= \frac{2^n}{(n-1)} [1 + \frac{n-1}{n-2} \cdot \frac{1}{2} + \cdots$$

$$+ \frac{n-1}{3} \cdot \frac{1}{2^{n-4}} + \frac{n-1}{2} \cdot \frac{1}{2^{n-3}}]$$

$$= \frac{2^n}{(n-1)} \sum_{k=0}^{n-3} \frac{n-1}{n-(k+1)} \cdot \frac{1}{2^k}.$$

It is easily checked that $\frac{n-1}{n-(k+1)} \leq k+1$ when $0 \leq k \leq n-3$, so

$$\sum_{k=2}^{n-1} 2^{k+1} \Delta \log k < \frac{2^n}{(n-1)} \sum_{k=0}^{n-3} \frac{k+1}{2^k}, \quad (5.2)$$

and the last sum is bounded by the ratio test. Using Eq. (5.2) in Eq. (5.1), we finally have

$$\sum_{k=2}^{n-1} 2^k \log k = 2^n \log n \left[1 + \mathcal{O}\left(\frac{1}{(n-1)\log n}\right)\right], \quad (n \to \infty).$$

In this example, the asymptotic behavior of the sum is given, not by the largest term, but rather by twice the largest term.

For a sum in which the terms are more slowly varying, the Euler summation formula (Theorem 2.10) is a valuable tool in establishing the asymptotic behavior. The following example is of particular importance since it gives a good approximation for $n!$ when n is large.

Example 5.5. Consider the sum $\sum_{k=1}^{n} \log k$. From Theorem 2.10, we have, for $i \geq 1$,

$$\sum_{k=1}^{n} \log k = n \log n - n + 1 + \frac{\log n}{2}$$

$$+ \sum_{j=1}^{i} \frac{B_{2j}}{(2j)!} \left[\frac{(2j-2)!}{n^{2j-1}} - (2j-2)!\right]$$

$$+ \frac{1}{(2i)!} \int_{1}^{n} \frac{(2i-1)!}{x^{2i}} B_{2i}(x - [x]) dx.$$

Simplifying and rearranging, we have

$$\sum_{k=1}^{n} \log k = n \log n - n + \frac{\log n}{2} + \sum_{j=1}^{i} \frac{B_{2j}}{(2j)(2j-1)} \frac{1}{n^{2j-1}} \quad (5.3)$$

$$+ \left\{1 - \sum_{j=1}^{i} \frac{B_{2j}}{(2j)(2j-1)} + \frac{1}{2i} \int_{1}^{\infty} \frac{B_{2i}(x - [x])}{x^{2i}} dx\right\}$$

$$- \frac{1}{2i} \int_{n}^{\infty} \frac{B_{2i}(x - [x])}{x^{2i}} dx.$$

5.2 Asymptotic Analysis of Sums

The expression in braces is independent of n; let's give it the name $\gamma(i)$. Now by Eq. (5.3)

$$\gamma(i+1) - \gamma(i) = \int_n^\infty \left(\frac{B_{2i+2}(x-[x])}{(2i+2)x^{2i+2}} - \frac{B_{2i}(x-[x])}{2ix^{2i}} \right) dx$$

$$- \frac{B_{2(i+1)}}{(2i+2)(2i+1)} \frac{1}{n^{2i+1}}$$

$$\to 0 \quad \text{as} \quad n \to \infty,$$

so γ is independent of i as well.

Equation (5.3) yields asymptotic estimates of $\sum_{k=1}^n \log k$ for each $i = 1, 2, \cdots$. For $i = 2$, we have

$$\sum_{k=1}^n \log k = n \log n - n + \frac{\log n}{2} + \frac{1}{12n} + \gamma + \mathcal{O}\left(\frac{1}{n^3}\right), \quad (n \to \infty). \quad (5.4)$$

Exponentiation gives

$$n! = e^\gamma \left(\frac{n}{e}\right)^n \sqrt{n} \, e^{\frac{1}{12n} + \mathcal{O}(\frac{1}{n^3})}, \quad (n \to \infty),$$

or by Taylor's formula

$$n! = e^\gamma \left(\frac{n}{e}\right)^n \sqrt{n} \left(1 + \frac{1}{12n} + \frac{1}{288n^2} + \mathcal{O}\left(\frac{1}{n^3}\right)\right), \quad (n \to \infty). \quad (5.5)$$

It remains to compute γ in Eqs. (5.4) and (5.5). We will make use of Wallis' formula

$$\frac{\pi}{2} = \lim_{n \to \infty} \left[\frac{2 \cdot 4 \cdots 2n}{1 \cdot 3 \cdots (2n-1)} \right]^2 \frac{1}{2n+1}$$

(see Exercise 5.18). Now

$$\left[\frac{2 \cdot 4 \cdots 2n}{1 \cdot 3 \cdots (2n-1)} \right]^2 \frac{1}{2n+1} = \frac{2^{4n}(n!)^4}{((2n)!)^2} \frac{1}{2n+1},$$

and from Eq. (5.5)

$$\frac{1}{2n+1}\frac{2^{4n}(n!)^4}{((2n)!)^2} \sim \frac{2^{4n}}{2n+1}\frac{e^{4\gamma}(\frac{n}{e})^{4n}n^2}{e^{2\gamma}(\frac{2n}{e})^{4n}2n}$$

$$\sim \frac{e^{2\gamma}}{4}, \quad (n \to \infty),$$

so by Wallis' formula

$$\frac{e^{2\gamma}}{4} = \frac{\pi}{2},$$

and finally

$$e^\gamma = \sqrt{2\pi}.$$

Then Eq. (5.5) is

$$n! = \sqrt{2\pi n}\left(\frac{n}{e}\right)^n\left(1 + \frac{1}{12n} + \frac{1}{288n^2} + \mathcal{O}\left(\frac{1}{n^3}\right)\right), \quad (n \to \infty). \quad (5.6)$$

Equation (5.6) is called Stirling's formula, and it gives a good estimate for $n!$, even for moderate values of n. The asymptotic series

$$1 + \frac{1}{12n} + \frac{1}{288n^2} + \cdots$$

turns out to be a divergent series. It can be shown that the best approximation for $n!$ is obtained if the series is truncated after the smallest term. However, a few terms suffice for most calculations. If $n = 5$, then $n! = 120$, while

$$\sqrt{2 \cdot \pi \cdot 5}\left(\frac{5}{e}\right)^5 \simeq 118.019$$

5.2 Asymptotic Analysis of Sums

and

$$\sqrt{2\cdot\pi\cdot 5}\left(\frac{5}{e}\right)^5\left(1+\frac{1}{12\cdot 5}\right) \simeq 119.986.$$

The numbers generated by Eq. (5.6) are extremely large when n is large, so Eq. (5.4) is preferable for such computations.

Using the methods of Example 5.5 together with additional properties of the gamma function, it can be shown that Stirling's formula is also valid for the gamma function

$$\Gamma(t+1) = \sqrt{2\pi t}\left(\frac{t}{e}\right)^t\left(1+\frac{1}{12t}+\frac{1}{288t^2}+\mathcal{O}\left(\frac{1}{t^3}\right)\right), \quad (t\to\infty). \quad (5.7)$$

(See DeBruijn [54].)

We saw in Chapter 3 that the solutions of some linear equations can be represented in terms of the gamma function. Stirling's formula can then be used to obtain asymptotic information.

Example 5.6. Consider the first order linear equation

$$t\Delta u(t) - \frac{1}{2}u(t) = 0.$$

We rearrange this equation to read

$$u(t+1) = \frac{t+\frac{1}{2}}{t}u(t)$$

and apply the method of Example 3.3 to find the general solution

$$u(t) = C\frac{\Gamma(t+\frac{1}{2})}{\Gamma(t)}.$$

By Eq. (5.7),

$$u(t) \sim C \frac{\sqrt{2\pi(t+\frac{1}{2})}\left(\frac{t+\frac{1}{2}}{e}\right)^{t+\frac{1}{2}}}{\sqrt{2\pi t}(\frac{t}{e})^t}, \quad (t \to \infty).$$

Simplifying and using the fact that $\lim_{t\to\infty}\left(\frac{t+\frac{1}{2}}{t}\right)^t = \sqrt{e}$, we obtain

$$u(t) \sim Ct^{\frac{1}{2}}, \quad (t \to \infty).$$

This relation motivates the choice of the factorial series representation

$$u(t) = \sum_{k=0}^{\infty} a_k t^{(-k+\frac{1}{2})}$$

in Exercise 3.85 since the lead term is

$$a_0 t^{(\frac{1}{2})} = a_0 \frac{\Gamma(t+1)}{\Gamma(t+\frac{1}{2})} \sim a_0 t^{\frac{1}{2}}, \quad (t \to \infty).$$

In conclusion, here is another useful formula that is a special case of the Euler summation formula:

$$\sum_{k=1}^{n} \frac{1}{k} = \log n + \gamma + \frac{1}{2n} - \frac{1}{12n^2} + \mathcal{O}\left(\frac{1}{n^4}\right), \tag{5.8}$$

where

$$\gamma = \lim_{n\to\infty}\left(1 + \frac{1}{2} + \cdots + \frac{1}{n} - \log n\right) \simeq 0.577$$

is Euler's constant (see Exercises 5.24 and 5.25).

The asymptotic analysis of sums is discussed in the books by De-Bruijn [54] and Olver [190].

5.3 Linear Equations

This section is an introduction to the study of the asymptotic behavior of solutions of homogeneous linear equations. If the equation has constant coefficients, then the asymptotic behavior is available from an expression for the exact solution. Nevertheless, we begin by considering this case since there is an aspect of the analysis that carries over to the more general setting.

Let $u(t)$ be any nontrivial solution of the equation

$$u(t+2) + p_1 u(t+1) + p_0 u(t) = 0,$$

where p_0, p_1 are constants and the characteristic roots λ_1, λ_2 satisfy $|\lambda_1| > |\lambda_2|$. Then $u(t) = a\lambda_1^t + b\lambda_2^t$ for some constants a, b. If $a \neq 0$, then

$$\frac{u(t+1)}{u(t)} = \frac{a\lambda_1^{t+1} + b\lambda_2^{t+1}}{a\lambda_1^t + b\lambda_2^t}$$

$$= \frac{\lambda_1 + \frac{b}{a}\lambda_1 \left(\frac{\lambda_2}{\lambda_1}\right)^{t+1}}{1 + \frac{b}{a}\left(\frac{\lambda_2}{\lambda_1}\right)^t} \to \lambda_1, \quad (t \to \infty).$$

If $a = 0$, then

$$\frac{u(t+1)}{u(t)} = \frac{b\lambda_2^{t+1}}{b\lambda_2^t} = \lambda_2,$$

so in any case the ratio $\frac{u(t+1)}{u(t)}$ converges to a root of the characteristic equation as t goes to infinity.

If $|\lambda_1| = |\lambda_2|$, then this property may fail. The equation

$$u(t+2) - u(t) = 0$$

has characteristic roots $\lambda = \pm 1$ (so $|\lambda_1| = |\lambda_2|$), and for the solution $u(t) = 2 + (-1)^t$ we find

$$\frac{u(t+1)}{u(t)} = \frac{2 + (-1)^{t+1}}{2 + (-1)^t}.$$

This expression produces a sequence that alternates between 3 and $\frac{1}{3}$.

A fundamental result in the analysis of asymptotic behavior is a theorem due to H. Poincaré, which states that certain linear homogeneous equations have the property of convergence of ratios of successive values.

Definition 5.4. A homogeneous linear equation

$$u(t+n) + p_{n-1}(t)u(t+n-1) + \cdots + p_0(t)u(t) = 0 \qquad (5.9)$$

is said to be of "Poincaré type" if $\lim_{t\to\infty} p_k(t) = p_k$ for $k = 0, 1, \cdots, n-1$ (i.e., if the coefficient functions converge to constants as t goes to infinity).

Theorem 5.1. (Poincaré's Theorem) If Eq. (5.9) is of Poincaré type and if the roots $\lambda_1, \cdots, \lambda_n$ of $\lambda^n + p_{n-1}\lambda^{n-1} + \cdots + p_0 = 0$ satisfy $|\lambda_1| > |\lambda_2| > \cdots > |\lambda_n|$, then every nontrivial solution u of Eq. (5.9) satisifes

$$\lim_{t\to\infty} \frac{u(t+1)}{u(t)} = \lambda_i$$

for some i.

Since the proof of Poincaré's Theorem is rather technical, we delay giving it until the end of this section.

Poincaré's Theorem leaves unanswered a natural question: is it true that for every characteristic root λ_i there is a solution $u(t)$ so that $\frac{u(t+1)}{u(t)} \to \lambda_i$, $(t \to \infty)$? The following result due to O. Perron gives an affirmative response.

Theorem 5.2. (Perron's Theorem) In addition to the assumptions of Theorem 5.1, suppose that $p_0(t) \neq 0$ for each t. Then there are n independent solutions $u_1, \cdots, u_n$ of Eq. (5.9) that satisfy

$$\lim_{t\to\infty} \frac{u_i(t+1)}{u_i(t)} = \lambda_i, \quad (i = 1, \cdots, n).$$

See [180] for a proof of Perron's Theorem.

Example 5.7. $(t+2)u(t+2) - (t+3)u(t+1) + 2u(t) = 0$.

Dividing through by $t+2$, we obtain

$$u(t+2) - \frac{t+3}{t+2}u(t+1) + \frac{2}{t+2}u(t) = 0,$$

5.3 Linear Equations

and the equation is of Poincaré type since $\frac{t+3}{t+2} \to 1$ and $\frac{2}{t+2} \to 0$ as $t \to \infty$. The associated characteristic equation is $\lambda^2 - \lambda = 0$, so $\lambda_1 = 1$ and $\lambda_2 = 0$. By Perron's Theorem, there are independent solutions u_1, u_2 so that $\frac{u_1(t+1)}{u_1(t)} \to 1$, $\frac{u_2(t+1)}{u_2(t)} \to 0$ as $t \to \infty$.

For most purposes, we would like to have information about the asymptotic behavior of the solutions themselves. Knowing the limiting value of $\frac{u(t+1)}{u(t)}$ gives partial information but does not immediately yield an asymptotic approximation for $u(t)$. For example, some of the functions that satisfy $\lim_{t \to \infty} \frac{u(t+1)}{u(t)} = 1$ are $u(t) = 5, t, 3t^2 + 12, t^{67}, e^{\sqrt{t}}, e^{-\sqrt{t}}, \frac{1}{t^3 - 7}, \log t$, etc.

Theorem 5.3. Suppose $\frac{u(t+1)}{u(t)} \to \lambda$ $(t \to \infty)$.
  a) If $\lambda \neq 0$, then $u(t) = \pm \lambda^t e^{z(t)}$ with $z(t) \ll t$, $(t \to \infty)$.
  b) If $\lambda = 0$, then $|u(t)| = e^{-z(t)}$ with $z(t) \gg t$, $(t \to \infty)$.

Proof.
  a) Let $v(t) = \left| \frac{u(t)}{\lambda^t} \right|$. Then

$$\frac{v(t+1)}{v(t)} = \frac{\left|\frac{u(t+1)}{\lambda^{t+1}}\right|}{\left|\frac{u(t)}{\lambda^t}\right|} = \left| \frac{1}{\lambda} \frac{u(t+1)}{u(t)} \right| \to 1, \quad (t \to \infty).$$

Since $v(t)$ is positive for t sufficiently large, we can let $z(t) = \log v(t)$. Then

$$z(t+1) - z(t) = \log \frac{v(t+1)}{v(t)} \to 0, \quad (t \to \infty).$$

Let $\epsilon > 0$, and choose m so that $|z(t+1) - z(t)| < \epsilon$ for all $t > m$. For $t > m$,

$$|z(t) - z(m)| \leq \sum_{k=m+1}^{t} |z(k) - z(k-1)| < \epsilon(t-m),$$

so

$$|z(t)| < \epsilon(t-m) + |z(m)|$$

or

$$\left|\frac{z(t)}{t}\right| < \epsilon\left(1 - \frac{m}{t}\right) + \left|\frac{z(m)}{t}\right|$$
$$< 2\epsilon$$

for t sufficiently large. Since $\epsilon > 0$ was arbitrary, $z(t) \ll t$, $(t \to \infty)$, and the proof of (a) is complete.

b) The proof is left as an exercise. ∎

If $\lambda = 0$, then Theorem 5.3(b) implies that $u(t)$ must tend to zero faster than e^{-ct} for every positive constant c. For $\lambda > 0$, Theorem 5.3(a) is equivalent to the statement that $(\lambda - \delta)^t \ll |u(t)| \ll (\lambda + \delta)^t$, $(t \to \infty)$ for each small $\delta > 0$ (see Exercise 5.30).

Example 5.7. (continued) In Example 3.23, we obtained one solution of

$$(t+2)u(t+2) - (t+3)u(t+1) + 2u(t) = 0,$$

namely $u(t) = \frac{2^t}{t!}$. Note that $\frac{u(t+1)}{u(t)} = \frac{2}{t+1} \to 0$ as $t \to \infty$, so we can take $u_2(t) = \frac{2^t}{t!}$.

Let's try to produce more information about $u_1(t)$. We know that

$$\frac{u_1(t+1)}{u_1(t)} = 1 + \varphi(t), \tag{5.10}$$

where $\varphi(t) \to 0$ as $t \to \infty$. Writing the difference equation in the form

$$(t+2)\frac{u_1(t+2)}{u_1(t+1)} - (t+3) + 2\frac{u_1(t)}{u_1(t+1)} = 0$$

and substituting Eq. (5.10), we have

$$(t+2)(1 + \varphi(t+1)) - (t+3) + \frac{2}{1+\varphi(t)} = 0.$$

5.3 Linear Equations

By the Mean Value Theorem (applied to the function $\frac{2}{1+u}$)

$$\frac{2}{1+\varphi(t)} = 2 + \mathcal{O}(\varphi(t)), \quad (t \to \infty),$$

so we have

$$(t+2)(1+\varphi(t+1)) - (t+3) + 2 + \mathcal{O}(\varphi(t)) = 0.$$

Rearranging,

$$\varphi(t+1) = -\frac{1}{t+2} + \mathcal{O}\left(\frac{\varphi(t)}{t}\right), \quad (t \to \infty).$$

We conclude that

$$\varphi(t) = -\frac{1}{t+1} + \mathcal{O}\left(\frac{1}{t^2}\right), \quad (t \to \infty).$$

Substitute this last expression into Eq. (5.10) to obtain

$$u_1(t+1) = \frac{t}{t+1}\left(1 + \mathcal{O}\left(\frac{1}{t^2}\right)\right) u_1(t).$$

As in Section 3.1, we solve this equation by iteration, beginning with a value $t = t_0$ so that $u_1(t_0) \neq 0$ and $1 + \mathcal{O}(\frac{1}{t^2}) > 0$ for $t \geq t_0$:

$$u_1(t) = \prod_{s=t_0}^{t-1} \frac{s}{s+1} \prod_{s=t_0}^{t-1}\left(1 + \mathcal{O}\left(\frac{1}{s^2}\right)\right) u_1(t_0)$$

$$= \frac{t_0}{t} u_1(t_0) \prod_{s=t_0}^{t-1}\left(1 + \mathcal{O}\left(\frac{1}{s^2}\right)\right).$$

In order to complete the calculation, we need the following theorem.

Theorem 5.4. Assume both $\sum_{s=t_0}^{\infty} a_s$ and $\sum_{s=t_0}^{\infty} a_s^2$ converge and $1 + a_s > 0$ for $s \geq t_0$. Then $\lim_{t \to \infty} \prod_{s=t}^{t-1}(1 + a_s)$ exists and is equal to a positive constant.

The proof is outlined in Exercise 5.34.

Returning to our calculation, we see that $\lim_{t\to\infty} tu_1(t) = C \neq 0$, so we finally have

$$u_1(t) \sim \frac{C}{t}, \quad (t \to \infty).$$

Frequently, an equation that is not of Poincaré type can be converted to one of Poincaré type by a change of variable.

Example 5.8. $u(t+2) - (t+1)u(t+1) + u(t) = 0$.

If this equation has a solution that increases rapidly as t increases, then the terms $u(t+2)$ and $(t+1)u(t+1)$ will increase more rapidly than the term $u(t)$, so

$$u(t+2) \sim (t+1)u(t+1), \quad (t \to \infty).$$

This relation suggests that $u(t)$ may grow like $(t-1)!$ Consequently, we factor off this behavior by making the change of variable

$$u(t) = (t-1)!v(t).$$

The resulting equation for v is

$$v(t+2) - v(t+1) + \frac{v(t)}{t(t+1)} = 0,$$

which is of Poincaré type with characteristic roots $\lambda = 0, 1$. As in the previous example, set

$$\frac{v(t+1)}{v(t)} = 1 + \varphi(t),$$

where $\varphi(t) \to 0$ as $t \to \infty$, and substitution yields an equation for φ:

$$\varphi(t+1) + \frac{1}{t(t+1)}\frac{1}{1+\varphi(t)} = 0.$$

5.3 Linear Equations

Since $\frac{1}{1+\varphi(t)} = 1 + \mathcal{O}(\varphi(t))$, $(t \to \infty)$,

$$\varphi(t+1) = -\frac{1}{t(t+1)}\left(1 + \mathcal{O}(\varphi(t))\right), \quad (t \to \infty),$$

so

$$\varphi(t) = \mathcal{O}\left(\frac{1}{t^2}\right), \quad (t \to \infty).$$

Then

$$v(t+1) = \left(1 + \mathcal{O}\left(\frac{1}{t^2}\right)\right) v(t),$$

and Theorem 5.4 implies

$$v(t) \sim C, \quad (t \to \infty),$$

for some constant C. Finally, we have

$$u_1(t) \sim C(t-1)!, \quad (t \to \infty).$$

Next, set

$$\frac{v(t+1)}{v(t)} = \psi(t),$$

where $\psi(t) \to 0$ as $t \to \infty$. Then ψ satisfies

$$\psi(t) = \frac{1}{t(t+1)} + \psi(t)\psi(t+1),$$

or

$$\psi(t) = \frac{1}{t(t+1)}\left(1 + \mathcal{O}\left(\frac{1}{t^2}\right)\right), \quad (t \to \infty).$$

It follows that

$$v(t+1) = \frac{1}{t(t+1)}\left(1 + \mathcal{O}\left(\frac{1}{t^2}\right)\right)v(t), \quad (t \to \infty).$$

Iteration and Theorem 5.4 yield a constant D so that

$$v(t) \sim \frac{D}{t!(t-1)!}, \quad (t \to \infty),$$

and we obtain a second solution $u_2(t)$ that satisfies

$$u_2(t) \sim \frac{D}{t!}, \quad (t \to \infty).$$

We present a final example that is somewhat more complicated than the previous ones.

Example 5.9. $u(t+2) - 3tu(t+1) + 2t^2 u(t) = 0.$

If we seek a rapidly increasing solution, then it is not clear in this case that any term is asymptotically smaller than the others. In fact, a growth rate of $t!$ would roughly balance the size of the three terms. Let

$$u(t) = t!v(t).$$

Then $v(t)$ satisfies

$$v(t+2) - \left(3 - \frac{6}{t+2}\right)v(t+1) + 2\left(1 - \frac{3t+2}{(t+1)(t+2)}\right)v(t) = 0,$$

which is of Poincaré type. By Perron's Theorem, there are independent solutions v_1, v_2 so that

$$\frac{v_1(t+1)}{v_1(t)} \to 1, \quad \frac{v_2(t+1)}{v_2(t)} \to 2,$$

as $t \to \infty$.

5.3 Linear Equations

Let $\frac{v_1(t+1)}{v_1(t)} = 1 + \varphi(t)$ so that $\lim_{t\to\infty} \varphi(t) = 0$. A short computation leads to

$$\varphi(t+1) - 2\varphi(t) = -\frac{2}{(t+1)(t+2)} + \mathcal{O}\left(\frac{\varphi(t)}{t}\right) + \mathcal{O}\left(\varphi^2(t)\right), \quad (t \to \infty).$$

If we call the righthand side of the preceding equation $r(t)$, then the general solution is (by Theorem 3.1)

$$\varphi(t) = 2^{t-1} \left(C + \sum_{s=1}^{t-1} \frac{r(s)}{2^s} \right).$$

To satisfy the condition $\lim_{t\to\infty} \varphi(t) = 0$, choose $C = -\sum_{s=1}^{\infty} \frac{r(s)}{2^s}$; then

$$\varphi(t) = -\sum_{s=t}^{\infty} \frac{r(s)}{2^{s-t+1}}$$

and

$$|\varphi(t)| \le \max_{s \ge t} |r(s)| \sum_{s=t}^{\infty} \frac{1}{2^{s-t+1}}$$

or

$$|\varphi(t)| \le \max_{s \ge t} |r(s)|.$$

It follows that $\varphi(t) = \mathcal{O}(\frac{1}{t^2})$, $(t \to \infty)$, so by Theorem 5.4, $v_1(t) \sim C_1$, $(t \to \infty)$. A solution $u_1(t)$ of the original equation then satisfies

$$u_1(t) \sim C_1 t!, \quad (t \to \infty).$$

Now set

$$\frac{v_2(t+1)}{v_2(t)} = 2 + \psi(t),$$

with $\psi(t) \ll 1$ $(t \to \infty)$. We find

$$\psi(t+1) - \frac{\psi(t)}{2} = \frac{3t+4}{(t+1)(t+2)} + \mathcal{O}\left(\psi^2(t)\right), \quad (t \to \infty).$$

Since

$$\frac{3t+4}{(t+1)(t+2)} = \frac{3}{t} + \mathcal{O}\left(\frac{1}{t^2}\right), \quad (t \to \infty),$$

the general solution is

$$\psi(t) = 2^{1-t}\left[C + \sum_{s=1}^{t-1} 2^s \left(\frac{3}{s} + \mathcal{O}\left(\frac{1}{s^2}\right) + \mathcal{O}(\psi^2(s))\right)\right], \quad (t \to \infty).$$

From Exercise 5.17,

$$\sum_{s=1}^{t-1} 2^s \left(\frac{3}{s}\right) = \frac{3}{t} 2^t \left[1 + \mathcal{O}\left(\frac{1}{t}\right)\right], \quad (t \to \infty).$$

Since

$$\sum_{s=1}^{t-1} 2^s \left(\frac{1}{s^2}\right) = \mathcal{O}\left(\frac{2^t}{t^2}\right), \quad (t \to \infty),$$

we have

$$\psi(t) = \frac{6}{t} + \mathcal{O}\left(\frac{1}{t^2}\right), \quad (t \to \infty).$$

Then v_2 satisfies

$$v_2(t+1) = 2\left(\frac{t+3}{t}\right)\left[1 + \mathcal{O}\left(\frac{1}{t^2}\right)\right] v_2(t),$$

5.3 Linear Equations

and we find by iteration

$$v_2(t) \sim C_2 t^3 2^{t-1}, \quad (t \to \infty),$$

so finally

$$u_2(t) \sim C_2 2^{t-1}(t+3)!, \quad (t \to \infty).$$

For a summary of what is known about the asymptotic behavior of solutions of linear difference equations see the appendix of Wimp [248]. Other useful references are Batchelder [22], Birkhoff and Trjitzinksy [26], Culmer and Harris [53], and Immink [135].

Proof of Poincaré's Theorem.

Since the main ideas of the proof are evident in the case $n = 2$, we consider that case only and write Eq. (5.9) in the form

$$u(t+2) + (a + \alpha(t))\, u(t+1) + (b + \beta(t))\, u(t) = 0, \qquad (5.11)$$

where $\alpha(t)$, $\beta(t) \to 0$ as $t \to \infty$. Recall that the roots λ_1, λ_2 of the characteristic equation $\lambda^2 + a\lambda + b = 0$ satisfy $|\lambda_1| > |\lambda_2|$.

Let $u(t)$ be a nontrivial solution of Eq. (5.11) and let $x(t)$, $y(t)$ be chosen to satisfy

$$\begin{aligned} x(t) + y(t) &= u(t), \\ \lambda_1 x(t) + \lambda_2 y(t) &= u(t+1). \end{aligned} \qquad (5.12)$$

The system (5.12) has for each t a unique nontrivial solution since

$$\det \begin{bmatrix} 1 & 1 \\ \lambda_1 & \lambda_2 \end{bmatrix} = \lambda_2 - \lambda_1 \neq 0$$

and either $u(t)$ or $u(t+1)$ is not zero.

Using Eqs. (5.11) and (5.12), we arrive at the system

$$\begin{aligned} x(t+1) =\ & \lambda_1 x(t) + (\lambda_2 - \lambda_1)^{-1} \left[(\lambda_1 \alpha(t) + \beta(t))\, x(t) \right. \qquad (5.13) \\ & \left. + (\lambda_2 \alpha(t) + \beta(t))\, y(t) \right], \end{aligned}$$

$$y(t+1) = \lambda_2 y(t) + (\lambda_1 - \lambda_2)^{-1} \left[(\lambda_2 \alpha(t) + \beta(t)) y(t) \right. \quad (5.14)$$
$$\left. + (\lambda_1 \alpha(t) + \beta(t)) x(t) \right].$$

(See Exercise 5.37.)

Choose $\epsilon > 0$ small enough that $\frac{|\lambda_2| + \epsilon}{|\lambda_1| - \epsilon} < 1$, and choose N so large that

$$|\lambda_1 - \lambda_2|^{-1} |\lambda_i \alpha(t) + \beta(t)| < \frac{\epsilon}{2}, \quad (i = 1, 2)$$

if $t \geq N$.

Let $t \geq N$ and suppose $|x(t)| \geq |y(t)|$. From Eq. (5.13)

$$\begin{aligned} |x(t+1)| &\geq |\lambda_1| |x(t)| - \frac{\epsilon}{2} (|x(t)| + |y(t)|) \\ &\geq (|\lambda_1| - \epsilon) |x(t)|. \end{aligned}$$

From Eq. (5.14)

$$\begin{aligned} |y(t+1)| &\leq |\lambda_2| |y(t)| + \frac{\epsilon}{2} (|y(t)| + |x(t)|) \\ &\leq (|\lambda_2| + \epsilon) |x(t)|. \end{aligned}$$

Taking a ratio of these inequalities, we have

$$\left| \frac{y(t+1)}{x(t+1)} \right| \leq \frac{|\lambda_2| + \epsilon}{|\lambda_1| - \epsilon} < 1,$$

so $|x(t+1)| > |y(t+1)|$, and inductively we conclude $|x(s)| > |y(s)|$ for all $s > t$. Consequently, there is a number $M \geq N$ so that either

$$|x(t)| > |y(t)| \text{ for } t \geq M \quad (5.15)$$

or

$$|y(t)| > |x(t)| \text{ for } t \geq M. \quad (5.16)$$

5.3 Linear Equations

Suppose that Eq. (5.15) is true. There is a number r in $[0,1]$ (the "limit superior") so that for each $\delta > 0$

$$\left|\frac{y(t)}{x(t)}\right| < r + \delta \qquad (5.17)$$

for sufficiently large t, and

$$\left|\frac{y(t)}{x(t)}\right| > r - \delta \qquad (5.18)$$

for infinitely many values of t.
From Eqs. (5.14) and (5.13),

$$|y(t+1)| \leq |\lambda_2||y(t)| + \epsilon|x(t)|,$$
$$|x(t+1)| \geq |\lambda_1||x(t)| - \epsilon|x(t)|,$$

for $t \geq M$, so by Eq. (5.18)

$$r - \delta < \left|\frac{y(t+1)}{x(t+1)}\right| \leq \frac{|\lambda_2|\left|\frac{y(t)}{x(t)}\right| + \epsilon}{|\lambda_1| - \epsilon}$$

for infinitely many values of t. By Eq. (5.17)

$$r - \delta < \frac{|\lambda_2|(r+\delta) + \epsilon}{|\lambda_1| - \epsilon}$$

or

$$r < \frac{\delta(|\lambda_1| + |\lambda_2| - \epsilon) + \epsilon}{|\lambda_1| - |\lambda_2| - \epsilon}.$$

Since ϵ and δ may be chosen as small as we like, it follows that $r = 0$. Thus if Eq. (5.15) is true, then $\lim_{t \to \infty} \frac{y(t)}{x(t)} = 0$. From Eq. (5.12), $\frac{u(t)}{x(t)} \to 1$ and $\frac{u(t+1)}{x(t)} \to \lambda_1$ as $t \to \infty$. Then $\frac{u(t+1)}{u(t)} \to \lambda_1$ as $t \to \infty$.
  In a similar manner, one can show that Eq. (5.16) implies $\lim_{t \to \infty} \frac{x(t)}{y(t)} = 0$ and $\lim_{t \to \infty} \frac{u(t+1)}{u(t)} = \lambda_2$. ∎

5.4 Nonlinear Equations

In Section 4.3, we considered stability questions for the equation

$$u(t+1) = f(u(t)). \tag{5.19}$$

These equations are of great importance in numerical analysis, especially for approximating roots for nonlinear equations. For example, Newton's method for solving $g(y) = 0$ is of this type with $f(u) = u - \frac{g(u)}{g'(u)}$. We will analyze more closely the asymptotic behavior of solutions of Eq. (5.19) as $t \to \infty$ in this section. For simplicity, f will be a real-valued function of a real variable. Also, we assume $f(0) = 0$ for most of the discussion since taking zero to be the fixed point simplifies calculations and the fixed point can always be translated to zero.

We begin by examining the case that $0 < |f'(0)| < 1$. If f' is continuous near zero, then there is a $\delta > 0$ and an $0 < \alpha < 1$ so that $|f'(u)| \leq \alpha$ for $|u| \leq \delta$. As in the proof of Theorem 4.8,

$$|u(t)| \leq \alpha^t |u(0)| \quad (t \geq 0) \tag{5.20}$$

for every solution u of Eq. (5.19) with $|u(0)| \leq \delta$, and $u(t) \to 0$ as $t \to \infty$.

From Taylor's formula,

$$f(u) = f'(0)u + \frac{f''(c)}{2}u^2,$$

where c is between 0 and u, so

$$u(t+1) = f'(0)\left[1 + \frac{f''(c)u(t)}{2f'(0)}\right]u(t). \tag{5.21}$$

If $f(u) = 0$ for some $0 < |u| \leq \delta$, then it is possible for a nontrivial solution to reach zero in a finite number of iterations. For solutions that do not reach zero in finitely many steps, inequality (5.20), Theorem 5.4, and iteration of (5.21) yield

$$u(t) \sim Cu(0)\left(f'(0)\right)^t, \quad (t \to \infty),$$

5.4 Nonlinear Equations

where $C \neq 0$ varies with $u(0)$. It is convenient to use the convention $u(t) \sim 0$, $(t \to \infty)$ when u reaches zero in finitely many steps so that the preceding asymptotic relation applies to all solutions u with $|u(0)| < \delta$. We have proved the following theorem:

Theorem 5.5. Assume $f''(u)$ is bounded near $u = 0$, $f(0) = 0$, $0 < |f'(0)| < 1$ and $|f'(y)| < 1$ for $|u| \leq \delta$. For each solution u of Eq. (5.19) with $|u(0)| \leq \delta$,

$$u(t) \sim Cu(0)\left(f'(0)\right)^t, \quad (t \to \infty),$$

where C depends on $u(0)$.

For a given initial value $u(0)$, the constant C can be computed as accurately as desired by iteration.

Example 5.10. $u(t+1) = 0.5u(t)(1 + u(t))$.

For this equation, $f(u) = 0.5u(1+u)$, $f'(0) = 0.5$, and $|f'(u)| < 1$ for $-1.5 < u < 0.5$. First, let $u(0) = 0.2$ and $c(t) = \frac{u(t)}{.2(0.5)^t}$. Then $c(t)$ satisfies the equation

$$c(t+1) = c(t)\left(1 + 0.2(0.5)^t c(t)\right).$$

By iteration, we find $\lim_{t \to \infty} c(t) \simeq 1.54$, so by Theorem 5.5 the approximation

$$u(t) \simeq (1.54)(0.2)(0.5)^t$$

is accurate to about three significant digits for large t.

If we begin instead with $u(0) = -0.2$, we find that $C \simeq 0.69$.

The type of convergence described in Theorem 5.5 is called "linear convergence" since $u(t+1) \sim f'(0)u(t)$, $(t \to \infty)$.

Example 5.11. The modified Newton's method for solving $g(v) = 0$ is given by

$$v(t+1) = v(t) - \frac{g(v(t))}{g'(v(0))}, \quad (t \geq 0). \tag{5.22}$$

The advantage of this method over Newton's method is that the derivative of g is computed only once at the initial point. Let z be the desired root, i.e., let $g(z) = 0$. We make the change of variables $u = v - z$ to translate the root to zero. Equation (5.22) becomes

$$u(t+1) = u(t) - \frac{g(u(t)+z)}{g'(u(0)+z)}$$
$$\equiv f(u(t)).$$

Then $f(0) = 0$ and $f'(u) = 1 - \frac{g'(u+z)}{g'(u(0)+z)}$. If $g'(z) \neq 0$ and g' is continuous, we can choose $\delta > 0$ so that

$$\left| 1 - \frac{g'(u+z)}{g'(u(0)+z)} \right| < 1$$

for $|u|, |u(0)| < \delta$. By Theorem 5.5, we expect to have linear convergence of Eq. (5.22) to the root z of $g(0) = 0$ if the initial point $v(0)$ satisfies $|v(0) - z| < \delta$.

Next, suppose $f'(0) = 0$. In this case, solutions of Eq. (5.19) that start near zero will exhibit convergence to zero that is more rapid than linear convergence. If $f''(0) \neq 0$, Taylor's formula yields

$$u(t+1) = \frac{f''(0)}{2} u^2(t) \left[1 + \frac{f'''(c)}{3f''(0)} u(t) \right],$$

for some c between $u(t)$ and 0. Again Eq. (5.20) holds if $|u(0)| \leq \delta$, so

$$\beta(t) \equiv \frac{f'''(c)}{3f''(0)} u(t) = \mathcal{O}(\alpha^t), \quad (t \to \infty), \tag{5.23}$$

where $0 < \alpha < 1$.

Either $u(t)$ goes to zero in a finite number of steps or

$$\log |u(t+1)| = 2 \log |u(t)| + \log \left| \frac{f''(0)}{2} (1 + \beta(t)) \right|.$$

5.4 Nonlinear Equations

This is a first order linear equation with solution

$$\log |u(t)| = 2^t \left[\log |u(0)| + \sum_{s=0}^{t-1} 2^{-s-1} \log \left| \frac{f''(0)}{2} (1+\beta(s)) \right| \right].$$

Let

$$D = \log |u(0)| + \sum_{s=0}^{\infty} 2^{-s-1} \log \left| \frac{f''(0)}{2} (1+\beta(s)) \right|.$$

Then

$$\log |u(t)| = 2^t \left[D - \sum_{s=t}^{\infty} 2^{-s-1} \log \left| \frac{f''(0)}{2} (1+\beta(s)) \right| \right],$$

and

$$|u(t)| = \frac{2}{|f''(0)|} (Cu(0))^{2^t} e^{-2^t \sum_{s=t}^{\infty} 2^{-s-1} \log |1+\beta(s)|},$$

where $C = \frac{e^D}{|u(0)|}$. Using Eq. (5.23), it can be shown that

$$\lim_{t \to \infty} 2^t \sum_{s=t}^{\infty} 2^{-s-1} \log |1 + \beta(s)| = 0$$

(see Exercise (5.43)), so finally

$$u(t) \sim \frac{2}{f''(0)} (Cu(0))^{2^t}, \quad (t \to \infty). \qquad (5.24)$$

In a similar way, we can prove the following theorem:

Theorem 5.6. Assume for some $m \geq 2$ that $f^{(m+1)}$ exists near zero, $f(0) = f'(0) = \cdots = f^{(m-1)}(0) = 0$ and $f^{(m)}(0) \neq 0$. Let $|f'(u)| < 1$ for

$|u| \leq \delta$. For each solution u of Eq. (5.19) with $|u(0)| \leq \delta$,

$$|u(t)| \sim \left(\frac{m!}{|f^{(m)}(0)|}\right)^{\frac{1}{m-1}} (C|u(0)|)^{m^t}, \quad (t \to \infty),$$

where C depends on $u(0)$.

Note that the proof for the case $m = 2$ shows that C is a bounded function of $u(0)$. Furthermore, we must have $|Cu(0)| < 1$ when $|u(0)| \leq \delta$ since Eq. (5.20) implies that $u(t) \to 0$, $(t \to \infty)$. We are again using the convention $u(t) \sim 0$, $(t \to \infty)$, for the case that u reaches zero in finitely many steps.

A short calculation gives

$$\left|\frac{u(t+1)}{(u(t))^m}\right| \sim \frac{|f^{(m)}(0)|}{m!}, \quad (t \to \infty).$$

Consequently, the type of convergence described by Theorem 5.6 is called "convergence of order m." The terms "quadratic convergence" for $m = 2$ and "cubic convergence" for $m = 3$ are also commonly used.

Example 5.12. Consider Newton's method

$$v(t+1) = v(t) - \frac{g(v(t))}{g'(v(t))}, \quad (t \geq 0),$$

for approximating a root z of $g(v) = 0$. Let $u = v - z$. Then

$$u(t+1) = u(t) - \frac{g(u(t)+z)}{g'(u(t)+z)} \equiv f(u(t)).$$

We have $f(0) = 0$ and

$$f'(u) = \frac{g(u+z)g''(u+z)}{g'^2(u+z)},$$

so $f'(0) = 0$. Now $f''(0) = \frac{g''(z)}{g'(z)}$ and Newton's method will converge quadratically if $g'(z)$ and $g''(z)$ are not zero. By Theorem 5.6 and Eq. (5.24)

5.4 Nonlinear Equations

$$v(t) - z \sim \frac{2g'(z)}{g''(z)} \left(C(v(0) - z)\right)^{2^t}, \quad (t \to \infty),$$

provided that

$$\left|\frac{g(v)g''(v)}{g'^2(v)}\right| < 1$$

for $|v - z| \leq |v(0) - z|$. For a given equation $g(v) = 0$, the staircase method provides an elementary means of determining for which initial estimates $v(0)$ Newton's method will converge (see Exercise 5.46).

It is also of interest to obtain asymptotic approximations of solutions of Eq. (5.19) that diverge to infinitely as $t \to \infty$. As the following example shows, the analysis can sometimes be carried out by using the substitution $u = \frac{1}{v}$. The variable v will converge to zero as $t \to \infty$, and its rate of convergence may be given by one of the previous theorems. An asymptotic estimate for u is then obtained by inversion.

Example 5.13. Consider again the equation of Example 5.10,

$$u(t+1) = 0.5u(t)(1 + u(t)).$$

Letting $v = \frac{1}{u}$, we have

$$v(t+1) = \frac{2v^2(t)}{v(t) + 1} \equiv f(v(t)).$$

Now $f(0) = f'(0) = 0$ and $f''(0) = 4$. From Eq. (5.24),

$$v(t) \sim .5\left(Cv(0)\right)^{2^t}, \quad (t \to \infty),$$

if $v(0)$ is sufficiently near zero. It follows that

$$u(t) \sim 2\left(Du(0)\right)^{2^t}, \quad (t \to \infty),$$

if $u(0)$ is sufficiently large, where D depends on $u(0)$.

In the next example, we consider a family of equations with solutions that diverge to infinity more slowly.

Example 5.14. $u(t+1) = u(t) + a + g(u(t))$.

Assume a is a positive constant and $g(u) = \mathcal{O}\left(\frac{1}{u^b}\right)$, $(u \to \infty)$, for some $b > 0$. Choose u_0 large enough so that $g(u) \geq -\frac{a}{2}$ for $u \geq u_0 > 0$. then $u(t+1) - u(t) \geq \frac{a}{2}$, $(t \geq 0)$, if $u(0) \geq u_0$, and it follows by iteration that $u(t) > \frac{at}{2}$, $(t \geq 0)$.

Now we have

$$|g(u(t))| \leq \frac{M}{u^b} \leq \frac{2^b M}{a^b t^b}, \quad (t \geq 0),$$

for some $M > 0$. Since

$$\Delta u(t) = a + \mathcal{O}(t^{-b}), \quad (t \to \infty),$$

summation yields

$$u(t) = at + \mathcal{O}(\sum t^{-b}), \quad (t \to \infty).$$

By the Integral Test (or the Euler summation formula),

$$\sum t^{-b} = \begin{cases} \mathcal{O}(1) & \text{if} \quad b > 1, \\ \mathcal{O}(\log t) & \text{if} \quad b = 1, \\ \mathcal{O}(t^{1-b}) & \text{if} \quad 0 < b < 1. \end{cases}$$

Our final example makes use of Example 5.14 in analyzing the rate of convergence of solutions of Eq. (5.19) to zero in the case that $f'(0) = 1$.

Example 5.15. $u(t+1) = u(t)\left(1 - u^2(t)\right)$.

Even though the derivative of $u(1-u^2)$ at $u = 0$ is one, it is evident from a staircase diagram that solutions with initial values near zero will converge to zero as $t \to \infty$.

5.4 Nonlinear Equations

Let $u = \frac{1}{\sqrt{v}}$. The equation for v is

$$\begin{aligned} v(t+1) &= \frac{v^3(t)}{(v(t)-1)^2} \\ &= v(t) + 2 + \frac{3v(t)-2}{(v(t)-1)^2}. \end{aligned}$$

From Example 5.14,

$$v(t) = 2t + \mathcal{O}(\log t), \quad (t \to \infty).$$

Then

$$\begin{aligned} u(t) &= (2t + \mathcal{O}(\log t))^{-\frac{1}{2}}, \quad (t \to \infty) \\ &= \frac{1}{\sqrt{2t}} \left(1 + \mathcal{O}\left(\frac{\log t}{2t}\right) \right)^{-\frac{1}{2}}, \quad (t \to \infty). \end{aligned}$$

By the Mean Value Theorem,

$$(1+w)^{-\frac{1}{2}} = 1 - \frac{1}{2}(1+c)^{-\frac{3}{2}} w$$

for some c between 0 and w. Finally

$$u(t) = \frac{1}{\sqrt{2t}} \left(1 + \mathcal{O}\left(\frac{\log t}{t}\right) \right), \quad (t \to \infty).$$

DeBruijn [54] gives a number of methods for computing additional terms in asymptotic approximations of solutions of nonlinear difference equations of first order.

Exercises

Section 5.1

5.1 Show that "$\sim$" is an "equivalence relation," i.e.,
  (a) $f(t) \sim f(t)$, $(t \to \infty)$;
  (b) $f(t) \sim g(t)$ implies $g(t) \sim f(t)$, $(t \to \infty)$;
  (c) $f(t) \sim g(t)$ and $g(t) \sim h(t)$ imply $f(t) \sim h(t)$, $(t \to \infty)$.

5.2 Verify the following asymptotic relations:
  (a) $\frac{1}{t^2+2t-7} \sim \frac{1}{t^2}$, $(t \to \infty)$;
  (b) $\cosh t \sim \frac{e^t}{2}$, $(t \to \infty)$;
  (c) $\sin \frac{3}{t} \sim \frac{3}{t}$, $(t \to \infty)$;
  (d) $\tan^{-1} t \sim \frac{\pi}{2}$, $(t \to \infty)$.

5.3 Verify
  (a) $t^2 \log t \ll t^3$, $(t \to \infty)$;
  (b) $\log \log t \ll \log t$, $(t \to \infty)$;
  (c) $t^4 \ll e^t$, $(t \to \infty)$;
  (d) $\tan(\frac{1}{t^2}) \ll \frac{10}{t}$, $(t \to \infty)$.

5.4 Show that if $u(t) = \mathcal{O}(w(t))$ and $v(t) = \mathcal{O}(w(t))$, $(t \to \infty)$, then for any constants C and D we have $Cu(t) + Dv(t) = \mathcal{O}(w(t))$, $(t \to \infty)$. Does it also follow that $u(t)v(t) = \mathcal{O}(w(t))$, $(t \to \infty)$?

5.5 Verify
  (a) $5x^2 \sin 3x = \mathcal{O}(x^2)$, $(x \to \infty)$;
  (b) $\frac{1}{x-2} = \frac{1}{x}\left[1 + \mathcal{O}(\frac{1}{x})\right]$, $(x \to \infty)$;
  (c) $\frac{1}{x-2} = \frac{1}{x}\left[1 + \frac{2}{x} + \mathcal{O}(\frac{1}{x^2})\right]$, $(x \to \infty)$.

5.6
  (a) Show that $\sqrt{t^2+1} = t[1 + \frac{1}{2t^2} + \mathcal{O}(\frac{1}{t^4})]$, $(t \to \infty)$.
  (b) Use the equation in part (a) to estimate $\sqrt{101}$ and $\sqrt{10001}$.
  (c) What are the relative errors in the estimates in part (b)?

5.7 Prove that if f and g are continuous and have convergent integrals on $[1, \infty)$, then $f(t) \sim g(t)$, $(t \to \infty)$, implies that $\int_t^\infty f(x)dx \sim \int_t^\infty g(x)dx$, $(t \to \infty)$.

5.8 Give estimates of
  (a) $\int_1^\infty \frac{e^{-3t}}{t^{50}} dt$;   (b) $\int_1^\infty \frac{e^{-50t}}{t^3} dt$.

Exercises

5.9 Find bounds on the errors for the estimates in Exercise 5.8.

5.10 Use integration by parts to show

$$\int_0^\infty \frac{e^{-t}}{x+t}\,dt = \frac{1}{x}\left(1 - \frac{1}{x} + \mathcal{O}\left(\frac{1}{x^2}\right)\right), \quad (x \to \infty).$$

5.11 Complete the whole asymptotic series for the integral in Exercise 5.10.

Section 5.2

5.12 Show that the asymptotic estimate in Example 5.2 can be improved to

$$y_n = (n-1)!\left[y_1 + e - 1 - \frac{1}{n!} + \mathcal{O}\left(\frac{1}{(n+1)!}\right)\right], \quad (n \to \infty).$$

5.13 Use Taylor's formula to obtain an asymptotic estimate for

$$\sum_{k=1}^{n-1} \frac{1}{(2k)!}, \quad (n \to \infty).$$

5.14 Given that $\sum_{k=1}^\infty \frac{(-1)^{k+1}}{k^2} = \frac{\pi^2}{12}$, show that

$$\sum_{k=1}^{n-1} \frac{(-1)^{k+1}}{k^2} = \frac{\pi^2}{12} + \mathcal{O}\left(\frac{1}{n^2}\right), \quad (n \to \infty).$$

5.15 Verify
  (a) $\sum_{k=1}^\infty k! = n![1 + \mathcal{O}(\frac{1}{n})], \quad (n \to \infty)$;
  (b) $\sum_{k=1}^n k! = n![1 + \frac{1}{n} + \mathcal{O}(\frac{1}{n^2})], \quad (n \to \infty)$.

5.16 Generalize the calculation in Example 5.4 to find an asymptotic approximation for

$$\sum_{k=2}^{n-1} a^k \log k$$

as $n \to \infty$ if $a > 1$.

5.17 Use summation by parts to show

$$\sum_{k=1}^{n-1} \frac{2^k}{k} = \frac{2^n}{n}\left[1 + \mathcal{O}\left(\frac{1}{n}\right)\right], \quad (n \to \infty).$$

5.18 Verify Wallis' formula:

$$\frac{\pi}{2} = \lim_{n\to\infty} \left[\frac{2\cdot 4 \cdots 2n}{1\cdot 3 \cdots (2n-1)}\right]^2 \frac{1}{2n+1}.$$

(Hint: first show that

$$\int_0^{\frac{\pi}{2}} \sin^{2n-1} x\, dx = \frac{2\cdot 4 \cdots (2n-2)}{1\cdot 3 \cdots (2n-1)}$$

and

$$\int_0^{\frac{\pi}{2}} \sin^{2n} x\, dx = \frac{1\cdot 3 \cdots (2n-1)}{2\cdot 4 \cdots (2n)} \frac{\pi}{2}.$$

Next, integrate the inequalities $\sin^{2n+1} x \leq \sin^{2n} x \leq \sin^{2n-1} x$, which hold on the interval $[0, \frac{\pi}{2}]$.)

5.19 What is the relative error if Eq. (5.6) is used to compute 6! ?

5.20 Find an asymptotic estimate like Eq. (5.4) for $\sum_{k=1}^{n} \log(2k+1)$.

5.21 Find the next term in the asymptotic series in Eq. (5.6).

5.22 Use Stirling's formula to determine the asymptotic behavior of the

Exercises 245

solutions of the equation $u(t+1) = \frac{2t^3}{(t+1)^2} u(t)$ as $t \to \infty$.

5.23 Show that $t^{(r)} \sim t^r$, $(t \to \infty)$.

5.24 Show $\lim_{n\to\infty} \left(\sum_{k=1}^n \frac{1}{k} - \log n\right)$ exists by using Euler's summation formula with $m = 1$.

5.25 Use Euler's summation formula with $m = 2$ to verify Eq. (5.8).

Section 5.3

5.26 If the second order equation with constant coefficients

$$u(t+2) + p_1 u(t+1) + p_0 u(t) = 0$$

has a double characteristic root $\lambda_1 = \lambda_2 = \lambda$, then show that each nontrivial solution $u(t)$ satisfies $\lim_{t\to\infty} \frac{u(t+1)}{u(t)} = \lambda$.

5.27 For the equation $u(t+2) - (1 + \frac{(-1)^t}{t+1})u(t) = 0$, show that $\lim_{t\to\infty} \frac{u(t+1)}{u(t)}$ fails to exist for every solution $u(t)$. (Hint: use iteration.)

5.28 What information does Perron's Theorem give about the following equations?
  (a) $(t+2)u(t+2) - (2t+3)xu(t+1) + (t+1)u(t) = 0$
  (b) $t(t+1)\Delta^2 u(t) - 2u(t) = 0$
  (c) $(2t^2 - 1)u(t+3) + (2t^2 + t)u(t+2) - 4t^2 u(t+1) + u(t) = 0$.
(Note: x is a parameter in part (a).)

5.29 Prove Theorem 5.3(b)

5.30 Show that if $\lim_{t\to\infty} \frac{u(t+1)}{u(t)} = \lambda > 0$, then for each δ in $(0, \lambda)$, $(\lambda - \delta)^t \ll |u(t)| \ll (\lambda + \delta)^t$, $(t \to \infty)$.

5.31 If we did not know $u_2(t) = \frac{2^t}{t!}$ in Example 5.7, show how we could obtain $u_2(t) \sim C\frac{2^t}{t!}$, $(t \to \infty)$. (Hint: let $\frac{u_2(t+1)}{u_2(t)} = \psi(t)$ and show that $\psi(t) = \frac{2(t+6)}{(t+3)(t+4)}\left(1 + \mathcal{O}(\frac{1}{t^2})\right)$ $(t \to \infty)$.)

5.32 Show that $(t+1)u(t+2) - (t+4)u(t+1) + u(t) = 0$ has solutions u_1, u_2 satisfying $u_1(t) \sim t^2$, $u_2(t) \sim \frac{1}{(t+2)!}$, $(t \to \infty)$.

5.33 Use the results of Example 3.23 and Example 5.7 to deduce $\sum_{k=0}^{t-1} \frac{k!}{2^{k+1}} \sim \frac{(t-1)!}{2^t}$, $(t \to \infty)$.

5.34 Prove Theorem 5.4. (Hint: by Taylor's Theorem, $|\log(1+a_s) - a_s| \leq a_s^2$ if $|a_s| < \frac{1}{2}$. Then $\sum_{s=t_0}^{\infty} (\log(1+a_s) - a_s)$ converges, so $\sum_{s=t_0}^{\infty} \log(1+a_s)$ converges.)

5.35 The Bessel functions $J_t(x)$ and $Y_t(x)$ satisfy the equation

$$u(t+2) - \frac{2}{x}(t+1)u(t+1) + u(t) = 0.$$

Using the procedure of Example 5.8, show that there are solutions u_1, u_2 that satisfy $u_1(t) \sim \frac{C}{t!}(\frac{x}{2})^t$ and $u_2(t) \sim D(t-1)!(\frac{2}{x})^t$, $(t \to \infty)$.

5.36 Show that $u(t+2) - (t+1)u(t+1) + (t+1)u(t) = 0$ has solutions u_1, u_2 so that $u_1(t) \sim C(t-2)!$, $u_2(t) \sim Dt$, $(t \to \infty)$.

5.37 Verify Eq. (5.13).

5.38 Investigate the asymptotic behavior of the solutions of

$$t^2 u(t+2) - 3tu(t+1) + 2u(t) = 0$$

as $t \to \infty$.

Section 5.4

5.39 Find all the solutions of $u(t+1) = 0.5u(t)(1 - u(t))$ that reach zero in three or fewer steps.

5.40 Find the asymptotic behavior of solutions of

$$u(t+1) = 0.5(u^3(t) - 3u^2(t) + 4u(t))$$

that start near a stable fixed point.

5.41 For the equation

$$u(t+1) = \frac{u^2(t) + 3u(t) + 1}{u^2(t) + 4},$$

show that $u(t) - 1 \sim Cu(0)(\frac{3}{5})^t$, $(t \to \infty)$ if $u(0)$ is near 1. Estimate C if $u(0) = 2$.

Exercises

5.42 (a) Write down a modified Newton's method for computing $\sqrt{2}$ and observe the linear convergence by computing five iterations.

(b) Write down Newton's method for computing $\sqrt{2}$ and observe the quadratic convergence by computing four iterations.

5.43 Show that

$$\lim_{t \to \infty} 2^t \sum_{s=t}^{\infty} 2^{-s-1} \log|1 + \beta(s)| = 0,$$

if β satisfies Eq. (5.23). (Hint: use the Mean Value Theorem to show $\log(1 + \beta(s)) = \mathcal{O}(\alpha^s)$, $(s \to \infty)$.)

5.44 Compute an asymptotic approximation for solutions of

$$u(t+1) = \frac{u(t)(u^2(t) + 3)}{3u^2(t) + 1}$$

that have initial values near $u = 1$.

5.45 Show that certain solutions of

$$u(t+1) = \frac{u(t)(u^2(t) + 3)}{3u^2(t) + a}, \quad (a > 0),$$

exhibit cubic convergence to $\sqrt{a}$.

5.46 Use the staircase method to show

(a) solutions u of

$$u(t+1) = \frac{u^2(t) + a}{2u(t)}$$

converge to $\sqrt{a}$ if $u(0) > 0$ and to $-\sqrt{a}$ if $u(0) < 0$;

(b) all solutions of

$$u(t+1) = \frac{2u^3(t) + 1}{3u^2(t) + 1}$$

converge to the unique solution of $u^3 + u - 1 = 0$.

5.47 Consider the solution of $\tan u = u$ in the interval $(\pi, \frac{3\pi}{2})$. Find an interval of initial values so that Newton's method will converge to this solution.

5.48 Let $u(t+1) = u^4(t) + 2u(t)$, $(t \geq 0)$.
  (a) Show $u(t) \to \infty$ as $t \to \infty$ if $u(0) > 0$.
  (b) Give an asymptotic approximation for $u(t)$ as $t \to \infty$ if $u(0) > 0$.

5.49 Consider $u(t+1) = u(t) + \frac{1}{u(t)}$.
  (a) Show that $u(t) \to \infty$ as $t \to \infty$ if $u(0) > 0$.
  (b) Find an asymptotic approximation for $u(t)$. (Hint: let $u = \sqrt{v}$.)

5.50 Find the asymptotic behavior as $t \to \infty$ of solutions of

$$u(t+1) = u(t) - u^2(t) + u^3(t)$$

that converge to zero.

5.51 Compute an asymptotic approximation for solutions of $u(t+1) = \sin u(t)$ with initial values near zero.

Chapter 6
The Self-Adjoint Second Order Linear Equation

6.1 Introduction

In this section we will introduce the second order self-adjoint difference equation, which is the main topic of this chapter. We will show which second order linear difference equations can be put in the self-adjoint form and establish a number of useful identities.

A very important equation in applied mathematics is the second order linear self-adjoint differential equation

$$(P(x)z'(x))' + Q(x)z(x) = 0, \qquad (6.1)$$

where we assume $P(x) > 0$ in $[c,d]$ and $P(x)$, $Q(x)$ are continuous on $[c,d]$. Let us show that Eq. (6.1) is related to a self-adjoint difference equation. Now for small $h = \frac{d-c}{n}$,

$$z'(x) \approx \frac{z(x) - z(x-h)}{h}$$

and

$$(P(x)z'(x))' \approx \frac{1}{h}\left\{\frac{P(x+h)[z(x+h) - z(x)]}{h} - \frac{P(x)[z(x) - z(x-h)]}{h}\right\},$$

so

$$(P(x)z'(x))' \approx \frac{1}{h^2}\{P(x+h)z(x+h)$$
$$-[P(x+h) + P(x)]z(x) + P(x)z(x-h)\}.$$

250   Chapter 6. The Self-Adjoint Second Order Linear Equation

Let

$$x = c + th,$$

where t is a discrete variable taking on the integer values $0 \leq t \leq n$, and let

$$y(t) = z(c + th),$$

where $z(x)$ is a solution of Eq. (6.1) on $[c, d]$. Then

$$P(c + (t+1)h)\, z\,(c + (t+1)h)$$
$$- \;[P(c + (t+1)h) + P(c + th)]\, z(c + th)$$
$$+ \;P(c + th)z\,(c + (t-1)h) \;+\; h^2 Q(c + th) z(c + th) \approx 0.$$

If we set

$$p(t-1) = P(c + th),$$
$$q(t) = h^2 Q(c + th),$$

for $1 \leq t \leq n$ and $1 \leq t \leq n-1$ respectively, then

$$p(t)y(t+1) - (p(t) + p(t-1))\, y(t) + p(t-1)y(t-1) + q(t)y(t) \approx 0.$$

Finally, we can write this in the form

$$\Delta\,(p(t-1)\Delta y(t-1)) + q(t)y(t) \;\approx\; 0,$$

$1 \leq t \leq n-1$. Note that $y(t)$ is defined for $0 \leq t \leq n$.

The linear second order self-adjoint difference equation is then defined to be

$$\Delta\,(p(t-1)\Delta y(t-1)) + q(t)y(t) \;=\; 0, \qquad (6.2)$$

where we assume that $p(t)$ is defined and positive on the set of integers $[a, b+1] \equiv \{a, a+1, \cdots, b+1\}$ and $q(t)$ is defined on the set of integers

6.1 Introduction

$[a+1, b+1]$. We can also write Eq. (6.2) in the form

$$p(t)y(t+1) + c(t)y(t) + p(t-1)y(t-1) = 0, \qquad (6.3)$$

where

$$c(t) = q(t) - p(t) - p(t-1) \qquad (6.4)$$

for t in $[a+1, b+1]$.

Since Eq. (6.3) can be solved uniquely for $y(t+1)$ and $y(t-1)$, the initial value problem (6.3),

$$y(t_0) = A,$$
$$y(t_0 + 1) = B,$$

where t_0 is in $[a, b+1]$, A, B are constants, has a unique solution defined in all of $[a, b+2]$ ($\equiv \{a, a+1, \cdots, b+2\}$). Of course, a similar statement is true for the corresponding nonhomogeneous equation.

Note that any equation written in the form of Eq. (6.3), where $p(t) > 0$ on $[a, b+1]$, can be written in the self-adjoint form of Eq. (6.2) by taking

$$q(t) = c(t) + p(t) + p(t-1). \qquad (6.5)$$

Example 6.1. Write

$$2^t y(t+1) + (\sin t - 3 \cdot 2^{t-1})y(t) + 2^{t-1}y(t-1) = 0$$

in the self-adjoint form.

Here $p(t) = 2^t$ and $c(t) = \sin t - 3 \cdot 2^{t-1}$. Hence by Eq. (6.5)

$$q(t) = \sin t - 3 \cdot 2^{t-1} + 2^t + 2^{t-1}$$
$$= \sin t.$$

252   Chapter 6. The Self-Adjoint Second Order Linear Equation

Then the self-adjoint form of this equation is

$$\Delta\left(2^{t-1}\Delta y(t-1)\right) + \sin t\, y(t) = 0.$$

Actually any equation of the form

$$\alpha(t)y(t+1) + \beta(t)y(t) + \gamma(t)y(t-1) = 0, \tag{6.6}$$

where $\alpha(t) > 0$ on $[a, b+1]$, $\gamma(t) > 0$ on $[a+1, b+1]$, can be written in the self-adjoint form of Eq. (6.2). To see this, multiply both sides of Eq. (6.6) by a positive function $h(t)$ to be chosen later to obtain

$$\alpha(t)h(t)y(t+1) + \beta(t)h(t)y(t) + \gamma(t)h(t)y(t-1) = 0.$$

This would be of the form of Eq. (6.3), which we know we can write in self-adjoint form, provided

$$\alpha(t)h(t) = p(t)$$
$$\gamma(t)h(t) = p(t-1).$$

Consequently, we want to pick a positive function $h(t)$ so that

$$\alpha(t)h(t) = \gamma(t+1)h(t+1)$$

or

$$h(t+1) = \frac{\alpha(t)}{\gamma(t+1)}h(t)$$

for t in $[a, b]$. Then

$$h(t) = A \prod_{s=a}^{t-1} \frac{\alpha(s)}{\gamma(s+1)},$$

6.1 Introduction

where A is any positive constant. If we choose

$$p(t) = A\alpha(t) \prod_{s=a}^{t-1} \frac{\alpha(s)}{\gamma(s+1)},$$

and by Eq. (6.5)

$$q(t) = \beta(t)h(t) + p(t) + p(t-1),$$

then we have that Eq. (6.6) is equivalent to Eq. (6.2).

Example 6.2. Write the difference equation

$$(t-1)y(t+1) + \left(\frac{t^2}{\Gamma(t-1)} - t\right)y(t) + y(t-1) = 0, \qquad (6.7)$$

$t \geq 2$, in self-adjoint form.

Take

$$h(t) = \prod_{s=2}^{t-1}(s-1)$$
$$= (t-2)! = \Gamma(t-1).$$

Then

$$p(t) = (t-1)\Gamma(t-1) = \Gamma(t),$$

and

$$q(t) = \left(\frac{t^2}{\Gamma(t-1)} - t\right)\Gamma(t-1) + \Gamma(t) + \Gamma(t-1)$$
$$= t^2 - t\Gamma(t-1) + (t-1)\Gamma(t-1) + \Gamma(t-1)$$
$$= t^2.$$

254 Chapter 6. The Self-Adjoint Second Order Linear Equation

A self-adjoint form of Eq. (6.7) is

$$\Delta\left[\Gamma(t-1)\Delta y(t-1)\right] + t^2 y(t) = 0.$$

Let $y(t)$, $z(t)$ be solutions of Eq. (6.2) in $[a, b+2]$. Recall that in Chapter 2 we defined the Casoratian of $y(t)$ and $z(t)$ by

$$w(t) = w[y(t), z(t)] = \begin{vmatrix} y(t) & z(t) \\ y(t+1) & z(t+1) \end{vmatrix}$$
$$= \begin{vmatrix} y(t) & z(t) \\ \Delta y(t) & \Delta z(t) \end{vmatrix}.$$

Define a linear operator L on the set of functions y defined on $[a, b+2]$ by

$$Ly(t) = \Delta\left(p(t-1)\Delta y(t-1)\right) + q(t)y(t)$$

for t in $[a+1, b+1]$.

Theorem 6.1. (Lagrange Identity) If $y(t)$ and $z(t)$ are defined on $[a, b+2]$, then

$$z(t)Ly(t) - y(t)Lz(t) = \Delta\{p(t-1)w[z(t-1), y(t-1)]\}$$

for t in $[a+1, b+1]$.

Proof. For t in $[a+1, b+1]$ consider

$$\begin{aligned}
z(t)Ly(t) &= z(t)\Delta\left[p(t-1)\Delta y(t-1)\right] + z(t)q(t)y(t) \\
&= \Delta\left[z(t-1)p(t-1)\Delta y(t-1)\right] \\
&\quad - (\Delta z(t-1))\,p(t-1)\Delta y(t-1) \\
&\quad + z(t)q(t)y(t) \\
&= \Delta\left[z(t-1)p(t-1)\Delta y(t-1) - y(t-1)p(t-1)\Delta z(t-1)\right] \\
&\quad + y(t)\Delta\left(p(t-1)\Delta z(t-1)\right) + y(t)q(t)z(t) \\
&= \Delta\{p(t-1)w[z(t-1), y(t-1)]\} + y(t)Lz(t),
\end{aligned}$$

which gives the desired result. ∎

6.1 Introduction

By summing both sides of the Lagrange Identity from $a+1$ to $b+1$ we get the following corollary.

Corollary 6.1. (Green's Theorem) Assume $y(t)$ and $z(t)$ are defined on $[a, b+2]$. Then

$$\sum_{t=a+1}^{b+1} z(t)Ly(t) - \sum_{t=a+1}^{b+1} y(t)Lz(t) = \{p(t)w[z(t), y(t)]\}_a^{b+1}.$$

Corollary 6.2. (Liouville's Formula) If $y(t)$ and $z(t)$ are solutions of Eq. (6.2), then

$$w[y(t), z(t)] = \frac{C}{p(t)}$$

for t in $[a, b+1]$, where C is a constant.

Proof. By the Lagrange Identity

$$\Delta\{p(t-1)w[y(t-1), z(t-1)]\} = 0$$

for t in $[a+1, b+1]$. Hence

$$p(t-1)w[y(t-1), z(t-1)] = C,$$

where C is a constant, for t in $[a+1, b+2]$. Therefore

$$w[y(t), z(t)] = \frac{C}{p(t)}$$

for t in $[a, b+1]$.

It follows from Corollary 6.2 that if $y(t)$ and $z(t)$ are solutions of Eq. (6.2), then either $w[y(t), z(t)] = 0$ for all t in $[a, b+1]$ ($y(t), z(t)$ are linearly dependent on $[a, b+2]$) or $w[y(t), z(t)]$ is of one sign ($y(t)$ and $z(t)$ are linearly independent on $[a, b+2]$). See also Theorem 3.4.

Theorem 6.2. (Polya Factorization) Assume $z(t)$ is a solution of Eq. (6.2) with $z(t) > 0$ on $[a, b+2]$. Then there exist functions $\varrho_i(t)$, $i = 1, 2$, with

$\varrho_1(t) > 0$ on $[a, b+2]$, $\varrho_2(t) > 0$ on $[a+1, b+2]$, such that for any function $y(t)$ defined on $[a, b+2]$, for t in $[a+1, b+1]$,

$$Ly(t) = \varrho_1(t)\Delta\left[\varrho_2(t)\Delta\left(\varrho_1(t-1)y(t-1)\right)\right].$$

Proof. Since $z(t)$ is a positive solution of Eq. (6.2) we have by the Lagrange identity that

$$Ly(t) = \frac{1}{z(t)}\Delta\left\{p(t-1)w[z(t-1), y(t-1)]\right\}$$

for t in $[a+1, b+1]$. By Theorem 2.1(e)

$$\Delta\left\{\frac{y(t-1)}{z(t-1)}\right\} = \frac{z(t-1)\Delta y(t-1) - y(t-1)\Delta z(t-1)}{z(t-1)z(t)}$$
$$= \frac{w[z(t-1), y(t-1)]}{z(t-1)z(t)}.$$

Then

$$Ly(t) = \frac{1}{z(t)}\Delta\left[p(t-1)z(t-1)z(t)\Delta\left(\frac{y(t-1)}{z(t-1)}\right)\right] \quad (6.8)$$

$$\varrho_1(t) = \frac{1}{z(t)} > 0, \quad (t \text{ in } [a, b+2])$$
$$\varrho_2(t) = p(t-1)z(t-1)z(t) > 0, \quad (t \text{ in } [a+1, b+2])$$

so
$$Ly(t) = \varrho_1(t)\Delta\left[\varrho_2(t)\Delta\left(\varrho_1(t-1)y(t-1)\right)\right],$$

as desired. ∎

Example 6.3. Find a Polya factorization for

$$y(t+1) - 6y(t) + 8y(t-1) = 0. \quad (6.9)$$

6.1 Introduction

The characteristic equation is $(\lambda - 2)(\lambda - 4) = 0$. Hence $z(t) = 2^t$ is a positive solution of this equation. We can write this equation in self-adjoint form with $p(t) = \left(\frac{1}{8}\right)^t$. From Eq. (6.8) we obtain the Polya factorization of Eq. (6.9),

$$2^{-t}\Delta\left[\left(\frac{1}{8}\right)^{t-1} 2^{t-1} 2^t \Delta\left(\frac{y(t-1)}{2^{t-1}}\right)\right] = 0$$

or

$$\Delta\left[2^{-t}\Delta\left(2^{-t}y(t-1)\right)\right] = 0.$$

Check that 2^t and 4^t are solutions of this last equation.

Definition 6.1. The Cauchy function $y(t,s)$, defined for $a \leq t \leq b+2$, $a+1 \leq s \leq b+1$, is defined as the function that for each fixed s in $[a+1, b+1]$ is the solution of the initial value problem (6.2), $y(s,s) = 0$, $y(s+1,s) = \frac{1}{p(s)}$.

Example 6.4 Find the Cauchy function for $\Delta^2 y(t-1) = 0$.

A general solution of $\Delta^2 y(t-1) = 0$ is $y(t) = A + Bt$. Using $y(s,s) = 0$, $y(s+1,s) = \frac{1}{p(s)} = 1$ we have that the Cauchy function is

$$y(t,s) = t - s.$$

Theorem 6.3. If $u_1(t)$, $u_2(t)$ are linearly independent solutions of Eq. (6.2), then the Cauchy function for Eq. (6.2) is given by

$$y(t,s) = \frac{\begin{vmatrix} u_1(s) & u_2(s) \\ u_1(t) & u_2(t) \end{vmatrix}}{p(s)\begin{vmatrix} u_1(s) & u_2(s) \\ u_1(s+1) & u_2(s+1) \end{vmatrix}}, \quad (6.10)$$

$a \leq t \leq b+2$, $a+1 \leq s \leq b+1$.

Proof. Since $u_1(t)$, $u_2(t)$ are linearly independent, $w[u_1(t), u_2(t)] \neq 0$ for t in $[a, b+1]$. Hence Eq. (6.10) is well defined. Note that by expanding $y(t,s)$ in (6.10) by the second row in the numerator we have that for each fixed s in $[a+1, b+1]$, $y(t,s)$ is a linear combination of $u_1(t)$ and $u_2(t)$ and so is a solution of Eq. (6.2). Clearly $y(s,s) = 0$ and $y(s+1,s) = \frac{1}{p(s)}$. ∎

Example 6.5. Use Theorem 6.3 to find the Cauchy function $y(t,s)$ for $\Delta^2 y(t-1) = 0$.
Take $u_1(t) = 1$ and $u_2(t) = t$, then

$$y(t,s) = \frac{\begin{vmatrix} 1 & s \\ 1 & t \end{vmatrix}}{\begin{vmatrix} 1 & s \\ 1 & s+1 \end{vmatrix}} = t - s.$$

The next theorem shows how the Cauchy function is used to solve initial value problems.

Theorem 6.4. (Variation of constants formula) The solution of the initial value problem

$$Ly(t) = h(t), \quad (t \text{ in } [a+1, b+1])$$
$$y(a) = 0,$$
$$y(a+1) = 0,$$

is given by

$$y(t) = \sum_{s=a+1}^{t} y(t,s) h(s) \qquad (6.11)$$

for t in $[a, b+2]$, where $y(t,s)$ is the Cauchy function for $Ly(t) = 0$. (Here if $t = b+2$, then the term $y(b+2, b+2)h(b+2)$ is understood to be zero.)

Proof. Let $y(t)$ be given by Eq. (6.11). By convention $y(a) = 0$. Also

$$y(a+1) = y(a+1, a+1)h(a+1) = 0,$$
$$y(a+2) = y(a+2, a+1)h(a+1) + y(a+2, a+2)h(a+2)$$
$$= \frac{h(a+1)}{p(a+1)}.$$

It follows that $y(t)$ satisfies $Ly(t) = h(t)$ for $t = a+1$.

6.1 Introduction

Now assume $a + 2 \le t \le b + 1$. Then

$$\begin{aligned}
Ly(t) &= p(t-1)y(t-1) + c(t)y(t) + p(t)y(t+1) \\
&= \sum_{s=a+1}^{t-1} p(t-1)y(t-1,s)h(s) + \sum_{s=a+1}^{t} c(t)y(t,s)h(s) \\
&\quad + \sum_{s=a+1}^{t+1} p(t)y(t+1,s)h(s) \\
&= \sum_{s=a+1}^{t-1} Ly(t,s)h(s) + c(t)y(t,t)h(t) \\
&\quad + p(t)y(t+1,t)h(t) + p(t)y(t+1,t+1)h(t+1) \\
&= h(t). \quad \blacksquare
\end{aligned}$$

Corollary 6.3. The solution of the initial value problem

$$\begin{aligned}
Ly(t) &= h(t), \quad (t \text{ in } [a+1, b+1]) \\
y(a) &= A, \\
y(a+1) &= B
\end{aligned}$$

is given by

$$y(t) = u(t) + \sum_{s=a+1}^{t} y(t,s)h(s),$$

where $y(t,s)$ is the Cauchy function for $Ly(t) = 0$ and $u(t)$ is the solution of the initial value problem $Lu(t) = 0$, $u(a) = A$, $u(a+1) = B$.

Proof. Since $u(t)$ is a solution of $Lu(t) = 0$ and $\sum_{s=a+1}^{t} y(t,s)h(s)$ is a solution of $Ly(t) = h(t)$, we have that

$$y(t) = u(t) + \sum_{s=a+1}^{t} y(t,s)h(s)$$

is a solution of $Ly(t) = h(t)$. Further, $y(a) = u(a) = A$ and $y(a+1) = u(a+1) = B$. $\blacksquare$

Example 6.6. Use the variation of constants formula to solve the initial value problem

$$\Delta^2 y(t-1) = t,$$
$$y(0) = y(1) = 0.$$

By Theorem 6.4 and Example 6.4, the solution $y(t)$ of this initial value problem is given by

$$y(t) = \sum_{s=1}^{t}(t-s)s$$
$$= t\sum_{s=1}^{t} s - \sum_{s=1}^{t} s^2.$$

Here we could use summation formulas from calculus, but instead we use factorial polynomials.

$$y(t) = t\sum_{s=1}^{t} s^{(1)} - \sum_{s=1}^{t}\left[s^{(2)} + s^{(1)}\right]$$
$$= t\left[\frac{s^{(2)}}{2}\right]_1^{t+1} - \left[\frac{s^{(3)}}{3} + \frac{s^{(2)}}{2}\right]_1^{t+1}$$
$$= t\frac{(t+1)^{(2)}}{2} - \frac{(t+1)^{(3)}}{3} - \frac{(t+1)^{(2)}}{2}$$
$$= \left(\frac{1}{2}t^3 + \frac{1}{2}t^2\right) - \left(\frac{1}{3}t^3 - \frac{1}{3}t\right) - \left(\frac{1}{2}t^2 + \frac{1}{2}t\right)$$
$$= \frac{1}{6}t^3 - \frac{1}{6}t.$$

6.2 Sturmian Theory

In this section we introduce the very important concept of generalized zero, which is due to Philip Hartman [114]. This concept provides a mechanism for obtaining fundamental results about second order self-adjoint equations and also represents the best approach for extending these

6.2 Sturmian Theory

results to higher order equations. Our first objective is to present the Sturm separation theorem for the self-adjoint difference equation (6.2). It is believed that Sturm actually proved the Sturm separation theorem for difference equations before he proved the corresponding result for differential equations. In contrast to the differential equations case, the Sturm separation theorem does not hold for all second order homogeneous difference equations. The important concept of disconjugacy will be introduced in this section, and we will see that it is very important in proving comparison theorems.

The following simple lemma shows that there is no nontrivial solution of Eq. (6.2) with $y(t_0) = 0$ and $y(t_0 - 1)y(t_0 + 1) \geq 0$, $t_0 > a$. In some sense this lemma says that nontrivial solutions of Eq. (6.2) can have only "simple" zeros.

Lemma 6.1. If $y(t)$ is a nontrivial solution of Eq. (6.2) such that $y(t_0) = 0$, $a < t_0 < b + 2$, then $y(t_0 - 1)y(t_0 + 1) < 0$.

Proof. Since $y(t)$ is a solution of Eq. (6.2) with $y(t_0) = 0$, $a < t_0 < b + 2$, we obtain from Eq. (6.3)

$$p(t_0)y(t_0 + 1) = -p(t_0 - 1)y(t_0 - 1).$$

Since $y(t_0 + 1)$, $y(t_0 - 1) \neq 0$ and $p(t) > 0$, it follows that

$$y(t_0 - 1)y(t_0 + 1) < 0. \blacksquare$$

Lemma 6.1 is fundamental in the Sturmian theory for difference equations as in the proof of the next theorem. Because of Lemma 6.1 we can define the generalized zero of a solution of Eq. (6.2) as follows.

Definition 6.2. We say that a solution $y(t)$ of Eq. (6.2) has a generalized zero at t_0 provided $y(t_0) = 0$ if $t_0 = a$ and if $t_0 > a$ either $y(t_0) = 0$ or $y(t_0 - 1)y(t_0) < 0$.

Theorem 6.5. (Sturm Separation Theorem) Two linearly independent solutions of Eq. (6.2) can not have a common zero. If a nontrivial solution of Eq. (6.2) has a zero at t_1 and a generalized zero at $t_2 > t_1$, then any second linearly independent solution has a generalized zero in $(t_1, t_2]$. If a nontrivial solution of Eq. (6.2) has a generalized zero at t_1 and a generalized zero at $t_2 > t_1$, then any second linearly independent solution has a generalized zero in $[t_1, t_2]$.

262 Chapter 6. The Self-Adjoint Second Order Linear Equation

Proof. Assume two solutions $y(t)$, $z(t)$ of Eq. (6.2) have a common zero at t_0. Then the Casoratian $w[y(t), z(t)]$ is zero at t_0 and hence $y(t)$ and $z(t)$ are linearly dependent.

Next assume $y(t)$ is a nontrivial solution of Eq. (6.2) with a zero at t_1 and a generalized zero at t_2. Without loss of generality $t_2 > t_1 + 1$ is the first generalized zero of $y(t)$ to the right of t_1, $y(t) > 0$ in (t_1, t_2), and $y(t_2) \leq 0$. Assume $z(t)$ is a second linearly independent solution with no generalized zeros in $(t_1, t_2]$. Without loss of generality, $z(t) > 0$ on $[t_1, t_2]$. Pick a constant $T > 0$ such that there is a $t_0 \in (t_1, t_2)$ such that $z(t_0) = Ty(t_0)$ but $z(t) \geq Ty(t)$ on $[t_1, t_2]$. Then $u(t) = z(t) - Ty(t)$ is a nontrivial solution with $u(t_0) = 0$, $u(t_0 - 1)u(t_0 + 1) \geq 0$, $t_0 > a$, which contradicts Lemma 6.1.

The last statement in this theorem is left as an exercise. ∎

The next example shows that the Sturm Separation Theorem does not hold for all second order linear homogeneous difference equations.

Example 6.7. Show that the conclusions of Theorem 6.5 do not hold for the Fibonocci difference equation

$$y(t+1) - y(t) - y(t-1) = 0.$$

The characteristic equation is $\lambda^2 - \lambda - 1 = 0$. Hence characteristic values are $\frac{1}{2} \pm \frac{\sqrt{5}}{2}$. Take $y(t) = \left(\frac{1}{2} - \frac{\sqrt{5}}{2}\right)^t$ and $z(t) = \left(\frac{1}{2} + \frac{\sqrt{5}}{2}\right)^t$. Note that $y(t)$ has a generalized zero at every integer while $z(t) > 0$ for all t. This of course does not contradict Theorem 6.5 because this equation can not be written in self-adjoint form.

In Theorem 6.5 it was noted that two linearly independent solutions can not have a common zero. The following example shows that this is not true for generalized zeros.

Example 6.8. The difference equation

$$y(t+1) + 2y(t) + 2y(t-1) = 0,$$

which can be put in self-adjoint form, has $y(t) = 2^{\frac{t}{2}} \sin \frac{3\pi}{4} t$, $z(t) = 2^{\frac{t}{2}} \cos \frac{3\pi}{4} t$ as linearly independent solutions. Note that both of these solutions have a generalized zero at $t = 2$.

6.2 Sturmian Theory

Definition 6.3. We say that the difference equation (6.2) is "disconjugate" on $[a, b+2]$ provided no nontrivial solution of (6.2) has two or more generalized zeros on $[a, b+2]$.

Of course in any interval $[a, b+2]$ there is a nontrivial solution with at least one generalized zero.

Example 6.9. The difference equation

$$y(t+1) - \sqrt{3}y(t) + y(t-1) = 0$$

is disconjugate on any interval of length less than six.

This follows from the fact that any solution of this equation is of the form $y(t) = A\sin\left(\frac{\pi t}{6} + B\right)$.

Example 6.10. The difference equation

$$y(t+2) - 7y(t+1) + 12y(t) = 0$$

is disconjugate on any interval.

Using Theorem 6.4 we can prove the following comparison theorem.

Theorem 6.6. Assume $Ly(t) = 0$ is disconjugate on $[a, b+2]$ and $u(t)$, $v(t)$ satisfy

$$Lu(t) \geq Lv(t), \quad (t \text{ in } [a+1, b+1])$$
$$u(a) = v(a),$$
$$u(a+1) = v(a+1).$$

Then $u(t) \geq v(t)$ on $[a, b+2]$.

Proof. Set

$$w(t) = u(t) - v(t).$$

Then

$$\begin{aligned} h(t) &\equiv Lw(t) \\ &= Lu(t) - Lv(t) \\ &\geq 0, \quad (t \text{ in } [a+1, b+1]). \end{aligned}$$

Hence $w(t)$ solves the initial value problem

$$Lw(t) = h(t),$$
$$w(a) = 0,$$
$$w(a+1) = 0.$$

By the variation of constants formula

$$w(t) = \sum_{s=a+1}^{t} y(t,s)h(s),$$

where $y(t,s)$ is the Cauchy function for $Ly(t) = 0$. Since $Ly(t) = 0$ is disconjugate and $y(s,s) = 0$, $y(s+1,s) = \frac{1}{p(s)} > 0$, we have that $y(t,s) > 0$ for $s+1 \le t \le b+2$. It follows that $w(t) \ge 0$ on $[a, b+2]$, which gives us the desired result. ∎

Example 6.11. Find bounds on the solution $y(t)$ of the initial value problem

$$\Delta^2 y(t-1) = \frac{2}{1+t^2}, \quad (t \ge 1)$$
$$y(0) = 0,$$
$$y(1) = 1.$$

Let $v(t)$ be the solution of the initial value problem $\Delta^2 v(t-1) = 0$, $v(0) = 0$, $v(1) = 1$ and $u(t)$ be the solution of the initial value problem $\Delta^2 u(t-1) = 2$, $u(0) = 0$, $u(1) = 1$. Since $Ly(t) \equiv \Delta^2 y(t-1) = 0$ is disconjugate on any interval,

$$Lu(t) = 2 \ge Ly(t) = \frac{2}{1+t^2} \ge 0 = Lv(t)$$

for $t \ge 1$, and

$$u(0) = y(0) = v(0),$$
$$u(1) = y(1) = v(1),$$

6.2 Sturmian Theory

we have by Theorem 6.6 that

$$u(t) \geq y(t) \geq v(t)$$

for $t \geq 0$. But it is easy to show that $v(t) = t$ and $u(t) = t^2$. Hence $y(t)$ satisfies

$$t^2 \geq y(t) \geq t$$

for $t = 0, 1, 2, \cdots$.

Consider the boundary value problem (BVP)

$$\Delta^2 y(t-1) + 2y(t) = 0,$$
$$y(0) = A, \; y(2) = B.$$

If $A = B = 0$, this boundary value problem has infinitely many solutions. If $A = 0$, $B \neq 0$, it has no solutions. We show in the following theorem that with the assumption of disconjugacy this type of boundary value problem has a unique solution.

Theorem 6.7. If $Ly(t) = 0$ is disconjugate on $[a, b+2]$, then the boundary value problem

$$Ly(t) = h(t),$$
$$y(t_1) = A, \; y(t_2) = B,$$

where $a \leq t_1 < t_2 \leq b+2$, A, B constants, has a unique solution.

Proof. Let $y_1(t)$, $y_2(t)$ be linearly independent solutions of $Ly(t) = 0$ and let $y_p(t)$ be a particular solution of $Ly(t) = h(t)$; then a general solution of $Ly(t) = h(t)$ is

$$y(t) = c_1 y_1(t) + c_2 y_2(t) + y_p(t).$$

The boundary conditions lead to the system of equations

$$c_1 y_1(t_1) + c_2 y_2(t_2) = A - y_p(t_1),$$
$$c_1 y_1(t_2) + c_2 y_2(t_2) = B - y_p(t_2).$$

This system of equations has a unique solution if and only if

$$\begin{vmatrix} y_1(t_1) & y_2(t_1) \\ y_1(t_2) & y_2(t_2) \end{vmatrix} \neq 0.$$

Assume

$$\begin{vmatrix} y_1(t_1) & y_2(t_1) \\ y_1(t_2) & y_2(t_2) \end{vmatrix} = 0.$$

Then there are constants d_1, d_2, not both zero, such that the nontrivial solution

$$y(t) = d_1 y_1(t) + d_2 y_2(t)$$

satisfies

$$y(t_1) = y(t_2) = 0.$$

This contradicts the disconjugacy of $Ly(t) = 0$ on $[a, b+2]$, and the proof is complete. ∎

6.3 Green's Functions

In this section we introduce the Green's function for a two point conjugate boundary value problem. We do this in such a way that other boundary value problems for second order problems (see Exercises 6.15-6.18 for the development of the Green's function for a focal boundary value problem) and Green's functions for n^{th} order equations are analogous. It will follow that under certain conditions the solution of a nonhomogeneous boundary value problem can be expressed in terms of Green's functions. The Green's function will also be used to prove an important comparison theorem for conjugate BVPs.

By Theorem 6.7 if $Ly(t) = 0$ is disconjugate on $[a, b+2]$, then the boundary value problem

$$Ly(t) = h(t), \quad (t \text{ in } [a+1, b+1]) \tag{6.12}$$

6.3 Green's Functions

$$y(a) = 0, \qquad (6.13)$$
$$y(b+2) = 0 \qquad (6.14)$$

has a unique solution $y(t)$. We would like to have a formula like the variation of constants formula for $y(t)$. Let us begin by proving several results concerning what we will later define as the Green's function $G(t,s)$ for the BVP $Ly(t) = 0$, (6.13), (6.14).

First assume there is a function $G(t,s)$ that satisfies the following:
a) $G(t,s)$ is defined for $a \leq t \leq b+2$, $a+1 \leq s \leq b+1$;
b) $LG(t,s) = \delta_{ts}$ for $a+1 \leq t \leq b+1$, $a+1 \leq s \leq b+1$, where δ_{ts} is the Kronecker delta ($\delta_{ts} = 0$ if $t \neq s$, $\delta_{ts} = 1$ if $t = s$);
c) $G(a,s) = G(b+2,s) = 0$, $a+1 \leq s \leq b+1$.
Set

$$y(t) = \sum_{s=a+1}^{b+1} G(t,s)h(s);$$

then we claim $y(t)$ satisfies Eqs. (6.12)–(6.14). First by (c),

$$y(a) = \sum_{s=a+1}^{b+1} G(a,s)h(s) = 0$$

and

$$y(b+2) = \sum_{a=s+1}^{b+1} G(b+2,s)h(s) = 0,$$

so Eqs. (6.13) and (6.14) hold. Next note that

$$Ly(t) = \sum_{s=a+1}^{b+1} LG(t,s)h(s)$$
$$= \sum_{s=a+1}^{b+1} \delta_{ts} h(s)$$
$$= h(t), \quad (a+1 \leq t \leq b+1).$$

Thus we have shown that if there is a function $G(t,s)$ satisfying (a)-(c), then

$$y(t) = \sum_{s=a+1}^{b+1} G(t,s)h(s)$$

satisfies the BVP (6.12)-(6.14).

We now show that if $Ly(t) = 0$ is disconjugate on $[a, b+2]$, then there is a function $G(t,s)$ satisfying (a)-(c). To this end let $y_1(t)$ be the solution of the IVP (6.2), $y_1(a) = 0$, $y_1(a+1) = 1$, and let $y(t,s)$ be the Cauchy function for $Ly(t) = 0$. Define $G(t,s)$ for $a \leq t \leq b+2$, $a+1 \leq s \leq b+1$, by

$$G(t,s) = \begin{cases} -\frac{y(b+2,s)y_1(t)}{y_1(b+2)}, & t \leq s \\ y(t,s) - \frac{y(b+2,s)y_1(t)}{y_1(b+2)}, & s \leq t. \end{cases} \quad (6.15)$$

Since $Ly(t) = 0$ is disconjugate on $[a, b+2]$, $y_1(b+2) > 0$, so we are not dividing by zero in the definition of $G(t,s)$. Also note that since $y(s,s) = 0$ it is okay to write $t \leq s$ and $s \leq t$ in the definition of $G(t,s)$.

Since

$$G(a,s) = -\frac{y(b+2,s)y_1(a)}{y_1(b+2)} = 0$$

and

$$G(b+2,s) = y(b+2,s) - \frac{y(b+2,s)y_1(b+2)}{y_1(b+2)} = 0,$$

we have that $G(t,s)$ satisfies (c).

Next we show that $G(t,s)$ satisfies (b). If $t \geq s+1$, then

$$LG(t,s) = Ly(t,s) - \frac{y(b+2,s)}{y_1(b+2)}Ly_1(t)$$
$$= 0 = \delta_{ts}.$$

6.3 Green's Functions

If $t \leq s - 1$, then

$$\begin{aligned} LG(t,s) &= -\frac{y(b+2,s)}{y_1(b+2)} Ly_1(t) \\ &= 0 = \delta_{ts}. \end{aligned}$$

Finally, if $t = s$, then

$$\begin{aligned} LG(s,s) &= p(s)G(s+1,s) + c(s)G(s,s) + p(s-1)G(s-1,s) \\ &= p(s)y(s+1,s) - \frac{y(b+2,s)}{y_1(b+2)} Ly_1(s) \\ &= 1 = \delta_{ts}. \end{aligned}$$

Hence $G(t,s)$ satisfies (a)-(c).

Let us show that if $Ly(t) = 0$ is disconjugate on $[a, b+2]$, then there is a unique function satisfying (a)-(c). We know $G(t,s)$ defined by Eq (6.15) satisfies (i)-(iii). Assume $H(t,s)$ satisfies (a)-(c). Fix s in $[a+1, b+1]$ and set

$$y(t) = G(t,s) - H(t,s).$$

It follows from (b) that $y(t)$ is a solution of $Ly(t) = 0$ on $[a, b+2]$. By (c) $y(a) = 0$, $y(b+2) = 0$. Since $Ly(t) = 0$ is disconjugate on $[a, b+2]$ we must have $y(t) = 0$ on $[a, b+2]$. Since s in $[a+1, b+2]$ is arbitrary, it follows that $G(t,s) \equiv H(t,s)$ for $a \leq t \leq b+2$, $a+1 \leq s \leq b+1$.

Definition 6.4. If $Ly(t) = 0$ is disconjugate on $[a, b+2]$, then we define the Green's function for the BVP (6.12), (6.13), (6.14) to be the unique function $G(t,s)$ satisfying (a)-(c).

Theorem 6.8. If $Ly(t) = 0$ is disconjugate on $[a, b+2]$, then the unique solution of

$$\begin{aligned} Ly(t) &= h(t), \\ y(a) &= 0 = y(b+2) \end{aligned}$$

270 Chapter 6. The Self-Adjoint Second Order Linear Equation

is given by

$$y(t) = \sum_{s=a+1}^{b+1} G(t,s)h(s).$$

Furthermore,

$$G(t,s) = \begin{cases} -\frac{y(b+2,s)}{y_1(b+2)}y_1(t), & t \leq s \\ y(t,s) - \frac{y(b+2,s)}{y_1(b+2)}y_1(t), & s \leq t \end{cases}$$

and $G(t,s) < 0$ on the square $a+1 \leq t, s \leq b+1$.

Proof. It remains to show that $G(t,s) < 0$ on the square $a+1 \leq t, s \leq b+1$. To see this, fix s in $[a+1, b+1]$. Since $Ly(t) = 0$ is disconjugate on $[a, b+2]$, $y_1(t) > 0$ for $a < t \leq b+2$ and $y(t,s) > 0$ for $s < t \leq b+2$. Hence

$$G(t,s) = -\frac{y(b+2,s)}{y_1(b+2)}y_1(t) < 0$$

for $a+1 \leq t \leq s$.

If $s \leq t \leq b+2$, then

$$G(t,s) = y(t,s) - \frac{y(b+2,s)}{y_1(b+2)}y_1(t),$$

which as a function of t is a solution of $Ly(t) = 0$ on $[a, b+2]$. Since $G(b+2, s) = 0$ and $G(s,s) < 0$, we have that

$$G(t,s) < 0 \quad (\text{for } s \leq t \leq b+1).$$

Since s in $[a+1, b+1]$ is arbitrary we get the desired result. ∎

Example 6.12. Find the Green's function for the BVP

$$\Delta^2 y(t-1) = 0$$
$$y(a) = 0 = y(b+2).$$

6.3 Green's Functions

By Example 6.4 the Cauchy function for $\Delta^2 y(t-1) = 0$ is $y(t,s) = t - s$. The solution $y_1(t)$ of the IVP $\Delta^2 y(t-1) = 0$, $y(a) = 0$, $y(a+1) = 1$ is $y_1(t) = t - a$. By Theorem 6.8 we have for $t \leq s$ that

$$G(t,s) = -\frac{y(b+2,s)}{y_1(b+2)} y_1(t)$$

$$= -\frac{(b+2-s)(t-a)}{b+2-a}.$$

If $s \leq t$, then

$$G(t,s) = y(t,s) - \frac{y(b+2,s)}{y_1(b+2)} y_1(t)$$

$$= (t-s) - \frac{(b+2-s)(t-a)}{b+2-a}$$

$$= -\frac{(s-a)(b+2-t)}{b+2-a}.$$

Thus

$$G(t,s) = \begin{cases} -\frac{(b+2-s)(t-a)}{b+2-a}, & t \leq s \\ -\frac{(b+2-t)(s-a)}{b+2-a}, & s \leq t. \end{cases}$$

Note that $G(t,s) < 0$ on the square $a+1 \leq t, s \leq b+1$ as guaranteed by Theorem 6.8. Note also that $G(t,s)$ is symmetric on the square $a+1 \leq t, s \leq b+1$.

Example 6.13. Use the appropriate Green's function to solve the BVP

$$\Delta^2 y(t-1) = 12,$$
$$y(0) = 0 = y(6).$$

Chapter 6. The Self-Adjoint Second Order Linear Equation

By Theorem 6.8, the solution of this BVP is given by

$$y(t) = \sum_{s=1}^{5} G(t,s)12.$$

By Example 6.12

$$G(t,s) = \begin{cases} -\frac{(6-s)(t)}{6}, & t \leq s \\ -\frac{(6-t)(s)}{6}, & s \leq t. \end{cases}$$

Therefore

$$y(t) = 2\sum_{s=1}^{t}(t-6)s + 2\sum_{s=t+1}^{5} t(s-6)$$

$$= 2(t-6)\frac{t(t+1)}{2} + 2t\left[\frac{s^{(2)}}{2} - 6s\right]_{t+1}^{6}$$

$$= t^3 - 5t^2 - 6t + 2t\left[\left(\frac{30}{2} - 36\right) - \left(\frac{(t+1)t}{2} - 6(t+1)\right)\right]$$

$$= t^3 - 5t^2 - 6t - t^3 + 11t^2 - 30t$$

$$= 6t^2 - 36t.$$

The proof of the following corollary is left as an exercise.

Corollary 6.4. If $Ly(t) = 0$ is disconjugate on $[a, b+2]$, then the unique solution of the boundary value problem

$$Ly(t) = h(t)$$
$$y(a) = A, \quad y(b+2) = B$$

6.3 Green's Functions

is given by

$$y(t) = u(t) + \sum_{s=a+1}^{b+1} G(t,s)h(s),$$

where $G(t,s)$ is the Green's function for the BVP $Ly(t) = 0$, $y(a) = 0 = y(b+2)$ and $u(t)$ is the solution of the BVP $Lu(t) = 0$, $u(a) = A$, $u(b+2) = B$.

Example 6.14. Solve the BVP

$$\Delta^2 y(t-1) = 12,$$
$$y(0) = 1,$$
$$y(6) = 7.$$

By Corollary 6.4 and Example 6.13, the desired solution is

$$y(t) = u(t) + 6t^2 - 36t,$$

where $u(t)$ is the solution of the BVP $\Delta^2 y(t-1) = 0$, $u(0) = 1$, $u(6) = 7$. It follows that $u(t) = 1 + t$ and consequently

$$y(t) = 6t^2 - 35t + 1.$$

Theorem 6.9. Assume $Ly(t) = 0$ is disconjugate on $[a, b+2]$ and assume $u(t)$, $v(t)$ satisfy

$$Lu(t) \leq Lv(t), \quad (t \text{ in } [a+1, b+1]),$$
$$u(a) \geq v(a),$$
$$u(b+2) \geq v(b+2),$$

then $u(t) \geq v(t)$ on $[a, b+2]$.

Proof. Set $w(t) = u(t) - v(t)$ for t in $[a, b+2]$. Then

$$h(t) \equiv Lw(t) = Lu(t) - Lv(t) \leq 0$$

for t in $[a+1, b+1]$. If $A \equiv u(a) - v(a) \geq 0$, $B \equiv u(b+2) - v(b+2) \geq 0$, then $w(t)$ solves the BVP

$$Lw(t) = h(t),$$
$$w(a) = A,$$
$$w(b+2) = B.$$

By Corollary 6.4,

$$w(t) = y(t) + \sum_{s=a+1}^{b+1} G(t,s) h(s), \tag{6.16}$$

where $G(t,s)$ is the Green's function for the BVP $Ly(t) = 0$, $y(a) = 0 = y(b+2)$ and $y(t)$ is the solution of the BVP $Ly(t) = 0$, $y(a) = A$, $y(b+2) = B$. By Theorem 6.8, $G(t,s) \leq 0$. Since $Ly(t) = 0$ is disconjugate on $[a, b+2]$ and $y(a) \geq 0$, $y(b+2) \geq 0$, it follows that $y(t) \geq 0$. By Eq. (6.16) $w(t) \geq 0$, which gives the desired result. ∎

6.4 Disconjugacy

In Section 6.2 we introduced the concept of disconjugacy and showed that disconjugacy is important in obtaining a comparison result for solutions of initial value problems (Theorem 6.6) and an existence and uniqueness result for solutions of boundary value problems (Theorem 6.7). The existence and uniqueness of the Green's functions in Section 6.3 also relied on disconjugacy. Theorem 6.9 used disconjugacy to establish a comparison result for solutions of a two point boundary value problem.

This section is a continuation of the study of disconjugacy. We will develop several criteria for disconjugacy and discover further consequences of disconjugacy. In Chapter 8 the role of disconjugacy in the discrete calculus of variations will be examined. In particular, we will see that disconjugacy is equivalent to the positive definiteness of a certain quadratic functional.

Theorem 6.10. The difference equation $Ly(t) = 0$ is disconjugate on $[a, b+2]$ if and only if there is a positive solution of $Ly(t) = 0$ on $[a, b+2]$.

6.4 Disconjugacy

Proof. Assume $Ly(t) = 0$ is disconjugate on $[a, b+2]$. Let $u(t)$, $v(t)$ be solutions of $Ly(t) = 0$ satisfying

$$u(a) = 0, \quad u(a+1) = 1,$$
$$v(b+1) = 1, \quad v(b+2) = 0.$$

By the disconjugacy $u(t) > 0$ on $[a+1, b+2]$ and $v(t) > 0$ on $[a, b+1]$. It follows that $y(t) = u(t) + v(t)$ is a positive solution of $Ly(t) = 0$.

Conversely assume $Ly(t) = 0$ has a positive solution on $[a, b+2]$. It follows from the Sturm separation theorem that no nontrivial solution has two generalized zeros in $[a, b+2]$. ∎

Corollary 6.5. The difference equation (6.2) is disconjugate on $[a, b+2]$ if and only if has a Polya factorization on $[a, b+2]$.

Proof. If Eq. (6.2) is disconjugate, then by Theorem 6.10 it has a positive solution. By Theorem 6.2, $Ly(t) = 0$ has a Polya factorization.

Conversely assume $Ly(t) = 0$ has the Polya factorization

$$\varrho_1(t)\Delta\{\varrho_2(t)\Delta(\varrho_1(t-1)y(t-1))\} = 0,$$

where $\varrho_1(t) > 0$ in $[a, b+2]$, $\varrho_2(t) > 0$ in $[a+1, b+2]$. It follows that $y(t) = \frac{1}{\varrho_1(t)}$ is a positive solution. By Theorem 6.10, $Ly(t) = 0$ is disconjugate on $[a, b+2]$. ∎

Define the k by k tridiagonal determinants $D_k(t)$, $a+1 \le t \le b+1$, $1 \le k \le b+2-t$, to be

$$\begin{vmatrix} c(t) & p(t) & 0 & 0 & \cdots & 0 \\ p(t) & c(t+1) & p(t+1) & 0 & \cdots & 0 \\ 0 & p(t+1) & c(t+2) & p(t+2) & \cdots & 0 \\ \vdots & \vdots & \ddots & \ddots & \ddots & \vdots \\ 0 & 0 & \cdots & p(t+k-3) & c(t+k-2) & p(t+k-2) \\ 0 & 0 & \cdots & 0 & p(t+k-2) & c(t+k-1) \end{vmatrix}$$

where $c(t)$ is given by Eq. (6.4).

276   Chapter 6. The Self-Adjoint Second Order Linear Equation

Theorem 6.11. The difference equation $Ly(t) = 0$ is disconjugate on $[a, b+2]$ if and only if the coefficients of $Ly(t) = 0$ satisfy

$$(-1)^k D_k(t) > 0 \qquad (6.17)$$

for $a+1 \leq t \leq b+1$, $1 \leq k \leq b+2-t$.

Proof. Assume $Ly(t) = 0$ is disconjugate on $[a, b+2]$. We will show that Eq. (6.17) holds for $1 \leq k \leq b-a+1$, $a+1 \leq t \leq b+2-k$ by induction on k.

For $k = 1$ we now show that $-D_1(t) = -c(t) > 0$ for $a+1 \leq t \leq b+1$. To this end, fix $t_0 \in [a+1, b+1]$ and let $y(t)$ be the solution of Eq. (6.2) satisfying $y(t_0 - 1) = 0$, $y(t_0) = 1$. Since $Ly(t_0) = 0$ we have from Eq. (6.3) that

$$p(t_0)y(t_0+1) + c(t_0)y(t_0) = 0,$$
$$c(t_0) = -p(t_0)y(t_0+1).$$

By the disconjugacy $y(t_0+1) > 0$, so $c(t_0) < 0$. Since $t_0 \in [a+1, b+1]$ is arbitrary, $c(t) < 0$ for $a+1 \leq t \leq b+1$, and the first step of the induction is complete. Now assume $1 < k \leq b-a+1$ and

$$(-1)^{k-1} D_{k-1}(t) > 0 \qquad (6.18)$$

for $a+1 \leq t \leq b+3-k$. We will use this induction hypothesis to show that Eq. (6.17) holds. Fix $t_1 \in [a+1, b+2-k]$ and let $u(t)$ be the solution of $Ly(t) = 0$, $u(t_1 - 1) = 0$, $u(t_1 + k) = 1$. (Why do we know there is such a solution?) Using these boundary conditions and $Ly(t) = 0$ for $t_1 \leq t \leq t_1 + k - 1$, we arrive at the equations

$$c(t_1)u(t_1) + p(t_1)u(t_1+1) = 0$$
$$p(t_1)u(t_1) + c(t_1+1)u(t_1+1) + p(t_1+1)u(t_1+2) = 0$$
$$\vdots$$
$$p(t_1+k-3)u(t_1+k-3) + c(t_1+k-2)u(t_1+k-2)$$
$$+ p(t_1+k-2)u(t_1+k-1) = 0$$
$$p(t_1+k-2)u(t_1+k-2) + c(t_1+k-1)u(t_1+k-1) = -p(t_1+k-1).$$

6.4 Disconjugacy

Note that the determinant of the coefficients is $D_k(t_1)$. It is left as an exercise to show that $D_k(t_1) \neq 0$. We can use Cramer's rule to solve the above system for $u(t_1 + k - 1)$ to obtain

$$u(t_1 + k - 1) = -\frac{p(t_1 + k - 1)D_{k-1}(t_1)}{D_k(t_1)}.$$

By the disconjugacy $u(t_1 + k - 1) > 0$, so using Eq. (6.18) we have $(-1)^k D_k(t_1) > 0$. Since $t_1 \in [a+1, b+2-k]$ is arbitrary, Eq. (6.17) holds for t in $[a+1, b+2-k]$.

The converse statement of this theorem is a special case of the following theorem.

Theorem 6.12. If $(-1)^k D_k(a+1) > 0$ for $1 \leq k \leq b-a+1$, then $Ly(t) = 0$ is disconjugate on $[a, b+2]$.

Proof. Let $u(t)$ be the solution of $Ly(t) = 0$ satisfying $u(a) = 0$, $u(a+1) = 1$. By the Sturm separation theorem it suffices to show that $u(t) > 0$ on $[a+1, b+2]$.

We will show that $u(a+k) > 0$ for $1 \leq k \leq b-a+2$ by induction on k. For $k = 1$, $u(a+1) = 1 > 0$. Assume $1 < k \leq b-a+2$ and that $u(a+k-1) > 0$. Using $Lu(t) = 0$, $a+1 \leq t \leq a+k-1$ and $u(a) = 0$ we get the $k-1$ equations

$$c(a+1)u(a+1) + p(a+1)u(a+2) = 0$$
$$p(a+1)u(a+1) + c(a+2)u(a+2) + p(a+2)u(a+3) = 0$$
$$\vdots$$
$$p(a+k-3)u(a+k-3) + c(a+k-2)u(a+k-2)$$
$$+ p(a+k-2)u(a+k-1) = 0$$
$$p(a+k-2)u(a+k-2) + c(a+k-1)u(a+k-1)$$
$$+ p(a+k-1)u(a+k) = 0.$$

By Cramer's rule (here $D_0(a+1) \equiv 1$)

$$u(a+k-1) = -\frac{p(a+k-1)u(a+k)D_{k-2}(a+1)}{D_{k-1}(a+1)}.$$

It follows that $u(a+k) > 0$, so by induction $u(t) > 0$ in $[a+1, b+2]$. Hence

278   Chapter 6. The Self-Adjoint Second Order Linear Equation

$Ly(t) = 0$ is disconjugate on $[a, b+2]$. ∎

We say $Ly(t) = 0$ is disconjugate on an infinite set of integers $[a, \infty)$ provided no nontrivial solution has two generalized zeros on $[a, \infty)$.

Example 6.15. Use Theorem 6.12 to show that $\Delta^2 y(t-1) = 0$ is disconjugate on $[0, \infty)$.

By Theorem 6.12 it suffices to show that $(-1)^k D_k(1) > 0$ for $k \geq 1$. Here $p(t) = 1$, $c(t) = -2$. Thus,

$$D_1(1) = -2,$$
$$D_2(1) = \begin{vmatrix} -2 & 1 \\ 1 & -2 \end{vmatrix} = 3.$$

Expanding $D_{k+2}(1)$ along the first row we get

$$D_{k+2}(1) = -2D_{k+1}(1) - D_k(1).$$

By solving the initial value problem

$$D_{k+2}(1) + 2D_{k+1}(1) + D_k(1) = 0,$$
$$D_1(1) = -2, D_2(1) = 3,$$

we obtain

$$D_k(1) = (-1)^k (k+1).$$

Then

$$(-1)^k D_k(1) > 0, \quad (k \geq 1),$$

and so by Theorem 6.12, $\Delta^2 y(t-1) = 0$ is disconjugate on $[0, \infty)$.

Theorem 6.13. If $Ly(t) = 0$ is disconjugate on $[a, b+2]$, then there are solutions $u(t), v(t)$ such that $u(t) > 0$, $v(t) > 0$ on $[a, b+2]$ and

(6.19) $$\begin{vmatrix} u(t_1) & v(t_1) \\ u(t_2) & v(t_2) \end{vmatrix} > 0$$

6.4 Disconjugacy

whenever $a \le t_1 < t_2 \le b+2$.

Proof. By the disconjugacy we have from Theorem 6.10 that there is a positive solution $u(t)$ on $[a, b+2]$. Let $y(t)$ be a solution of Eq. (6.2) such that $u(t)$, $y(t)$ are linearly independent. By Liouville's formula the Casoratian $w[u(t), y(t)]$ is of one sign on $[a, b+1]$. If necessary we can replace $y(t)$ by $-y(t)$ so we can assume that

$$w[u(t), y(t)] > 0$$

on $[a, b+1]$. Pick $C > 0$ sufficiently large so that

$$v(t) = y(t) + Cu(t) > 0$$

on $[a, b+2]$. Note that

$$w[u(t), v(t)] = w[u(t), y(t)] > 0 \qquad (6.20)$$

on $[a, b+1]$.

We will now show that Eq. (6.19) holds. To see this fix t_1 in $[a, b+1]$. We will show by induction on k that

$$\begin{vmatrix} u(t_1) & v(t_1) \\ u(t_1+k) & v(t_1+k) \end{vmatrix} > 0$$

for $1 \le k \le b+2-t_1$. For $k = 1$ this is true because of Eq. (6.20). Now assume $1 < k \le b+2-t_1$ and

$$\begin{vmatrix} u(t_1) & v(t_1) \\ u(t_1+k-1) & v(t_1+k-1) \end{vmatrix} > 0.$$

The boundary value problem $Lz(t) = 0$, $z(t_1) = 0$, $z(t_1+k-1) = 1$ has a unique solution $z(t)$ by Theorem 6.7. Since $z(t)$ is a linear combination of $u(t)$ and $v(t)$,

$$\begin{vmatrix} z(t_1) & u(t_1) & v(t_1) \\ z(t_1+k-1) & u(t_1+k-1) & v(t_1+k-1) \\ z(t_1+k) & u(t_1+k) & v(t_1+k) \end{vmatrix} = 0.$$

280 Chapter 6. The Self-Adjoint Second Order Linear Equation

Expanding along the first column we get

$$\begin{vmatrix} u(t_1) & v(t_1) \\ u(t_1+k) & v(t_1+k) \end{vmatrix} = z(t_1+k) \begin{vmatrix} u(t_1) & v(t_1) \\ u(t_1+k-1) & v(t_1+k-1) \end{vmatrix}.$$

By the disconjugacy $z(t_1 + k) > 0$, so

$$\begin{vmatrix} u(t_1) & v(t_1) \\ u(t_1+k) & v(t_1+k) \end{vmatrix} > 0.$$

It follows that Eq. (6.19) holds for $a \leq t_1 < t_2 \leq b+2$. ∎

Theorem 6.14. Assume $Ly(t) = 0$ is disconjugate on $[a, \infty)$. Then there exists a nontrivial solution $y_1(t)$ such that if $y_2(t)$ is any second linearly independent solution, then

$$\lim_{t \to \infty} \frac{y_1(t)}{y_2(t)} = 0.$$

Furthermore,

$$\sum^{\infty} \frac{1}{p(s)y_1(s)y_1(s+1)} = \infty,$$

$$\sum^{\infty} \frac{1}{p(s)y_2(s)y_2(s+1)} < \infty.$$

Also if $w_i(t) = \frac{p(t)\Delta y_i(t)}{y_i(t)}$, $i = 1, 2$, then $w_1(t) < w_2(t)$ for all sufficiently large t.

Proof. Let $u(t), v(t)$ be linearly independent solutions of $Ly(t) = 0$. Since $Ly(t) = 0$ is disconjugate on $[a, \infty)$, there is an integer $T \geq a$ so that $v(t)$ is of one sign for $t \geq T$. For $t \geq T$ consider

$$\Delta\left(\frac{u(t)}{v(t)}\right) = \frac{w[v(t), u(t)]}{v(t)v(t+1)}$$

$$= \frac{C}{p(t)v(t)v(t+1)},$$

6.4 Disconjugacy

where C is a constant, by Liouville's formula. It follows that $\frac{u(t)}{v(t)}$ is either increasing or decreasing for $t \geq T$. Let

$$\gamma = \lim_{t \to \infty} \frac{u(t)}{v(t)},$$

where $-\infty \leq \gamma \leq \infty$. If $\gamma = \pm\infty$, then by interchanging $u(t)$ and $v(t)$ we get that $\gamma = 0$. Then we may as well assume $-\infty < \gamma < \infty$. If $\gamma \neq 0$, we can replace the solution $u(t)$ by the solution $u(t) - \gamma v(t)$ to get

$$\lim_{t \to \infty} \frac{u(t) - \gamma v(t)}{v(t)} = \gamma - \gamma = 0.$$

We may assume

$$\lim_{t \to \infty} \frac{u(t)}{v(t)} = 0.$$

Set $y_1(t) = u(t)$. If $y_2(t)$ is a second linearly independent solution then $y_2(t) = \alpha y_1(t) + \beta v(t)$, $\beta \neq 0$. We have

$$\lim_{t \to \infty} \frac{y_1(t)}{y_2(t)} = \lim_{t \to \infty} \frac{u(t)}{\alpha u(t) + \beta v(t)} = 0.$$

Pick an integer T_1 sufficiently large so that $y_2(t)$ is of the same sign for $t \geq T_1$.

Now consider for $t \geq T_1$

$$\Delta\left(\frac{y_1(t)}{y_2(t)}\right) = \frac{w[y_2(t), y_1(t)]}{y_2(t)y_2(t+1)}$$
$$= \frac{D}{p(t)y_2(t)y_2(t+1)},$$

where D is a constant, by Liouville's formula. Summing both sides of this

282   Chapter 6. The Self-Adjoint Second Order Linear Equation

equation from T_1 to $t-1$ we get

$$\frac{y_1(t)}{y_2(t)} - \frac{y_1(T_1)}{y_2(T_1)} = D \sum_{s=T_1}^{t-1} \frac{1}{p(s)y_2(s)y_2(s+1)}.$$

It follows that

$$\sum_{s=T_1}^{\infty} \frac{1}{p(s)y_2(s)y_2(s+1)} < \infty.$$

Pick an integer T_2 so that $y_1(t)$ is of one sign for $t \geq T_2$. Similar to the above

$$\Delta\left(\frac{y_2(t)}{y_1(t)}\right) = \frac{w[y_1(t), y_2(t)]}{y_1(t)y_1(t+1)}$$
$$= -\frac{D}{p(t)y_1(t)y_1(t+1)}.$$

Summing both sides of this equation from T_2 to $t-1$ we get

$$\frac{y_2(t)}{y_1(t)} - \frac{y_2(T_2)}{y_1(T_2)} = -D \sum_{s=T_2}^{t-1} \frac{1}{p(s)y_1(s)y_1(s+1)}. \tag{6.21}$$

It follows that

$$\sum_{s=T_2}^{\infty} \frac{1}{p(s)y_1(s)y_1(s+1)} = \infty.$$

To prove the last statement in the theorem pick an integer T_3 so that both $y_1(t)$ and $y_2(t)$ are of one sign for $t \geq T_3$. Then with

$$w_i(t) \equiv \frac{p(t)\Delta y_i(t)}{y_i(t)},$$

6.4 Disconjugacy

$i = 1, 2, t \geq T_3$, we have that

$$\begin{aligned} w_1(t) - w_2(t) &= \frac{p(t)\Delta y_1(t)}{y_1(t)} - \frac{p(t)\Delta y_2(t)}{y_2(t)} \\ &= \frac{p(t)w[y_2(t), y_1(t)]}{y_1(t)y_2(t)} \\ &= \frac{D}{y_1(t)y_2(t)}. \end{aligned} \qquad (6.22)$$

Since $w_i(t)$ is not changed if we replace $y_i(t)$ by $-y_i(t)$ we can assume $y_i(t) > 0$, $i = 1, 2$, $t \geq T_3$. Using $\lim_{t \to \infty} \frac{y_2(t)}{y_1(t)} = \infty$ and Eq. (6.21) we get $D < 0$. It then follows from Eq. (6.22) that $w_1(t) < w_2(t)$ for $t \geq T_3$. ∎

A solution $y_1(t)$ as in Theorem 6.14 is called a recessive (first principal) solution of Eq. (6.2) at ∞. It is a "smallest" solution at ∞. A recessive solution is unique up to multiplication by a nonzero constant. A solution like $y_2(t)$ in Theorem 6.14 is called a dominant (second principal) solution of Eq. (6.2) at ∞. The existence of these solutions is important in the computation of special functions using difference equations. Later we will see that a recessive solution corresponds to a minimum solution in a neighborhood of ∞ of the Riccati equation associated with $Ly(t) = 0$.

Example 6.16. Find a recessive solution $y_1(t)$ and a dominant solution $y_2(t)$ of the disconjugate equation

$$y(t+1) - 6y(t) + 8y(t-1) = 0, \quad (t \geq 1),$$

and verify directly that the conclusions of Theorem 6.14 are true for these solutions.

The characteristic equation is

$$(\lambda - 2)(\lambda - 4) = 0.$$

Take $y_1(t) = 2^t$, $y_2(t) = 4^t$. Then

$$\lim_{t \to \infty} \frac{y_1(t)}{y_2(t)} = \lim_{t \to \infty} \left(\frac{1}{2}\right)^t = 0.$$

284   Chapter 6. The Self-Adjoint Second Order Linear Equation

If we write this equation in self-adjoint form, we get $p(t) = (\frac{1}{8})^t$, $q(t) = 3(\frac{1}{8})^t$. Hence

$$\sum_{t=0}^{\infty} \frac{1}{p(t)y_1(t)y_1(t+1)} = \frac{1}{2}\sum_{t=0}^{\infty} 2^t = \infty,$$

and

$$\sum_{t=0}^{\infty} \frac{1}{p(t)y_2(t)y_2(t+1)} = \frac{1}{4}\sum_{t=0}^{\infty} \left(\frac{1}{2}\right)^t < \infty.$$

Finally note that

$$w_1(t) = 8^{-t} < w_2(t) = 3 \cdot 8^{-t}, \quad (t \geq 0).$$

6.5 The Riccati Equation

In this section we introduce the Riccati equation associated with $Ly(t) = 0$. We will find that disconjugacy of $Ly(t) = 0$ is closely related to conditions on the associated Riccati equation. In Section 6.6 we will see that the Riccati equation is very important in oscillation theory.

Theorem 6.15. The difference equation $Ly(t) = 0$ has a solution of one sign on $[a, b+2]$ {on $[a, \infty)$} if and only the Riccati equation

$$Rz(t) \equiv \Delta z(t) + q(t) + \frac{z^2(t)}{z(t) + p(t-1)} = 0$$

has a solution $z(t)$ on $[a+1, b+2]$ {on $[a+1, \infty)$} with $z(t) + p(t-1) > 0$ on $[a+1, b+2]$ {on $[a+1, \infty)$}.

Proof. We will prove the theorem for the finite interval case. Assume $y(t)$ is a solution of $Ly(t) = 0$ of one sign on $[a, b+2]$.

6.5 The Riccati Equation

Set (the Riccati substitution)

$$z(t) = \frac{p(t-1)\Delta y(t-1)}{y(t-1)}, \quad (a+1 \le t \le b+2)$$

$$= p(t-1)\left[\frac{y(t)}{y(t-1)} - 1\right].$$

Then

$$z(t) + p(t-1) = p(t-1)\frac{y(t)}{y(t-1)} > 0$$

on $[a+1, b+2]$. To show that $z(t)$ satisfies the Riccati equation, consider

$$\Delta z(t) = \frac{1}{y(t)}\Delta\left[p(t-1)\Delta y(t-1)\right] + p(t-1)\Delta y(t-1)\Delta\left[\frac{1}{y(t-1)}\right]$$

$$= -q(t) + z(t)y(t-1)\left[\frac{1}{y(t)} - \frac{1}{y(t-1)}\right]$$

$$= -q(t) + z(t)\left[\frac{y(t-1)}{y(t)} - 1\right]$$

$$= -q(t) + z(t)\left[\frac{p(t-1)}{z(t) + p(t-1)} - 1\right]$$

$$= -q(t) - \frac{z^2(t)}{z(t) + p(t-1)}.$$

Hence $Rz(t) = 0$ for t in $[a+1, b+1]$.

Conversely assume $z(t)$ is a solution of the Riccati equation $Rz(t) = 0$ with $z(t) + p(t-1) > 0$ in $[a+1, b+2]$. Let $y(t)$ be the solution of the initial value problem

$$y(t) = \frac{z(t) + p(t-1)}{p(t-1)}y(t-1) \qquad (6.23)$$

$$y(a) = 1.$$

286 Chapter 6. The Self-Adjoint Second Order Linear Equation

It follows that $y(t) > 0$ on $[a, b+2]$. From Eq. (6.23) we have that

$$\Delta y(t-1) = \frac{z(t)y(t-1)}{p(t-1)}.$$

Hence

$$p(t-1)\Delta y(t-1) = z(t)y(t-1),$$

and so

$$\begin{aligned}
\Delta \left[p(t-1)\Delta y(t-1)\right] &= y(t)\Delta z(t) + z(t)\Delta y(t-1) \\
&= -q(t)y(t) - \frac{z^2(t)y(t)}{z(t)+p(t-1)} + \frac{z^2(t)y(t-1)}{p(t-1)} \\
&= -q(t)y(t)
\end{aligned}$$

by Eq. (6.23). Then $y(t)$ is a positive solution of $Ly(t) = 0$. ∎

Corollary 6.6. The Riccati difference equation has a solution $z(t)$ on $[a+1, b+2]$ {on $[a+1, \infty)$} with $z(t) + p(t-1) > 0$ on $[a+1, b+2]$ {on $[a+1, \infty)$} if and only if $Ly(t) = 0$ is disconjugate on $[a, b+2]$ {on $[a, \infty)$}.

Proof. This corollary follows from Theorem 6.15 and Theorem 6.10. ∎

Example 6.17.

Solve the Riccati equation

$$\Delta z(t) + 2\left(\frac{1}{6}\right)^t + \frac{z^2(t)}{z(t) + (\frac{1}{6})^{t-1}} = 0.$$

Here $q(t) = 2(\frac{1}{6})^t$, $p(t) = (\frac{1}{6})^t$. The associated self-adjoint equation is

$$\Delta\left[\left(\frac{1}{6}\right)^{t-1}\Delta y(t-1)\right] + 2\left(\frac{1}{6}\right)^t y(t) = 0,$$

6.5 The Riccati Equation

or after simplifying

$$y(t+1) - 5y(t) + 6y(t-1) = 0.$$

A general solution of this equation is

$$y(t) = A2^t + B3^t,$$

so

$$z(t) = \frac{p(t-1)\Delta y(t-1)}{y(t-1)}$$
$$= \left(\frac{1}{6}\right)^{t-1} \frac{A2^{t-1} + 2B3^{t-1}}{A2^{t-1} + B3^{t-1}}.$$

For $A \neq 0$ we get the solutions

$$z(t) = \left(\frac{1}{6}\right)^{t-1} \frac{1 + 2C(\frac{3}{2})^{t-1}}{1 + C(\frac{3}{2})^{t-1}}.$$

For $A = 0$ we have

$$z(t) = 2\left(\frac{1}{6}\right)^{t-1}.$$

Consider an initial value problem for the Riccati equation

$$Rz(t) = 0,$$
$$z(a+1) = z_0.$$

We can rewrite this as

$$z(t+1) = z(t) - q(t) - \frac{z^2(t)}{z(t) + p(t-1)}.$$

288　Chapter 6. The Self-Adjoint Second Order Linear Equation

Note that we want $z(a+1) + p(a) = z_0 + p(a) \neq 0$ so that z is defined at $a+2$. If we want to continue our solution to $t = a+3$ we also need that $z(a+2) + p(a+1) \neq 0$. Hence a solution of an initial value problem for a Riccati equation may not exist on the whole interval $[a+1, b+2]$. In the proof of the next theorem we will consider such an initial value problem.

Theorem 6.16. The difference equation $Ly(t) = 0$ is disconjugate on $[a, b+2]$ {on $[a, \infty)$} if and only if the Riccati inequality $Rw(t) \leq 0$ has a solution $w(t)$ on $[a+1, b+2]$ {on $[a+1, \infty)$} with $w(t) + p(t-1) > 0$ on $[a+1, b+2]$ {on $[a+1, \infty)$}.

Proof. If $Ly(t) = 0$ is disconjugate on $[a, b+2]$ then by Corollary 6.6 the Riccati equation (and hence the Riccati inequality) has a solution $w(t)$ on $[a+1, b+2]$ with $w(t) + p(t-1) > 0$ on $[a+1, b+2]$.

Conversely assume the Riccati inequality $Rw(t) \leq 0$ has a solution $w(t)$ on $[a+1, b+2]$ with $w(t) + p(t-1) > 0$ on $[a+1, b+2]$. Let $z(t)$ be the solution of the initial value problem $Rz(t) = 0$, $z(a+1) = w(a+1)$. We will show by mathematical induction that $z(t) \geq w(t)$ for $a+1 \leq t \leq b+2$. Since $z(t) \geq w(t)$ implies $z(t) + p(t-1) \geq w(t) + p(t-1) > 0$, we have simultaneously that $z(t)$ is a solution on $[a, b+2]$. For $t = a+1$ we have $z(a+1) = w(a+1)$. Assume $a+1 \leq t \leq b+1$ and $z(t) \geq w(t)$. We will use this to show that $z(t+1) \geq w(t+1)$. Now

$$0 \geq Rw(t) = \Delta w(t) + \frac{w^2(t)}{w(t) + p(t-1)} + q(t).$$

Hence

$$w(t+1) \leq \frac{p(t-1)w(t)}{w(t) + p(t-1)} - q(t).$$

Since $f(x) = \frac{cx}{x+c}$, $(c > 0)$, is an increasing function of x for $x + c > 0$,

$$w(t+1) \leq \frac{p(t-1)z(t)}{z(t) + p(t-1)} - q(t)$$
$$= z(t+1).$$

It follows that $Rz(t) = 0$ has a solution $z(t)$ on $[a+1, b+2]$ with $z(t) + p(t-1) > 0$ on $[a+1, b+2]$. ∎

6.5 The Riccati Equation

Corollary 6.7. If $q(t) \leq 0$ on $[a+1, b+1]$ {on $[a+1, \infty)$} then $Ly(t) = 0$ is disconjugate on $[a, b+2]$ {on $[a, \infty)$}.

Proof. If $q(t) \leq 0$, then $w(t) \equiv 0$ solves $Rw(t) \leq 0$ and $w(t) + p(t-1) > 0$. The result follows from Theorem 6.16. ∎

An important relation between the self-adjoint operator L and the Riccati operator R is given by the following lemma.

Lemma 6.2. If $y(t) > 0$ on $[a, b+2]$ and $z(t) = \frac{p(t-1)\Delta y(t-1)}{y(t-1)}$ on $[a+1, b+2]$, then $z(t) + p(t-1) > 0$ on $[a+1, b+2]$ and

$$Ly(t) = y(t)Rz(t) \tag{6.24}$$

for t in $[a+1, b+1]$.

Proof. By the quotient rule

$$y(t)\Delta z(t) = \frac{y(t-1)\Delta[p(t-1)\Delta y(t-1)] - p(t-1)[\Delta y(t-1)]^2}{y(t-1)}$$

$$= Ly(t) - q(t)y(t) - \left[\frac{p(t-1)\Delta y(t-1)}{y(t-1)}\right]^2 \frac{y(t-1)}{p(t-1)}.$$

Since $z(t) + p(t-1) = p(t-1)\frac{y(t)}{y(t-1)}$, $z(t) + p(t-1) > 0$ and

$$Ly(t) = y(t)\Delta z(t) + y(t)q(t) + z^2(t)\frac{y(t)}{z(t)+p(t-1)}.$$

Hence

$$Ly(t) = y(t)Rz(t)$$

for t in $[a+1, b+1]$. ∎

Now we can prove the following theorem.

Theorem 6.17. There is a function $y(t)$ with $y(t) > 0$ on $[a, b+2]$ and $Ly(t) \leq 0$, t in $[a+1, b+1]$ if and only if $Ly(t) = 0$ is disconjugate on $[a, b+2]$.

Proof. Set

$$w(t) = \frac{p(t-1)\Delta y(t-1)}{y(t-1)},$$

t in $[a+1, b+2]$. Then by Lemma 6.2

$$w(t) + p(t-1) > 0 \text{ in } [a+1, b+2]$$

and

$$Rw(t) = \frac{1}{y(t)} Ly(t)$$
$$\leq 0, \quad (a+1 \leq t \leq b+1).$$

Hence by Theorem 6.16, $Ly(t) = 0$ is disconjugate on $[a, b+2]$. The converse statement is left to the reader. ∎

Corollary 6.8. Assume there is a number $r > 0$ such that

$$p(t)r^2 + c(t)r + p(t-1) \leq 0$$

for t in $[a+1, b+1]$. Then $Ly(t) = 0$ is disconjugate on $[a, b+2]$.

Proof. Set

$$y(t) = r^t, \quad (a \leq t \leq b+2).$$

Then $y(t) > 0$ in $[a, b+2]$ and for t in $[a+1, b+1]$

$$\begin{aligned} Ly(t) &= p(t)r^{t+1} + c(t)r^t + p(t-1)r^{t-1} \\ &= r^{t-1}\left[p(t)r^2 + c(t)r + p(t-1)\right] \\ &\leq 0. \end{aligned}$$

By Theorem 6.17, $Ly(t) = 0$ is disconjugate on $[a, b+2]$. ∎

6.5 The Riccati Equation

Corollary 6.9. The difference equation

$$y(t+1) + \alpha(t)y(t) + \beta(t)y(t-1) = 0,$$

where $\beta(t) > 0$ in $[a+1, b+1]$, is disconjugate on $[a, b+2]$ provided there is a positive number r so that

$$r^2 + \alpha(t)r + \beta(t) \leq 0$$

for t in $[a+1, b+1]$.

Proof. Since $\beta(t) > 0$ on $[a+1, b+1]$ there is a positive function $p(t)$ so that

$$p(t)\left[y(t+1) + \alpha(t)y(t) + \beta(t)y(t-1)\right]$$
$$= p(t)y(t+1) + c(t)y(t) + p(t-1)y(t-1).$$

Then for t in $[a+1, b+1]$

$$p(t)r^2 + c(t)r + p(t-1)$$
$$= p(t)\left[r^2 + \alpha(t)r + \beta(t)\right] \leq 0.$$

By Corollary 6.8, $Ly(t) = 0$ is disconjugate on $[a, b+2]$. ∎

Example 6.18. Show that the equation

$$y(t+1) - y(t) + \left(\frac{1}{4} - \frac{1}{t^2}\right)y(t-1) = 0$$

is disconjugate on $[2, \infty)$.

Consider

$$h(r) = r^2 - r + \left(\frac{1}{4} - \frac{1}{t^2}\right).$$

Since $h(1/2) = -\frac{1}{t^2} \leq 0$ in $[1, \infty)$, we have by Corollary 6.9 that this difference equation is disconjugate on $[2, \infty)$.

Chapter 6. The Self-Adjoint Second Order Linear Equation

We can generalize Corollary 6.7 in the following manner.

Theorem 6.18. If $Ly(t) = 0$ is disconjugate on $[a, b+2]$ and $k(t) \leq 0$ for t in $[a+1, b+1]$, then

$$Ly(t) + k(t)y(t) = 0 \qquad (6.25)$$

is disconjugate on $[a, b+2]$.

Proof. Since $Ly(t) = 0$ is disconjugate on $[a, b+2]$ it has a Polya factorization. In particular

$$Ly(t) = \varrho_1(t)\Delta[\varrho_2(t)\Delta(\varrho_1(t-1)y(t-1))],$$

where

$$\varrho_1(t) > 0 \quad \text{in} \quad [a, b+2]$$
$$\varrho_2(t) > 0 \quad \text{in} \quad [a+1, b+2].$$

Equation (6.25) becomes

$$\varrho_1(t)\Delta\left\{\varrho_2(t)\left[\Delta\left(\varrho_1(t-1)y(t-1)\right)\right]\right\} + k(t)y(t) = 0.$$

Let $z(t) = \varrho_1(t)y(t)$ to get

$$\Delta\left[\varrho_2(t)\Delta z(t-1)\right] + \frac{k(t)}{\varrho_1^2(t)}z(t) = 0.$$

By Corollary 6.7 this last equation is disconjugate on $[a, b+2]$. Since $y(t)$ has a generalized zero at some point if and only if $z(t)$ does at the same point, we have that Eq. (6.25) is disconjugate on $[a, b+2]$. ∎

Now we can prove the following comparison theorem. In Chapter 8 we will give an improvement of this theorem.

Theorem 6.19. (Sturm Comparison Theorem) If $Ly(t) + q_1(t)y(t) = 0$ is disconjugate on $[a, b+2]$ and $q_2(t) \leq q_1(t)$ on $[a+1, b+1]$ then $Ly(t) + q_2(t)y(t) = 0$ is disconjugate on $[a, b+2]$.

Proof. The equation

$$Ly(t) + q_2(t)y(t) = 0$$

can be written in the form

$$My(t) + k(t)y(t) = 0,$$

where $My(t) \equiv Ly(t) + q_1(t)y(t) = 0$ is disconjugate on $[a, b+2]$ and

$$k(t) \equiv q_2(t) - q_1(t) \le 0$$

on $[a+1, b+1]$. Hence $Ly(t) + q_2(t)y(t) = 0$ is disconjugate on $[a, b+2]$ by Theorem 6.18. ∎

6.6 Oscillation

In this section we will be concerned with the self-adjoint difference equation

$$Ly(t) = \Delta[p(t-1)\Delta y(t-1)] + q(t)y(t) = 0,$$

where $p(t) > 0$ for integers $t \ge a$ and $q(t)$ is defined for integers $t \ge a+1$. We will define what it means for this equation to be oscillatory on $[a, \infty)$ and give several criteria for oscillation. The Riccati equation will be important in this development.

Definition 6.5. A nontrivial solution $y(t)$ of a second order linear homogeneous difference equation is said to be oscillatory on $[a, \infty)$ provided $y(t)$ has infinitely many generalized zeros on $[a, \infty)$. If a nontrivial solution is not oscillatory then it is said to be nonoscillatory. The difference equation $Ly(t) = 0$ is said to be oscillatory on $[a, \infty)$ provided it has a nontrivial oscillatory solution on $[a, \infty)$. If $Ly(t) = 0$ is not oscillatory on $[a, \infty)$, then we say $Ly(t) = 0$ is nonoscillatory on $[a, \infty)$.

Note that by the Sturm separation theorem if one nontrivial solution has infinitely many generalized zeros on $[a, \infty)$ then all nontrivial solutions have infinitely many generalized zeros on $[a, \infty)$.

Recall in Example 6.7 we saw that the Fibonacci difference equation

$y(t+1) - y(t) - y(t-1) = 0$ has an oscillatory solution $y(t) = \left(\frac{1}{2} - \frac{\sqrt{5}}{2}\right)^t$ and a nonoscillatory solution $z(t) = \left(\frac{1}{2} + \frac{\sqrt{5}}{2}\right)^t$ but the Fibonacci difference equation can not be put in self-adjoint form.

Example 6.19. Show that the difference equation

$$y(t+1) - \frac{t+7}{t+5} y(t) - \frac{t^2+1}{t^2+4} y(t-1) = 0$$

has an oscillatory and a nonoscillatory solution.
 Since

$$\lim_{t \to \infty} -\frac{t+7}{t+5} = -1$$

and

$$\lim_{t \to \infty} -\frac{t^2+1}{t^2+4} = -1,$$

we have that the equation is of Poincarè type. By Perron's theorem there are solutions $u(t), v(t)$ such that $\lim_{t \to \infty} \frac{u(t+1)}{u(t)} = \frac{1}{2} - \frac{\sqrt{5}}{2}$, $\lim_{t \to \infty} \frac{v(t+1)}{v(t)} = \frac{1}{2} + \frac{\sqrt{5}}{2}$, from which the desired result follows immediately.

Example 6.20. The difference equation

$$y(t+1) + 9y(t-1) = 0 \tag{6.26}$$

is oscillatory on $[0, \infty)$.
 The characteristic equation is $\lambda^2 + 9 = 0$, so the eigenvalues are $\lambda_1 = 3e^{i\frac{\pi}{2}}, \lambda_2 = 3e^{-i\frac{\pi}{2}}$. It follows that a general solution is

$$y(t) = A3^t \cos\frac{\pi}{2}t + B3^t \sin\frac{\pi}{2}t.$$

Hence Eq. (6.26) is oscillatory on $[0, \infty)$.
 Example 6.21 shows that we can have a nonoscillatory equation with as many generalized zeros as we want.

6.6 Oscillation

Example 6.21. The equation

$$y(t+1) + \alpha(t)y(t) + \beta(t)y(t-1) = 0,$$

where

$$\alpha(t) = \begin{cases} 0, & 1 \le t \le n \\ -2, & n+1 \le t \end{cases}$$

$$\beta(t) = \begin{cases} 9, & 1 \le t \le n \\ 1, & n+1 \le t \end{cases}$$

for $n \ge 3$ is nonoscillatory but not disconjugate on $[0, \infty)$.

Usually we can not actually solve the difference equation in question. We would like to develop theorems concerning the coefficients $p(t)$, $q(t)$ that will enable us to determine if the equation is oscillatory or nonoscillatory on $[a, \infty)$.

Theorem 6.20. If $p(t)$ is bounded above on $[a, \infty)$ and $Ly(t) = 0$ is nonoscillatory on $[a, \infty)$ then either $\sum_{t=a+1}^{\infty} q(t)$ exists as a finite number or $\sum_{t=a+1}^{\infty} q(t) = -\infty$.

Proof. Assume $Ly(t) = 0$ is nonoscillatory on $[a, \infty)$. Let $y(t)$ be a nontrivial solution. Then there is an integer $t_0 \ge a+1$ such that $y(t)$ is of one sign on $[t_0 - 1, \infty)$. Without loss of generality we can assume $y(t) > 0$ on $[t_0 - 1, \infty)$. Make the Riccati substitution

$$z(t) = \frac{p(t-1)\Delta y(t-1)}{y(t-1)}, \quad (t \ge t_0),$$

then $z(t) + p(t-1) > 0$ in $[t_0, \infty)$ and $z(t)$ satisfies the Riccati equation

$$\Delta z(t) = -q(t) - \frac{z^2(t)}{z(t) + p(t-1)}$$

on $[t_0, \infty)$. Summing both sides from t_0 to t we obtain

$$z(t+1) = z(t_0) - \sum_{s=t_0}^{t} q(s) - \sum_{s=t_0}^{t} \frac{z^2(s)}{z(s) + p(s-1)}. \tag{6.27}$$

296   Chapter 6. The Self-Adjoint Second Order Linear Equation

First assume $\sum_{s=t_0}^{\infty} \frac{z^2(s)}{z(s)+p(s-1)} < \infty$. Since $p(t)$ is bounded above on $[a,\infty)$, there is an $m > 0$ such that $p(t-1) \leq m$ on $[a+1,\infty)$. But $0 < z(t) + p(t-1) \leq z(t) + m$ implies

$$0 < \frac{z^2(t)}{z(t)+m} \leq \frac{z^2(t)}{z(t)+p(t-1)}.$$

By the comparison test for convergence of series $\sum_{t=t_0}^{\infty} \frac{z^2(t)}{z(t)+m}$ converges. But then $\lim_{t\to\infty} z(t) = 0$, so by Eq. (6.27) $\sum_{t=t_0}^{\infty} q(t)$ converges to a finite number.

Finally consider the case $\sum_{s=t_0}^{\infty} \frac{z^2(s)}{z(s)+p(s-1)} = \infty$. Since $-z(t+1) < p(t) \leq m$, Eq. (6.27) implies that $\sum_{t=t_0}^{\infty} q(t) = -\infty$. ∎

Theorem 6.20 gives us immediately the following two corollaries.

Corollary 6.10. If $p(t)$ is bounded above on $[a,\infty)$ and $\sum_{t=a+1}^{\infty} q(t) = \infty$, then $Ly(t) = 0$ is oscillatory on $[a,\infty)$.

Corollary 6.11. If $p(t)$ is bounded above on $[a,\infty)$ and

$$-\infty \leq \liminf_{t\to\infty} \sum_{s=a+1}^{t} q(s) < \limsup_{t\to\infty} \sum_{s=a+1}^{t} q(s) \leq \infty,$$

then $Ly(t) = 0$ is oscillatory on $[a,\infty)$.

Example 6.22. Show that the difference equation

$$\Delta^2 y(t-1) + \frac{1}{t} y(t) = 0, \quad (t \geq 1) \tag{6.28}$$

is oscillatory on $[0,\infty)$.

Here $p(t) = 1$, which is bounded on $[0,\infty)$. Since $\sum_{t=1}^{\infty} q(t) = \sum_{t=1}^{\infty} \frac{1}{t} = \infty$, we have by Corollary 6.10 that Eq. (6.28) is oscillatory on $[0,\infty)$.

Example 6.23. Show that the difference equation

$$\Delta^2 y(t-1) + (-1)^t \frac{1}{2} y(t) = 0, \quad (t \geq 1),$$

6.6 Oscillation

is oscillatory on $[0, \infty)$.

This follows from Corollary 6.11 because

$$-\frac{1}{2} = \liminf_{t\to\infty} \sum_{s=1}^{t}(-1)^s\frac{1}{2} < \limsup_{t\to\infty}\sum_{s=1}^{t}(-1)^s\frac{1}{2} = 0.$$

Theorem 6.21. If for all $t_0 \geq a+1$ there is a $t_1 \geq t_0$ such that

$$\limsup_{t\to\infty} \sum_{s=t_1}^{t} q(s) \geq 1, \tag{6.29}$$

then $\Delta^2 y(t-1) + q(t)y(t) = 0$ is oscillatory on $[a, \infty)$.

Proof. Assume $\Delta^2 y(t-1) + q(t)y(t) = 0$ is nonoscillatory on $[a, \infty)$. Then there is an integer $t_0 \geq a+1$ and a solution $y(t)$ such that $y(t) > 0$ for $t \geq t_0 - 1$. Make the Riccati substitution

$$z(t) = \frac{\Delta y(t-1)}{y(t-1)};$$

then $z(t) + p(t-1) = z(t) + 1 > 0$ for $t \geq t_0$ and

$$\Delta z(t) = -q(t) - \frac{z^2(t)}{z(t)+1}. \tag{6.30}$$

Pick $t_1 \geq t_0$ so that Eq. (6.29) holds. Summing both sides of Eq. (6.30) from t_1 to t we get that

$$z(t+1) - z(t_1) = -\sum_{s=t_1}^{t} q(s) - \sum_{s=t_1}^{t} \frac{z^2(s)}{z(s)+1}.$$

Hence

$$z(t+1) = \frac{z(t_1)}{1+z(t_1)} - \sum_{s=t_1}^{t} q(s) - \sum_{s=t_1+1}^{t} \frac{z^2(s)}{z(s)+1}. \tag{6.31}$$

If $\sum_{s=t_1+1}^{\infty} \frac{z^2(s)}{z(s)+1} = \infty$, then it is easy to get a contradiction from Eq. (6.31).

Now assume
$$\sum_{s=t_1+1}^{\infty} \frac{z^2(s)}{z(s)+1} < \infty.$$

But then $\lim_{t\to\infty} \frac{z^2(t)}{z(t)+1} = 0$ and consequently $\lim_{t\to\infty} z(t) = 0$. By Eq. (6.31)
$$-z(t+1) \geq -\frac{z(t_1)}{1+z(t_1)} + \sum_{s=t_1}^{t} q(s).$$

Hence
$$0 \geq -\frac{z(t_1)}{1+z(t_1)} + \limsup_{t\to\infty} \sum_{s=t_1}^{t} q(s)$$
$$> 0,$$

which is a contradiction. ∎

Example 6.24. Show that the difference equation
$$\Delta^2 y(t-1) + q(t)y(t) = 0, \quad \text{where}$$

$$\{q(t)\}_{t=a+1}^{\infty} = \{1, -\frac{1}{2}, -\frac{1}{2}, \frac{1}{3}, \frac{1}{3}, \frac{1}{3}, -\frac{1}{4}, -\frac{1}{4}, -\frac{1}{4}, -\frac{1}{4}, \ldots\}$$

is oscillatory on $[a, \infty)$.

Since for all $t_0 \geq a+1$ there is a $t_1 \geq t_0$ such that
$$\limsup_{t\to\infty} \sum_{s=t_1}^{t} q(s) = 1,$$

this difference equation is oscillatory by Theorem 6.21.

6.6 Oscillation

The following lemma will be used in the proof of Theorem 6.22 and will also be used to prove Theorem 8.10.

Lemma 6.3. Assume $z(t)$ is a solution of the Riccati equation $Rz(t) = 0$ on $[a+1, b+2]$ with $z(t) + p(t-1) > 0$ on $[a+1, b+2]$. If $u(t)$ is defined on $[a, b+2]$, then

$$\Delta\left[z(t)u^2(t-1)\right] = -\{q(t)u^2(t) - p(t-1)[\Delta u(t-1)]^2\} \qquad (6.32)$$
$$-\left(\frac{z(t)u(t)}{\sqrt{z(t)+p(t-1)}} - \sqrt{z(t)+p(t-1)}\Delta u(t-1)\right)^2$$

for t in $[a+1, b+1]$.

Proof. Let $z(t)$ and $u(t)$ be as in the statement of the lemma, and consider

$$\Delta\left(z(t)u^2(t-1)\right)$$
$$= u^2(t)\Delta z(t) + u(t)z(t)\Delta u(t-1) + z(t)u(t-1)\Delta u(t-1)$$
$$= u^2(t)\left[-q(t) - \frac{z^2(t)}{z(t)+p(t-1)}\right] + u(t)z(t)\Delta u(t-1) + z(t)u(t-1)\Delta u(t-1)$$
$$= -\left[q(t)u^2(t) - p(t-1)[\Delta u(t-1)]^2\right] - \frac{z^2(t)u^2(t)}{z(t)+p(t-1)}$$
$$\quad + u(t)z(t)\Delta u(t-1) + z(t)u(t-1)\Delta u(t-1) - p(t-1)[\Delta u(t-1)]^2$$
$$= -\left[q(t)u^2(t) - p(t-1)[\Delta u(t-1)]^2\right] - \frac{z^2(t)u^2(t)}{z(t)+p(t-1)}$$
$$\quad + 2u(t)z(t)\Delta u(t-1) - (z(t)+p(t-1))\left[\Delta u(t-1)\right]^2$$
$$= -\{q(t)u^2(t) - p(t-1)[\Delta u(t-1)]^2\}$$
$$\quad -\left(\frac{z(t)u(t)}{\sqrt{z(t)+p(t-1)}} - \sqrt{z(t)+p(t-1)}\Delta u(t-1)\right)^2. \qquad \blacksquare$$

Theorem 6.22. If $\sum_{t=a}^{\infty}\frac{1}{p(t)} = \infty$ and there is an integer $t_0 \geq a+1$ and a function $u(t) > 0$ in $[t_0, \infty)$ such that

$$\sum_{t=t_0}^{\infty}\{q(t)u^2(t) - p(t-1)[\Delta u(t-1)]^2\} = \infty, \qquad (6.33)$$

then $Ly(t) = 0$ is oscillatory on $[a, \infty)$.

Proof. Assume $Ly(t) = 0$ is nonoscillatory on $[a, \infty)$. Then there is a $t_1 \geq a$ such that $Ly(t) = 0$ is disconjugate on $[t_1, \infty)$. By Theorem 6.14 there is a dominant solution $y(t)$ and an integer $t_2 \geq t_1$ such that $y(t) > 0$ on $[t_2, \infty)$ and

$$\sum_{t=t_2}^{\infty} \frac{1}{p(t)y(t)y(t+1)} < \infty.$$

Let $T = \max\{t_2, t_1\} + 1$ and set for $t \geq T$

$$z(t) = \frac{p(t-1)\Delta y(t-1)}{y(t-1)}.$$

Then $z(t) + p(t-1) > 0$ in $[T, \infty)$ and $Rz(t) = 0$, $t \geq T$. By Lemma 6.3

$$\Delta\left[z(t)u^2(t-1)\right] = -\left\{q(t)u^2(t) - p(t-1)[\Delta u(t-1)]^2\right\}$$
$$- \left(\frac{z(t)u(t)}{\sqrt{z(t) + p(t-1)}} - \sqrt{z(t) + p(t-1)}\Delta u(t-1)\right)^2.$$

Summing both sides from T to t, we have that $\left[z(s)u^2(s-1)\right]_T^{t+1}$ is equal to

$$-\sum_{s=T}^{t}\{q(s)u^2(s) - p(s-1)[\Delta u(s-1)]^2\}$$

$$-\sum_{s=T}^{t}\left(\frac{z(s)u(s)}{\sqrt{z(s)+p(s-1)}} - \sqrt{z(s)+p(s-1)}\Delta u(s-1)\right)^2.$$

It follows from Eq. (6.33) that

$$\lim_{t \to \infty} z(t+1)u^2(t) = -\infty.$$

6.6 Oscillation

Then there is a $T_0 \geq T$ such that

$$z(t) < 0 \text{ for } t \geq T_0.$$

Hence

$$\frac{p(t-1)\Delta y(t-1)}{y(t-1)} < 0 \text{ for } t \geq T_0,$$

and so $\Delta y(t-1) < 0$ for $t \geq T_0$. Since $y(t)$ is decreasing for $t \geq T_0 - 1$ we have

$$\sum_{t=T_0}^{\infty} \frac{1}{p(t)} = y(T_0)y(T_0+1) \sum_{t=T_0}^{\infty} \frac{1}{p(t)y(T_0)y(T_0+1)}$$
$$< y(T_0)y(T_0+1) \sum_{t=T_0}^{\infty} \frac{1}{p(t)y(t)y(t+1)}$$
$$< \infty,$$

which is a contradiction. ∎

Corollary 6.12. If $\sum_{t=a}^{\infty} \frac{1}{p(t)} = \infty$ and $\sum_{t=a+1}^{\infty} q(t) = \infty$, then $Ly(t) = 0$ is oscillatory on $[a, \infty)$.

Proof. For $u(t) = 1$, $t \geq a$,

$$\sum_{t=a+1}^{\infty} \{q(t)u^2(t) - p(t-1)[\Delta u(t-1)]^2\} = \sum_{t=a+1}^{\infty} q(t) = \infty.$$

Hence by Theorem 6.22, $Ly(t) = 0$ is oscillatory on $[a, \infty)$. ∎

Example 6.25. Show that

$$\Delta\left[(t-1)\Delta y(t-1)\right] + \frac{1}{t}y(t) = 0 \qquad (6.34)$$

is oscillatory on $[2, \infty)$.

302   Chapter 6. The Self-Adjoint Second Order Linear Equation

Here $p(t) = t$, $q(t) = \frac{1}{t}$. Since

$$\sum_{t=2}^{\infty} \frac{1}{p(t)} = \infty = \sum_{t=3}^{\infty} q(t),$$

we have by Corollary 6.12 that Eq. (6.34) is oscillatory on $[a, \infty)$.

Example 6.26. Show that

$$\Delta\left[\left(\frac{1}{6}\right)^{t-1}\Delta y(t-1)\right] + \left(\frac{1}{5}\right)^{t} y(t) = 0 \qquad (6.35)$$

is oscillatory on $[0, \infty)$.

Let $u(t) = 5^t$; then $\Delta u(t) = 4 \cdot 5^t$. Hence

$$\sum_{t=1}^{\infty} \{q(t)u^2(t) - p(t-1)[\Delta u(t-1)]^2\}$$

$$= \sum_{t=1}^{\infty} \left\{\left(\frac{1}{5}\right)^{t} 5^{2t} - \left(\frac{1}{6}\right)^{t-1} 16 \cdot 5^{2t-2}\right\}$$

$$= \sum_{t=1}^{\infty} 5^t \left(1 - \frac{16}{5}\left(\frac{5}{6}\right)^{t-1}\right) = \infty.$$

By Theorem 6.22, equation Eq. (6.35) is oscillatory on $[0, \infty)$.

We now state without proof an oscillation theorem for a nonlinear difference equation. Oscillation in this case means every solution on $[a, \infty)$ has infinitely many generalized zeros. Nonoscillatory means there is at least one nontrivial solution with only a finite number of generalized zeros.

Theorem 6.23. If $q(t) \geq 0$, $t \geq a+1$, then the Emden-Fowler difference equation

$$\Delta^2 y(t-1) + q(t)y^{2n+1}(t) = 0, \quad (n \geq 1),$$

is oscillatory on $[a, \infty)$ if and only if $\sum_{t=a+1}^{\infty} tq(t) = \infty$.

6.6 Oscillation

Example 6.27. Show that the nonlinear difference equation

$$\Delta^2 y(t-1) + \frac{1}{t^2} y^3(t) = 0 \tag{6.36}$$

is oscillatory on $[0, \infty)$.

Here $n = 1$, $q(t) = \frac{1}{t^2}$, $t \geq 1$. Since

$$\sum_{t=1}^{\infty} t q(t) = \sum_{t=1}^{\infty} \frac{1}{t} = +\infty,$$

it follows from Theorem 6.23 that Eq. (6.36) is oscillatory on $[0, \infty)$.

For results concerning the oscillation of difference equations, the reader is referred to papers in the references by Bykov and Zivogladova [32], Bykov, Zivogladova and Sevcov [33], Chen and Erbe [40],[41], Chuanxi, Kuruklis and Ladas [48], Derr [55], Erbe and Zhang [70],[71], Gyori and Ladas [95],[96], Hinton and Lewis [125],[126], Hooker and Patula [131],[132],[133], Hooker, Kwong, and Patula [129],[130], Ladas [151],[152],[153], Ladas, Philos and Sficas [154],[155], Mingarelli [186], Patula [200],[201], Reid [227], Smith and Taylor [234], Szmanda [239],[240], and Wouk [250]. Analogs of some of the results in this chapter for higher order equations can be found in Cheng [43],[44], Eloe [64]-[67], Eloe and Henderson [69], Hankerson and Henderson [104], Hankerson and Peterson [105]-[111], Harris [112], Hartman [114], Hartman and Wintner [116], Henderson [117]-[120], Peil [203],[204], and Peterson [206]-[211]. Generalizations to matrix equations are contained in Ahlbrandt [8], Ahlbrandt and Hooker [9],[11],[12],[13], Chen and Erbe [40], Peil and Peterson [205], and Peterson and Ridenhour [212]-[216].

Exercises

Section 6.1

6.1 Each of the following are in the form of Eq. (6.3). Write each of these equations in the self-adjoint form of Eq. (6.2).
(a) $3^t y(t+1) + (e^t - 4 \cdot 3^{t-1})y(t) + 3^{t-1}y(t-1) = 0$.
(b) $(\cos \frac{1}{100}t)y(t+1) + 2^t y(t) + (\cos \frac{t-1}{100})y(t-1) = 0$.
(c) $y(t+1) + 2y(t) + y(t-1) = 0$.

6.2 Write each of the following equations in the self-adjoint form. In (a) and (d) use your factorization to solve the equation.
(a) $(t-1)y(t+1) - ty(t) + y(t-1) = 0$, $t \geq 3$.
(b) $(t-1)y(t+1) + (1-t)y(t) + y(t-1) = 0$, $t \geq 3$.
(c) $y(t+1) - 5y(t) + 6y(t-1) = 0$.
(d) $y(t+1) - 2y(t) + y(t-1) = 0$.

6.3 The standard solutions $z(\lambda) = J_\lambda(t)$ and $z(\lambda) = Y_\lambda(t)$ of the Bessel equation

$$t^2 y''(t) + ty'(t) + (t^2 - \lambda^2)y(t) = 0$$

satisfy the difference equation

$$z(\lambda + 1) - 2\lambda t^{-1} z(\lambda) + z(\lambda - 1) = 0.$$

Write this equation in self-adjoint form. If $u(m) = t^{-\lambda-m} z(\lambda + m)$, then verify that $u(m)$ satisfies

$$u(m+2) - 2(\lambda + m + 1)t^{-2} u(m+1) + t^{-2} u(m) = 0.$$

6.4 Find two linearly independent solutions $y(t)$ and $z(t)$ of each of the following:
(a) $y(t+1) - 5y(t) + 6y(t-1) = 0$,
(b) $y(t+1) - 2y(t) + y(t-1) = 0$,
(c) $\Delta^2 y(t-1) + 2y(t) = 0$.

Calculate $w[y(t), z(t)]$ directly and then check your answer using Liouville's formula.

6.5 Prove the form of the Lagrange identity for complex valued functions. Namely, if $y(t)$ and $z(t)$ are complex valued functions on $[a, b+2]$, then

$$\overline{z(t)}Ly(t) - y\overline{Lz(t)} = \Delta\{p(t-1)w[\overline{z(t-1)}, y(t-1)]\}$$

for t in $[a+1, b+1]$.

6.6 Find a Polya factorization for each of the following:
   (a) $y(t+1) - 2y(t) + y(t-1) = 0$,
   (b) $y(t+1) - 5y(t) + 6y(t-1) = 0$,
   (c) $y(t+1) - 4y(t) + 4y(t-1) = 0$,
   (d) $y(t+1) - 200y(t) + 10,000y(t-1) = 0$.

6.7 Find the Cauchy function $y(t, s)$ for each of the following:
   (a) $\Delta\left[\left(\frac{1}{6}\right)^{t-1} \Delta y(t-1)\right] + 2\left(\frac{1}{6}\right)^t y(t) = 0$,
   (b) $y(t+1) - 4y(t) + 4y(t-1) = 0$,
   (c) $\Delta^2 y(t-1) + 2y(t) = 0$.

6.8 Use the variation of constants formula to solve the following initial value problems:
   (a) $\Delta^2 y(t-1) = t$, $y(2) = y(3) = 0$,
   (b) $\Delta^2 y(t-1) = 3$, $y(0) = y(1) = 0$,
   (c) $\Delta\left[\left(\frac{1}{6}\right)^{t-1} \Delta y(t-1)\right] + 2\left(\frac{1}{6}\right)^t y(t) = 1$, $y(0) = y(1) = 0$,
   (d) $\Delta^2 y(t-1) = t^2$, $y(0) = y(1) = 0$.

6.9 Solve the initial value problem

$$\Delta^2 y(t-1) = t, \quad y(0) = 0, \quad y(1) = 1,$$

using Corollary 6.3.

Section 6.2

6.10 Solve the following difference equations and decide on what intervals these difference equations are disconjugate:
   (a) $y(t+1) - 9y(t) + 14y(t-1) = 0$,

(b) $y(t+1) - 2\sqrt{2}y(t) + 4y(t-1) = 0$.

6.11 Prove the last statement in Theorem 6.5.

6.12 Show that the existence of a solution $y(t)$ of $Ly(t) = 0$ with $y(t) \neq 0$ in $[a, b+2]$ does not in general imply $Ly(t) = 0$ is disconjugate on $[a, b+2]$.

6.13 Assume $Ly(t) = 0$ is disconjugate on $[a, b+2]$. Show that if $z(t)$ is a solution of the difference inequality $Lz(t) \geq 0$ on $[a, b+2]$ and $y(t)$ is a solution of $Ly(t) = 0$ in $[a, b+2]$ with $z(a) = y(a)$, $z(a+1) = y(a+1)$, then $y(t) \geq z(t)$ on $[a, b+2]$.

6.14 Show that if there is a solution of $Ly(t) = 0$ with $y(a) = 0$, $y(t) > 0$ in $[a+1, b+2]$, then $Ly(t) = 0$ is disconjugate on $[a, b+2]$.

Section 6.3

6.15 We say that $Ly(t) = 0$ is disfocal on $[a, b+2]$ if there is no nontrivial solution $y(t)$ such that $y(t)$ has a generalized zero at t_1 and $\Delta y(t)$ has a generalized zero at t_2 where $a \leq t_1 < t_2 \leq b+1$. Show that if $Ly(t) = 0$ is disfocal on $[a, b+2]$ then the boundary value problem $Ly(t) = h(t)$, $y(t_1) = A$, $\Delta y(t_1) = B$, $a \leq t_1 < t_2 \leq b+1$, has a unique solution. Show that $\Delta^2 y(t-1) = 0$ is disfocal on $[a, b+2]$.

6.16 Show that if $H(t, s)$ satisfies the properties
   (a) $H(t, s)$ is defined for $a \leq t \leq b+2$, $a+1 \leq s \leq b+1$,
   (b) $LH(t, s) = \delta_{ts}$ for $a \leq t \leq b+2$, $a+1 \leq s \leq b+1$,
   (c) $H(a, s) = 0 = \Delta H(b+1, s)$, $a+1 \leq s \leq b+1$,
then $y(t) = \sum_{s=a+1}^{b+1} H(t, s)h(s)$ solves the boundary value problem $Ly(t) = h(t)$, t in $[a+1, b+1]$, $y(a) = 0 = \Delta y(b+1)$.

6.17 Show that if $Ly(t) = 0$ is disfocal on $[a, b+2]$ then

$$H(t, s) = \begin{cases} -\frac{\Delta y(b+1,s)}{\Delta y_1(b+1)} y_1(t), & t \leq s \\ y(t, s) - \frac{\Delta y(b+1,s)}{\Delta y_1(b+1)} y_1(t), & s \leq t \end{cases}$$

satisfies (a)-(c) in Exercise 6.16 and $H(t, s) \leq 0$, $a \leq t \leq b+1$, $a+1 \leq s \leq b+1$.

6.18 Show that if $Ly(t) = 0$ is disfocal on $[a, b+2]$, then there is a unique function $H(t, s)$ satisfying properties (a)-(c) in Exercise 6.16. This function $H(t, s)$ is called the Green's function for the boundary value problem $Ly(t) = 0$, t in $[a+1, b+1]$, $y(a) = 0 = \Delta y(b+1)$.

Exercises

6.19 Find the Green's function for the boundary value problem $\Delta^2 y(t-1) = 0$, $y(a) = 0 = \Delta y(b+1)$. Show that $a - b - 2 \le H(t,s) \le 0$, $a \le t \le b+2$, $a+1 \le s \le b+1$, and $-1 \le \Delta H(t,s) \le 0$ for $a \le t \le b+1$, $a+1 \le s \le b+1$. (See the previous two exercises.)

6.20 Show that the Green's function $G(t,s)$ for the boundary value problem

$$\Delta^2 y(t-1) = 0, \quad y(a) = 0 = y(b+2),$$

satisfies

$$-\frac{b+2-a}{4} \le G(t,s) \le 0$$

for $a \le t \le b+2$, $a+1 \le s \le b+1$. Further show that

$$\sum_{s=a+1}^{b+1} |G(t,s)| \le \frac{(b+2-a)^2}{8}$$

for $a \le t \le b+2$.

6.21 Find the Green's function for the boundary value problem

$$\Delta\left[\left(\frac{1}{6}\right)^{t-1} \Delta y(t-1)\right] + 2\left(\frac{1}{6}\right)^t y(t) = 0,$$

$$y(0) = 0 = y(100).$$

6.22 Use the appropriate Green's function to solve the boundary value problem

$$\Delta^2 y(t-1) = t,$$
$$y(0) = 0 = y(8).$$

6.23 Prove Corollary 6.4.

Chapter 6. The Self-Adjoint Second Order Linear Equation

6.24 Use the appropriate Green's function to solve the boundary value problem
$$\Delta^2 y(t-1) = 10,$$
$$y(0) = 10, \quad y(10) = 70.$$

6.25 Use the appropriate Green's function to solve the boundary value problem
$$\Delta^2 y(t-1) = 8,$$
$$y(0) = 0, \quad y(8) = 4.$$

6.26 Use the appropriate Green's function (see Exercise 6.19) to solve the boundary value problem
$$\Delta^2 y(t-1) = 2,$$
$$y(0) = 0,$$
$$\Delta y(4) = 0.$$

Section 6.4

6.27 Show that $y(t+1) - (t+2)y(t) + 2ty(t-1) = 0$ is disconjugate on $[1, \infty)$.

6.28 Show that $D_k(t_1) \neq 0$ in the proof of Theorem 6.11.

Section 6.5

6.29 Show that the Riccati equation of this chapter $\Delta z(t) + q(t) + \frac{z^2(t)}{z(t)+p(t-1)} = 0$ can be written in the form of the Riccati equation $z(t+1)z(t) + \alpha(t)z(t+1) + \beta(t)z(t) + \gamma(t) = 0$ of Chapter 3.

6.30 For each of the following disconjugate equations on $[0, \infty)$, find a recessive solution $y_1(t)$ and a dominant solution $y_2(t)$ and verify directly that the conclusions of Theorem 6.14 hold for these two solutions.
    (a) $y(t+1) - 10y(t) + 25y(t-1) = 0$.
    (b) $2y(t+1) - 5y(t) + 2y(t-1) = 0$.

Exercises

6.31 Solve the Riccati equations

(a) $\Delta z(t) + 4\left(\frac{1}{9}\right)^t + \frac{z^2(t)}{z(t)+\left(\frac{1}{9}\right)^{t-1}} = 0,$

(b) $\Delta z(t) + \frac{z^2(t)}{z(t)+1} = 0,$

(c) $\Delta z(t) + 3\left(\frac{1}{8}\right)^t + \frac{z^2(t)}{z(t)+\left(\frac{1}{8}\right)^{t-1}} = 0.$

6.32 By setting $w(t) = (r-1)p(t-1)$, t in $[a+1, b+2]$, prove Corollary 6.8 directly from Theorem 6.17.

6.33 Use Theorem 6.17 to show that if $q(t) \leq 0$ on $[a+1, b+1]$, then $Ly(t) = 0$ is disconjugate on $[a, b+2]$.

6.34 Show that $y(t+1) - 3y(t) + (\frac{5}{4} - \sin t)y(t-1) = 0$ is disconjugate on $[1, \infty)$.

6.35 Show by use of Corollary 6.9 that

$$y(t+1) + \alpha y(t) + \beta y(t-1) = 0,$$

where $\beta > 0$, is disconjugate on any interval $[a, b+2]$ if $\alpha < 0$ and $\alpha^2 - 4\beta \geq 0$.

Section 6.6

6.36 Show that the difference equation $y(t+1) + y(t) - 6y(t-1) = 0$ has nontrivial oscillatory and nonoscillatory solutions.

6.37 (a) Show: if $\{t_n\}$ is a sequence of integers that is diverging to infinity and such that

$$q(t_n) > p(t_n) + p(t_n - 1),$$

then $Ly(t) = 0$ is oscillatory on $[a, \infty)$.

(b) Show that the difference equation $\Delta[(\frac{1}{2})^{t-1}\Delta y(t-1)] + 3(\frac{1}{2})^t y(t) = 0$ is oscillatory on $[0, \infty)$ by use of (a).

6.38 Show that the following difference equations are oscillatory:

(a) $\Delta^2 y(t-1) + \frac{1}{(t+1)\ln(t+1)} y(t) = 0$, $(t \geq 2)$,

(b) $\Delta^2 y(t-1) + Ay(t) = 0$, $(t \geq 1)$ $(A > 0)$,

(c) $\Delta^2 y(t-1) + [\sin(t+1) - \sin t]y(t) = 0$, $(t \geq 1)$,

Chapter 7
The Sturm-Liouville Problem

7.1 Introduction

In this chapter our main topic is the Sturm-Liouville difference equation

$$\Delta[p(t-1)\Delta y(t-1)] + [q(t) + \lambda r(t)]y(t) = 0. \qquad (7.1)$$

Here we assume $p(t)$ is defined and positive on the set of integers $[a, b+1] = \{a, a+1, \cdots, b+1\}$, $r(t)$ is defined and positive on $[a+1, b+1]$, $q(t)$ is defined and real valued on $[a+1, b+1]$, and λ is a parameter. At the outset we will consider the general linear homogeneous boundary conditions

$$Py \equiv a_{11}y(a) + a_{12}\Delta y(a) - b_{11}y(b+1) - b_{12}\Delta y(b+1) = 0,$$
$$Qy \equiv a_{21}y(a) + a_{22}\Delta y(a) - b_{21}y(b+1) - b_{22}\Delta y(b+1) = 0,$$

where the a's and b's are real constants. We assume that the boundary conditions $Py = 0$, $Qy = 0$ are not equivalent (that is, the vectors $[a_{11}, a_{12}, b_{11}, b_{12}]$, $[a_{21}, a_{22}, b_{21}, b_{22}]$ are linearly independent).

If $b_{11} = b_{12} = 0 = a_{21} = a_{22}$ we get the *separated boundary conditions*

$$\alpha y(a) + \beta \Delta y(a) = 0, \qquad (7.2)$$
$$\gamma y(b+1) + \delta \Delta y(b+1) = 0, \qquad (7.3)$$

where we assume that

$$\alpha^2 + \beta^2 \neq 0, \quad \gamma^2 + \delta^2 \neq 0. \qquad (7.4)$$

Definition 7.1. The boundary value problem (7.1)–(7.3) where Eq. (7.4) holds is called a Sturm-Liouville problem.

Another important special case of the boundary conditions $Py = 0$, $Qy = 0$ are the *periodic boundary conditions*

$$y(a) = y(b+1), \qquad (7.5)$$
$$\Delta y(a) = \Delta y(b+1). \qquad (7.6)$$

Definition 7.2. The boundary value problem (7.1), (7.5), (7.6), where we assume $p(a) = p(b+1)$, is called a periodic Sturm-Liouville problem.

Let us begin with some general definitions.

Definition 7.3. We say $\lambda = \lambda_0$ is an eigenvalue for the boundary value problem (7.1), $Py = 0$, $Qy = 0$, provided this boundary value problem for $\lambda = \lambda_0$ has a nontrivial solution $y_0(t)$. In such a case we say that $y_0(t)$ is an eigenfunction corresponding to the eigenvalue λ_0 and we say that the pair $(\lambda_0, y_0(t))$ is an eigenpair for the boundary value problem (7.1), $Py = 0$, $Qy = 0$. We say that an eigenvalue λ_0 is simple if there is only one linearly independent eigenfunction corresponding to λ_0. If an eigenvalue is not simple, then we say that it is a multiple eigenvalue.

Note that if $(\lambda_0, y_0(t))$ is an eigenpair for Eq. (7.1), $Py = 0$, $Qy = 0$, then $(\lambda_0, ky_0(t))$ for $k \neq 0$ is also an eigenpair for Eq. (7.1), $Py = 0$, $Qy = 0$. Later we will see that a Sturm-Liouville problem has only simple eigenvalues whereas a periodic Sturm-Liouville problem can have multiple eigenvalues.

Example 7.1. Find eigenpairs for the Sturm-Liouville problem

$$\Delta^2 y(t-1) + \lambda y(t) = 0, \qquad (7.7)$$
$$y(0) = 0, y(4) = 0.$$

The characteristic equation for Eq. (7.7) is

$$m^2 + (\lambda - 2)m + 1 = 0,$$

so

$$m = \frac{(2-\lambda) \pm \sqrt{(\lambda-2)^2 - 4}}{2}.$$

7.1 Introduction

If $|\lambda - 2| \geq 2$ it can be shown that there are no eigenvalues. Assume $|\lambda - 2| < 2$ and set

$$2 - \lambda = 2\cos\theta.$$

Then

$$m = \cos\theta \pm i\sin\theta = e^{\pm i\theta}.$$

Hence a general solution of Eq. (7.7) is

$$y(t) = A\cos\theta t + B\sin\theta t.$$

From the boundary conditions we have

$$y(0) = A = 0,$$
$$y(4) = B\sin 4\theta = 0.$$

Take

$$\theta_n = \frac{n\pi}{4}, \quad (n = 1, 2, 3);$$

then

$$\lambda_n = 2 - 2\cos\frac{n\pi}{4} \quad (n = 1, 2, 3).$$

Hence

$$(2 - \sqrt{2}, \sin\frac{\pi}{4}t), \quad (2, \sin\frac{\pi}{2}t), \quad (2 + \sqrt{2}, \sin\frac{3\pi}{4}t)$$

are eigenpairs for this Sturm-Liouville problem.

Note in the above example every eigenvalue is simple and the number of eigenvalues is the same as the number of integers in $[a+1, b+1] = [1, 3]$. We will see later that the number of eigenvalues for such a Sturm-Liouville problem is $b - a + 1$.

In the remainder of this chapter we will only consider boundary conditions of the form $Py = 0$, $Qy = 0$ where

$$p(b+1)\det A = p(a)\det B \tag{7.8}$$

and

$$A = \begin{bmatrix} a_{11} & a_{12} \\ a_{21} & a_{22} \end{bmatrix}, \quad B = \begin{bmatrix} b_{11} & b_{12} \\ b_{21} & b_{22} \end{bmatrix}.$$

Example 7.2. Show for a Sturm-Liouville problem that Eq. (7.8) is satisfied.

For the Sturm-Liouville problem

$$A = \begin{bmatrix} \alpha & \beta \\ 0 & 0 \end{bmatrix}, \quad B = \begin{bmatrix} 0 & 0 \\ \gamma & \delta \end{bmatrix}.$$

Since $\det A = 0 = \det B$, Eq. (7.8) holds.

Example 7.3. For the periodic Sturm-Liouville problem, Eq. (7.8) is satisfied.

For the periodic Sturm-Liouville problem

$$A = B = \begin{bmatrix} 1 & 0 \\ 0 & 1 \end{bmatrix}.$$

Hence $\det A = \det B = 1$. Since $p(t)$ is periodic with period $b+1-a$, Eq. (7.8) holds.

7.2 Finite Fourier Analysis

We now develop the basic results that are needed to define a finite Fourier series in terms of eigenfunctions for a Sturm-Liouville problem.

Definition 7.4. Let $y(t)$, $z(t)$ be complex valued functions defined on $[a+1, b+1]$; then we define the inner product $< \cdot, \cdot >$ by

$$< y, z > = \sum_{t=a+1}^{b+1} y(t)\overline{z(t)}.$$

7.2 Finite Fourier Analysis

It is easy to verify that this inner product satisfies the following:

a) $<y+z, w> = <y, w> + <z, w>$,
b) $<\alpha y, z> = \alpha <y, z>$,
c) $<y, z> = \overline{<z, y>}$,
d) $<y, y>> 0$ when y is not the trivial function on $[a+1, b+1]$.

Similarly if $r(t) > 0$ on $[a+1, b+1]$, then we can define an inner product with respect to (a weight function) $r(t)$ by

$$<y, z>_r = \sum_{t=a+1}^{b+1} r(t)y(t)\overline{z(t)}.$$

It can be shown that $<\cdot, \cdot>_r$ satisfies $(a) - (d)$ and, in addition, $<y, z>_r = <\sqrt{r}y, \sqrt{r}z>$.

Let

$$\mathcal{D} = \{\text{complex valued functions on } [a, b+2] : Py = 0 = Qy\}.$$

Definition 7.5. If $<Ly, z> = <y, Lz>$ for all y in $\mathcal{D}$, then we say that the boundary value problem (7.3), $Py = 0 = Qy$ is self-adjoint. (Here L is as in Chapter 6, namely $Ly(t) = \Delta[p(t-1)\Delta y(t-1)] + q(t)y(t)$.)

Note that Eq. (7.1) can be written in the form

$$Ly(t) + \lambda r(t)y(t) = 0. \tag{7.9}$$

Theorem 7.1. If Eq. (7.8) holds, then the boundary value problem (7.3), $Py = 0 = Qy$ is self-adjoint.

Proof. We will only prove this theorem in the special case that our boundary value problem is a Sturm-Liouville problem (see Exercise 7.4). Let y and z belong to $\mathcal{D}$. By the complex form of the Lagrange identity (Exercise 6.5)

$$<Ly, z> = <y, Lz> + \left\{p(t)w[\overline{z(t)}, y(t)]\right\}_a^{b+1}.$$

It suffices to show that y, z in $\mathcal{D}$ implies

$$\{p(t)w[\overline{z(t)}, y(t)]\}_a^{b+1} = 0.$$

We first show that $\{p(t)w[\overline{z(t)}, y(t)]\}_{t=a} = 0$.

Since

$$\alpha y(a) + \beta \Delta y(a) = 0,$$
$$\alpha \overline{z(a)} + \beta \overline{\Delta z(a)} = 0,$$

where α and β are not both zero because of Eq. (7.4),

$$\begin{vmatrix} y(a) & \Delta y(a) \\ \overline{z(a)} & \overline{\Delta z(a)} \end{vmatrix} = 0.$$

It follows that $\{p(t)w[\overline{z(t)}, y(t)]\}_{t=a} = 0$. Similarly using the boundary condition (7.3) we have that $\{p(t)w[\overline{z(t)}y(t)]\}_{t=b+1} = 0$, and the proof for the Sturm-Liouville problem case is complete. ∎

Theorem 7.2. If the boundary value problem (7.1), $Py = 0$, $Qy = 0$, is self-adjoint, then all eigenvalues are real. If λ_n, λ_m are distinct eigenvalues, then the corresponding eigenfunctions $y_n(t)$, $y_m(t)$ are orthogonal with respect to the weight function $r(t)$ on $[a+1, b+1]$, that is $< y_n, y_m >_r = 0$. For the Sturm-Liouville problem eigenvalues are simple.

Proof. Let $(\lambda_n, y_n(t))$, $(\lambda_m, y_m(t))$ be eigenpairs for the boundary value problem (7.1), $Py = 0$, $Qy = 0$. Since we have a self-adjoint boundary value problem and y_n, y_m are in $\mathcal{D}$,

$$< Ly_n, y_m > = < y_n, Ly_m >.$$

Using Eq. (7.9) with $\lambda = \lambda_n$, $\lambda = \lambda_m$, we have

$$< -\lambda_n r y_n, y_m > = < y_n, -\lambda_m r y_m >.$$

7.2 Finite Fourier Analysis

It follows that

$$(\lambda_n - \overline{\lambda}_m) < y_n, y_m >_r = 0.$$

If $m = n$, we get $\lambda_n = \overline{\lambda}_n$, so eigenvalues of self-adjoint boundary value problems are real. If $\lambda_n \neq \lambda_m$, then we obtain the orthogonality condition

$$< y_n, y_m >_r = 0.$$

Finally assume λ_0 is an eigenvalue for the Sturm-Liouville problem (7.1)–(7.3). Assume $y_0(t)$, $z_0(t)$ are eigenfunctions corresponding to the eigenvalue λ_0. Then

$$\alpha y_0(a) + \beta \Delta y_0(a) = 0,$$
$$\alpha z_0(a) + \beta \Delta z_0(a) = 0,$$

where by Eq. (7.4) α and β are not both zero. Hence

$$\begin{vmatrix} y_0(a) & \Delta y_0(a) \\ z_0(a) & \Delta z_0(a) \end{vmatrix} = 0$$

or

$$w\,[y_0(t), z_0(t)]\}_{t=a} = 0.$$

Since $y_0(t)$ and $z_0(t)$ are solutions of the same equation they must be linearly dependent. Hence all eigenvalues of a Sturm-Liouville problem are simple. ∎

Example 7.4. Show directly that eigenfunctions corresponding to distinct eigenvalues of the Sturm-Liouville problem

$$\Delta^2 y(t-1) + \lambda y(t) = 0,$$
$$y(0) = 0, \quad y(4) = 0$$

satisfy the orthogonality condition guaranteed by Theorem 7.2.

From Example 7.1 three linearly independent eigenfunctions are

$y_n(t) = \sin \frac{n\pi}{4} t$, $1 \leq n \leq 3$. Here $r(t) = 1$, so

$$< y_1, y_2 >_r = \sin \frac{\pi}{4} \sin \frac{\pi}{2} + \sin \frac{\pi}{2} \sin \pi + \sin \frac{3\pi}{4} \sin \frac{3\pi}{2}$$

$$= \frac{\sqrt{2}}{2} - \frac{\sqrt{2}}{2} = 0.$$

Similarly

$$< y_1, y_3 >_r = 0, \quad < y_2, y_3 >_r = 0.$$

Since eigenvalues of a self-adjoint boundary value problem (7.1), $Py = 0$, $Qy = 0$ are real it can be shown (see Exercise 7.7) that corresponding to each eigenvalue we can always pick a real eigenfunction. We will use this result in the proof of the next theorem.

Theorem 7.3. If λ is an eigenvalue of the Sturm-Liouville problem (7.1)–(7.3), $q(t) \leq 0$ on $[a+1, b+1]$, $\alpha\beta \leq 0$, and $\gamma\delta \geq 0$, then $\lambda \geq 0$. If, in addition, $q(t) > 0$ at two consecutive integers in $[a+1, b+1]$, then $\lambda > 0$.

Proof. Let λ be an eigenvalue with corresponding real eigenfunction $y(t)$ for the Sturm-Liouville problem (7.1)–(7.3).

Consider for $a + 1 \leq t \leq b + 1$,

$$y(t)[Ly(t) + \lambda r(t)y(t)] = 0,$$
$$y(t)\Delta[p(t-1)\Delta y(t-1)] + q(t)y^2(t) + \lambda r(t)y^2(t) = 0,$$
$$\Delta\{y(t-1)[p(t-1)\Delta y(t-1)]\} - p(t-1)[\Delta y(t-1)]^2$$
$$+ q(t)y^2(t) + \lambda r(t)y^2(t) = 0.$$

Summing both sides from $a + 1$ to $b + 1$ we have

$$y(b+1)p(b+1)\Delta y(b+1) - y(a)p(a)\Delta y(a)$$
$$- \sum_{t=a+1}^{b+1} p(t-1)[\Delta y(t-1)]^2 + \sum_{t=a+1}^{b+1} q(t)y^2(t) + \lambda \sum_{t=a+1}^{b+1} r(z)y^2(t) = 0.$$

7.2 Finite Fourier Analysis

It follows that

$$\lambda <y,y>_r \geq y(a)p(a)\Delta y(a) - y(b+1)p(b+1)\Delta y(b+1). \quad (7.10)$$

Since y satisfies (7.2),

$$\alpha y(a) + \beta \Delta y(a) = 0.$$

We claim $\alpha\beta \leq 0$ implies $y(a)p(a)\Delta y(a) \geq 0$. If $\alpha\beta = 0$ then $y(a)p(a)\Delta y(a) = 0$. Now assume $\alpha\beta < 0$, then

$$y(a)p(a)\Delta y(a) = -\frac{\beta p(a)[\Delta y(a)]^2}{\alpha} \geq 0.$$

Similarly we can use $\gamma\delta \geq 0$ and Eq. (7.3) to show that

$$-y(b+1)p(b+1)\Delta y(b+1) \geq 0.$$

From Eq. (7.10)

$$\lambda <y,y>_r \geq 0, \quad (7.11)$$

so $\lambda \geq 0$. If in addition $q(t) > 0$ at two consecutive integers in $[a+1,b+1]$ then the inequality in Eq. (7.10) and hence in Eq. (7.11) is strict. In this case we have $\lambda > 0$. ∎

Example 7.5. Show that all eigenvalues of the Sturm-Liouville problem

$$\Delta^2 y(t-1) + \left[-\frac{(t-50)^2}{t^2+1} + \lambda \frac{1}{t^2+1}\right] y(t) = 0,$$

$$y(0) = 0,$$
$$2y(100) + 3\Delta y(100) = 0$$

are positive.

Here $q(t) = -\frac{(t-50)^2}{t^2+1} \leq 0$ on $[1,100]$ and is strictly positive at two consecutive integers in $[1,100]$. Further $\alpha\beta = 0$ and $\gamma\delta = 6$. Hence by Theorem 7.3 all eigenvalues are positive.

We showed that eigenvalues for a Sturm-Liouville problem are simple. The following example shows that there are self-adjoint boundary value problems that have multiple eigenvalues.

Example 7.6. Show that $\lambda = 2$ is a multiple eigenvalue for the periodic Sturm-Liouville problem

$$\Delta^2 y(t-1) + \lambda y(t) = 0, \qquad (7.12)$$
$$y(0) = y(4),$$
$$\Delta y(0) = \Delta y(4).$$

For $\lambda = 2$ a general solution of Eq. (7.12) is

$$y(t) = A \cos \frac{\pi}{2} t + B \sin \frac{\pi}{2} t.$$

Note that all these solutions satisfy the boundary conditions. Hence there are two linearly independent eigenfunctions corresponding to $\lambda = 2$ and so $\lambda = 2$ is a multiple eigenvalue.

What follows is another method for finding the eigenvalues for the Sturm-Liouville problem (7.1)–(7.3) in the case that $\alpha \neq \beta$, $\gamma \neq \delta$.

First we write Eq. (7.1) in the form

$$p(t)y(t+1) + c(t)y(t) + p(t-1)y(t-1) = -\lambda r(t)y(t), \qquad (7.13)$$

where
$$c(t) = q(t) - p(t) - p(t-1).$$

Letting $t = a+1, a+2, \cdots, b+1$ in Eq. (7.13) and using the boundary conditions (7.2), (7.3), we obtain $N \equiv b - a + 1$ equations

$$\tilde{c}(a+1)y(a+1) + p(a+1)y(a+2) = -\lambda r(a+1)y(a+1),$$
$$p(a+1)y(a+1) + c(a+2)y(a+2)$$
$$\qquad + p(a+1)y(a+3) = -\lambda r(a+1)y(a+2),$$
$$\vdots$$
$$p(b)y(b) + \tilde{c}(b+1)y(b+1) = -\lambda r(b+1)y(b+1),$$

7.2 Finite Fourier Analysis

where

$$\tilde{c}(a+1) = c(a+1) + \frac{\beta p(a)y(a)}{\beta - \alpha}, \quad \tilde{c}(b+1) = c(b+1) + \frac{\delta p(b+1)y(b+2)}{\delta - \gamma}. \tag{7.14}$$

Note that $\tilde{c}(a+1) = c(a+1)$ if $\beta = 0$ and $\tilde{c}(b+1) = c(b+1)$ if $\delta = 0$.

We write this as the vector matrix equation

$$\mathcal{S}u = -\lambda R u, \tag{7.15}$$

where u is the column N vector

$$u = [y(a+1), y(a+2), \cdots, y(b+1)]^T,$$

$\mathcal{S}$ is the N by N tridiagonal matrix

$$\mathcal{S} = \begin{bmatrix} \tilde{c}(a+1) & p(a+1) & 0 & \cdots & 0 \\ p(a+1) & c(a+2) & p(a+2) & \cdots & 0 \\ 0 & \ddots & \ddots & \ddots & \vdots \\ \vdots & \cdots & p(b-1) & c(b) & p(b) \\ 0 & \cdots & 0 & p(b) & \tilde{c}(b+1) \end{bmatrix}$$

and R is the N by N diagonal matrix

$$R = \text{diag}\{r(a+1), r(a+2), \cdots, r(b+1)\}.$$

Because of the equivalence of the problem (7.13), (7.14) with the problem (7.15) it follows from matrix theory that the Sturm-Liouville problem (7.13), (7.14) has N linearly independent eigenfunctions with all eigenvalues real.

Let $\mu = -\lambda$, then we can write Eq. (7.15) in the form

$$R^{-1}\mathcal{S}u = \mu u.$$

It follows that (λ, y) is an eigenpair for (7.13), (7.14) if and only if $(-\lambda, u)$,

$u = [y(a+1), y(a+2), \cdots, y(b+1)]^T$ is an eigenpair for

$$R^{-1}\mathcal{S} = \begin{bmatrix} \frac{\tilde{c}(a+1)}{r(a+1)} & \frac{p(a+1)}{r(a+1)} & 0 & \cdots & 0 \\ \frac{p(a+1)}{r(a+2)} & \frac{c(a+2)}{r(a+2)} & \frac{p(a+2)}{r(a+2)} & \cdots & 0 \\ 0 & \ddots & \ddots & \ddots & \vdots \\ \vdots & \cdots & \cdots & \frac{c(b)}{r(b)} & \frac{p(b)}{r(b)} \\ 0 & \cdots & 0 & \frac{p(b)}{r(b+1)} & \frac{\tilde{c}(b+1)}{r(b+1)} \end{bmatrix}.$$

Example 7.7. Use $R^{-1}\mathcal{S}$ to find eigenpairs for the Sturm-Liouville problem

$$\Delta^2 y(t-1) + \lambda y(t) = 0,$$
$$y(0) = 0 = y(4).$$

Here $p(t-1) = 1$, $r(t) = 1$, $c(t) = -2$, $N = 3$, so

$$R^{-1}\mathcal{S} = \begin{bmatrix} -2 & 1 & 0 \\ 1 & -2 & 1 \\ 0 & 1 & -2 \end{bmatrix}.$$

The characteristic equation of $R^{-1}\mathcal{S}$ is

$$(\mu + 2)(\mu^2 + 4\mu + 2) = 0.$$

Hence $\mu = -2$, $-2 \pm \sqrt{-2}$ and consequently

$$\lambda = 2, \quad 2 - \sqrt{2}, \quad 2 + \sqrt{2}.$$

The corresponding eigenvectors are

$$[1, 0, -1]^T, \quad \left[\frac{\sqrt{2}}{2}, 1, \frac{\sqrt{2}}{2}\right]^T, \quad \left[\frac{\sqrt{2}}{2}, -1, \frac{\sqrt{2}}{2}\right]^T.$$

Compare these results to Example 7.1.

7.2 Finite Fourier Analysis

Let $N = b - a + 1$ be as in the previous section and let R^N be the set of all functions $y(t)$ defined on $[a+1, b+1]$. We call $y(a+1)$, $y(a+2)$, $\cdots$, $y(b+1)$ the components of $y(t)$. Assume $y_1(t), \cdots, y_N(t)$ are N linearly independent eigenfunctions of the Sturm-Liouville problem (7.1)–(7.3), $\alpha \neq \beta$. Given an arbitrary real valued function $\omega(t)$ defined in $[a+1, b+1]$ we would like to find constants $c_1, \cdots, c_N$ such that

$$\omega(t) = \sum_{k=1}^{N} c_k y_k(t) \tag{7.16}$$

for t in $[a+1, b+1]$. Note that

$$<\omega, y_k>_r = \left\langle \sum_{j=1}^{N} c_j y_j, y_k \right\rangle_r$$

$$= \sum_{j=1}^{N} c_j <y_j, y_k>_r$$

$$= c_k <y_k, y_k>_r$$

because of the orthogonality.

Hence for $1 \leq k \leq N$,

$$c_k = \frac{<\omega, y_k>_r}{<y_k, y_k>_r}. \tag{7.17}$$

The series (7.16), where c_k, $1 \leq k \leq N$, is given by Eq. (7.17), is called the Fourier series of $\omega(t)$; the coefficients c_k are called the Fourier coefficients of $\omega(t)$.

Example 7.8. Find the Fourier series of $\omega(t) = t^2 - 4t + 3$, $1 \leq t \leq 3$, with respect to the eigenfunctions of the Sturm-Liouville problem

$$\Delta^2 y(t-1) + \lambda y(t) = 0,$$
$$y(0) = 0 = y(4).$$

By Example 7.1, $y_k(t) = \sin \frac{k\pi}{4}$, $1 \leq k \leq 3$, are $N = 3$ linearly independent eigenfunctions of this Sturm-Liouville problem. From Eq. (7.17),

$$\begin{aligned} c_k &= \frac{<\omega, y_k>_1}{<y_k, y_k>_1} \\ &= \frac{\omega(1)y_k(1) + \omega(2)y_k(2) + \omega(3)y_k(3)}{y_k^2(1) + y_k^2(2) + y_k^2(3)} \\ &= -\frac{y_k(2)}{y_k^2(1) + y_k^2(2) + y_k^2(3)}. \end{aligned}$$

It follows that

$$c_1 = -\frac{1}{2}, \quad c_2 = 0, \quad c_3 = \frac{1}{2}.$$

Hence

$$\omega(t) = -\frac{1}{2}\sin\frac{\pi}{4} + \frac{1}{2}\sin\frac{3\pi t}{4}$$

for t in $[1, 3]$.

7.3 A Nonhomogeneous Problem

In this section we will be concerned with solving the boundary value problem

$$\begin{aligned} Ly(t) + \lambda_0 r(t)y(t) &= f(t), \quad (7.18) \\ \alpha y(a) + \beta \Delta y(a) &= 0, \\ \gamma y(b+1) + \delta \Delta y(b+1) &= 0, \end{aligned}$$

where Eq. (7.4) holds and λ_0 is an eigenvalue for the Sturm-Liouville problem (7.1)–(7.3). The main result of this section contains a necessary and sufficient condition for this boundary value problem to have a solution.

Let $y_1(t, \lambda)$, $y_2(t, \lambda)$ be the solutions of Eq. (7.1) satisfying the initial conditions

$$\begin{aligned} y_1(a, \lambda) &= 1, \quad \Delta y_1(a, \lambda) = 0 \\ y_2(a, \lambda) &= 0, \quad \Delta y_2(a, \lambda) = 1. \end{aligned}$$

7.3 A Nonhomogeneous Problem

Note that $y_1(t,\lambda)$, $y_2(t,\lambda)$ are linearly independent solutions of Eq. (7.1) and

$$y(t,\lambda) \equiv \beta y_1(t,\lambda) - \alpha y_2(t,\lambda)$$

is a nontrivial solution of Eq. (7.1) satisfying the boundary condition (7.2). Let Q be an operator corresponding to the boundary condition (7.3), that is,

$$Qy = \gamma y(b+1) + \delta \Delta y(b+1).$$

Set

$$\begin{aligned} g(\lambda) &= Qy(t,\lambda) \\ &= \gamma y(b+1,\lambda) + \delta \Delta y(b+1,\lambda) \end{aligned}$$

for $-\infty < \lambda < \infty$. Since eigenvalues of the Sturm-Liouville problem (7.1)–(7.3) are simple, $g(\lambda_0) = 0$ if and only if λ_0 is an eigenvalue of (7.1)–(7.3). Note that if λ_0 is an eigenvalue, then

$$y(t,\lambda_0) = \beta y_1(t,\lambda_0) - \alpha y_2(t,\lambda_0) \qquad (7.19)$$

is a corresponding eigenfunction.

The following lemma will be needed in the proof of the main existence theorem for Eqs. (7.18), (7.2), and (7.3).

Lemma 7.1. Let $y_1(t)$, $y_2(t)$ be solutions of $Ly = 0$ satisfying $y_1(a) = 1$, $\Delta y_1(a) = 0$, $y_2(a) = 0$, $\Delta y_2(a) = 1$; then the solution of the initial value problem $Ly = f$, $y(a) = 0$, $\Delta y(a) = -\frac{1}{p(a)} < f, y_1 >$ is given by

$$y(t) = -y_1(t) \sum_{s=a+1}^{t} \frac{y_2(s)f(s)}{p(a)} - y_2(t) \sum_{s=t+1}^{b+1} \frac{y_1(s)f(s)}{p(a)}$$

for t in $[a, b+2]$.

Proof. By our convention on sums and since $y_2(a) = 0$ we have that $y(a) = 0$. Note that $\Delta y(a) = y(a+1) = -\frac{1}{p(a)} <f, y_1>$.
Now assume $a + 1 \le t \le b+1$ and consider

$$y(t-1) = -y_1(t-1) \sum_{s=a+1}^{t-1} \frac{y_2(s)f(s)}{p(a)} - y_2(t-1) \sum_{s=t}^{b+1} \frac{y_1(s)f(s)}{p(a)}.$$

Then

$$\Delta y(t-1) = -y_1(t) \frac{y_2(t)f(t)}{p(a)} - \Delta y_1(t-1) \sum_{s=a+1}^{t-1} \frac{y_2(s)f(s)}{p(a)}$$
$$+ y_2(t) \frac{y_1(t)f(t)}{p(a)} - \Delta y_2(t-1) \sum_{s=t}^{b+1} \frac{y_1(s)f(s)}{p(a)}.$$

Hence

$$p(t-1)\Delta y(t-1) = -p(t-1)\Delta y_1(t-1) \sum_{s=a+1}^{t-1} \frac{y_2(s)f(s)}{p(a)}$$
$$- p(t-1)\Delta y_2(t-1) \sum_{s=t}^{b+1} \frac{y_1(s)f(s)}{p(a)}.$$

Therefore

$$\Delta[p(t-1)\Delta y(t-1)] = -p(t)\Delta y_1(t) \frac{y_2(t)f(t)}{p(a)} + p(t)\Delta y_2(t) \frac{y_1(t)f(t)}{p(a)}$$
$$- \Delta[p(t-1)\Delta y_1(t-1)] \sum_{s=a+1}^{t-1} \frac{y_2(s)f(s)}{p(s)}$$
$$- \Delta[p(t-1)\Delta y_2(t-1)] \sum_{s=t}^{b+1} \frac{y_1(s)f(s)}{p(a)}$$
$$= \frac{p(t)w[y_1(t), y_2(t)]}{p(a)} f(t) - q(t)y(t).$$

7.3 A Nonhomogeneous Problem

By Liouville's formula $p(t)w[y_1(t), y_2(t)] = C$, where C is a constant. Letting $t = a$ we get that $C = p(a)$. It follows that

$$Ly(t) = f(t), \quad (a+1 \leq t \leq b+1).$$ ■

Theorem 7.4. If $(\lambda_0, y_0(t))$ is an eigenpair for the Sturm-Liouville problem (7.1)–(7.3), then the boundary value problem

$$\begin{aligned} Ly(t) + \lambda_0 r(t) y(t) &= f(t), \\ \alpha y(a) + \beta \Delta y(a) &= 0, \\ \gamma y(b+1) + \delta \Delta y(b+1) &= 0, \end{aligned}$$

where Eq. (7.4) holds, has a solution if and only if $<f, y_0> = 0$.

Proof. Let $y_1(t) = y_1(t, \lambda_0)$, $y_2(t) = y_2(t, \lambda_0)$, then by Lemma 7.1 with $Ly(t) = 0$ replaced by $Ly(t) + \lambda_0 r(t) y(t) = 0$ a general solution of Eq. (7.18) is

$$y(t) = c_1 y_1(t) + c_2 y_2(t) - y_1(t) \sum_{s=a+1}^{t} \frac{y_2(s) f(s)}{p(a)} - y_2(t) \sum_{s=t+1}^{b+1} \frac{y_1(s) f(s)}{p(a)}.$$

Then $y(t)$ satisfies Eqs. (7.2), (7.3) if and only if c_1, c_2 satisfy the system of equations

$$\begin{aligned} \alpha c_1 + \beta c_2 &= \frac{\beta}{p(a)} <f, y_1>, \\ (Qy_1) c_1 + (Qy_2) c_2 &= \frac{1}{p(a)} Qy_1 <f, y_2>. \end{aligned}$$

Since λ_0 is an eigenvalue of the Sturm-Liouville problem (7.1)–(7.3),

$$\begin{vmatrix} \alpha & \beta \\ Qy_1 & Qy_2 \end{vmatrix} = 0. \tag{7.20}$$

Hence the above system has a solution c_1, c_2 if and only if the vectors

$$A = \left[\alpha, \beta, \frac{\beta}{p(a)} <f,y_1>\right],$$

$$B = \left[\mathcal{Q}y_1, \mathcal{Q}y_2, \frac{1}{p(a)}\mathcal{Q}y_1 <f,y_2>\right]$$

are linearly independent.

First consider the case $\alpha = 0$. Then $\beta \neq 0$ and by Eq. (7.20) $\mathcal{Q}y_1 = 0$ and so $B = (0, \mathcal{Q}y_2, 0)$. In this case by Eq. (7.19), βy_1 is an eigenfunction corresponding to λ_0. If $\mathcal{Q}y_2 = 0$, then $\mathcal{Q}y_1 = \mathcal{Q}y_2 = 0$ implies $\omega[y_1(t), y_2(t)] = 0$, which contradicts the fact that y_1 and y_2 are linearly independent. Hence $\mathcal{Q}y_2 \neq 0$. But then A and B are linearly dependent if and only if $<f, \beta y_1> = 0$ (if and only if $<f, y_0> = 0$).

Next consider the case $\alpha \neq 0$. By Eq. (7.20)

$$\alpha \mathcal{Q}y_2 - \beta \mathcal{Q}y_1 = 0,$$

or

$$\mathcal{Q}y_2 = \frac{\beta}{\alpha} \mathcal{Q}y_1. \tag{7.21}$$

Hence, in this case,

$$B = \frac{1}{\alpha}\mathcal{Q}y_1 \left[\alpha, \beta, \frac{\alpha}{p(a)} <f,y_2>\right].$$

Note that $\mathcal{Q}y_1 \neq 0$, for if $\mathcal{Q}y_1 = 0$, then by Eq. (7.21) $\mathcal{Q}y_2 = 0$, and earlier we saw this leads to a contradiction. Thus A and B are linearly dependent if and only if

$$\alpha <f, y_2> = \beta <f, y_1>.$$

But this is true if and only if

$$<f, \beta y_1 - \alpha y_2> = 0,$$

7.3 A Nonhomogeneous Problem

if and only if

$$< f, y_0 > = 0. \qquad \blacksquare$$

Example 7.9. Find necessary and sufficient conditions on f for the boundary value problem

$$\Delta^2 y(t-1) + 2y(t) = f(t),$$
$$y(0) = 0, \ y(4) = 0,$$

to have a solution.

By Example 7.1, $(\lambda_0, y_0(t)) = (2, \sin \frac{\pi}{2} t)$ is an eigenpair for the corresponding homogeneous problem

$$\Delta^2 y(t-1) + 2y(t) = 0,$$
$$y(t) = 0, \ y(4) = 0.$$

By Theorem 7.4 the nonhomogeneous boundary value problem has a solution if and only if

$$< f(t), \sin \frac{\pi}{2} t > = 0.$$

It follows that the nonhomogeneous boundary value problem has a solution if and only if

$$f(1) = f(3).$$

Theorem 7.5. The eigenvalues of the Sturm-Liouville problem (7.1)–(7.3) are precisely the zeros of $g(\lambda) = \gamma y(b+1, \lambda) + \delta \Delta y(b+1, \lambda)$, $-\infty < \lambda < \infty$. The zeros of $g(\lambda)$ are simple ($g(\lambda_0) = 0$ implies $g'(\lambda_0) \neq 0$).

Proof. The first statement was proved earlier. We now prove that $g(\lambda)$ has only simple zeros. To this end, set

$$z(t, \lambda) = \frac{d}{d\lambda} y(t, \lambda).$$

Since

$$\alpha y(a,\lambda) + \beta \Delta y(a,\lambda) = 0,$$

we obtain by differentiating with respect to λ that $z(t) = z(t,\lambda)$ satisfies Eq. (7.2). Since

$$Ly(t,\lambda) + \lambda r(t)y(t,\lambda) = 0,$$

we have in a similar way

$$Lz(t) + \lambda r(t)z(t) = -r(t)y(t,\lambda).$$

Assume $g(\lambda_0) = 0$. Then λ_0 is an eigenvalue of (7.1)–(7.3). Assume $g'(\lambda_0) = 0$; then

$$\gamma z(b+1,\lambda_0) + \delta \Delta z(b+1,\lambda_0) = 0.$$

We have shown that $z(t,\lambda_0)$ is a solution of the boundary value problem

$$Lz(t) + \lambda_0 r(t)z(t) = -r(t)y(t,\lambda_0),$$
$$\alpha z(a) + \beta \Delta z(a) = 0,$$
$$\gamma z(b+1) + \delta \Delta z(b+1) = 0.$$

It follows from Theorem 7.4 that

$$< r(t)y(t,\lambda_0), y(t,\lambda_0) > = 0.$$

But then $< y(t,\lambda_0), y(t,\lambda_0) >_r = 0$ so $y(t,\lambda_0) \equiv 0$ on $[a+1, b+1]$, which is a contradiction. ∎

Theorem 7.6. The Sturm-Liouville problem (7.1), $y(a) = 0 = y(b+2)$ has $N = b - a + 1$ eigenvalues $\lambda_1 < \lambda_2 < \cdots < \lambda_N$. If $(\lambda_i, y_i(t))$ are eigenpairs, $1 \leq i \leq N$, then $y_i(t)$ has $i - 1$ generalized zeros in $[a+1, b+1]$.

7.3 A Nonhomogeneous Problem

Proof. We will give just an outline of this proof. For more details the reader can see Fort [77]. Let $y(t, \lambda)$ be the solution of Eq. (7.1) satisfying the initial conditions $y(a, \lambda) = 0$, $y(a + 1, \lambda) = 1$. The eigenvalues of (7.1), $y(a) = 0 = y(b+2)$, are just the zeros of $y(b+2, \lambda)$, which are simple by the last theorem. Take λ' to be a sufficiently large negative number so that

$$\lambda' r(t) + q(t) \leq 0$$

on $[a+1, b+1]$. Then by Corollary 6.6, Eq. (7.1) is disconjugate on $[a, b+2]$. It follows that $y(t, \lambda) > 0$ in $[a+1, b+2]$. On the other hand we can see that by picking λ'' to be a sufficiently large positive number, then $y(t, \lambda'')$ satisfies

$$y(t, \lambda'') y(t+1, \lambda'') < 0$$

for $a+1 \leq t \leq b+1$. Note that $y(t, \lambda')$ has no generalized zeros in $[a+1, b+2]$ and $y(t, \lambda'')$ has $N+1$ generalized zeros in $[a+1, b+2]$. The eigenvalues of (7.1), $y(a) = 0 = y(b+2)$, are obtained by letting λ vary from λ' to λ''. The values $y(t, \lambda)$ for t in $[a+1, b+2]$ depend continuously on λ. Let λ_1 be the first value of λ in $[\lambda', \lambda'']$ such that $y(b+2, \lambda_1) = 0$. Because of the important Lemma 6.1, $y(t, \lambda_1) > 0$ on $[a+1, b+1]$. Hence an eigenfunction corresponding to λ_1 has zero generalized zeros in $[a+1, b+1]$. Increasing λ beyond λ_1 we get the next zero λ_2 of $y(b+2, \lambda)$. It can be shown (Fort[77]) that $y(t, \lambda_2)$ has exactly one generalized zero in $[a+1, b+1]$. Proceeding in this fashion we get that if λ_i is the ith zero of $y(b+2, \lambda)$ in $[\lambda', \lambda'']$, then $y(t, \lambda)$ has $i-1$ generalized zeros in $[a+1, b+1]$, $1 \leq i \leq N$. ∎

Next we will use Theorem 7.6 to get what is called Rayleigh's inequality. We will then give an example to show how Rayleigh's inequality can be used to give upper bounds for the smallest eigenvalue of (7.1), $y(a) = 0 = y(b+2)$.

Theorem 7.7. (Rayleigh's inequality) Let λ_1 be the smallest eigenvalue of (7.1), $y(a) = y(b+2) = 0$. Then

$$\lambda_1 \leq \frac{\sum_{t=a+1}^{b+2} p(t-1)[\Delta u(t-1)]^2 - \sum_{t=a+1}^{b+1} q(t) u^2(t)}{\sum_{t=a+1}^{b+1} r(t) u^2(t)},$$

where $u(t)$ is any nontrivial real-valued function defined on $[a, b+2]$ with

Chapter 7. The Sturm-Liouville Problem

$u(a) = u(b+2) = 0$. Furthermore, equality holds if and only if $u(t)$ is an eigenfunction corresponding to λ_1.

Proof. Let $\lambda = \lambda_1$ and assume $y(t)$ is an eigenfunction corresponding to λ. Then by Theorem 7.6 $y(t)$ has no generalized zeros in $[a+1, b+1]$ and hence is of constant sign there. Assume $u(t)$ is a real valued function defined on $[a, b+2]$ with $u(a) = u(b+2) = 0$ and consider for t in $[a+2, b+1]$

$$\Delta\left[\frac{u^2(t-1)}{y(t-1)} p(t-1)\Delta y(t-1)\right]$$

$$= \frac{u^2(t)}{y(t)}\Delta[p(t-1)\Delta y(t-1)] + p(t-1)\Delta y(t-1)\Delta\left\{\frac{u^2(t-1)}{y(t-1)}\right\}$$

$$= \frac{u^2(t)}{y(t)}[-\lambda r(t)y(t) - q(t)y(t)]$$

$$+ p(t-1)[y(t) - y(t-1)]\left[\frac{u^2(t)}{y(t)} - \frac{u^2(t-1)}{y(t-1)}\right]$$

$$= \lambda r(t)u^2(t) - q(t)u^2(t) + p(t-1)u^2(t)$$

$$+ p(t-1)u^2(t-1) - p(t-1)\frac{y(t)u^2(t-1)}{y(t-1)} - p(t-1)\frac{y(t-1)u^2(t)}{y(t)}$$

$$= -\lambda r(t)u^2(t) - q(t)u^2(t)$$

$$+ p(t-1)[u(t) - u(t-1)]^2 - p(t-1)y(t)y(t-1)\left(\frac{u(t)}{y(t)} - \frac{u(t-1)}{y(t-1)}\right)^2.$$

Summing both sides from $a+1$ to $b+1$ (we will only do the case where $a + 2 \leq b + 1$) we obtain

$$\frac{u^2(t-1)}{y(t-1)} p(t-1)\Delta y(t-1)\,]_{a+2}^{b+2}$$

$$= -\lambda \sum_{t=a+2}^{b+1} r(t)u^2(t) - \sum_{t=a+2}^{b+1} q(t)u^2(t)$$

$$+ \sum_{t=a+2}^{b+1} p(t-1)[\Delta u(t-1)]^2$$

$$- \sum_{t=a+2}^{b+1} p(t-1)y(t)y(t-1)\left[\Delta\left(\frac{u(t-1)}{y(t-1)}\right)\right]^2.$$

7.3 A Nonhomogeneous Problem

Since $y(b+2) = 0 = u(b+2)$, we have

$$\frac{u^2(a+1)p(a+1)\Delta y(a+1)}{y(a+1)}$$
$$= \lambda \sum_{t=a+2}^{b+1} r(t)u^2(t) + \sum_{t=a+2}^{b+1} q(t)u^2(t)$$
$$- \sum_{t=a+2}^{b+1} p(t-1)[\Delta u(t-1)]^2$$
$$+ \sum_{t=a+2}^{b+1} p(t-1)y(t)y(t-1)\left[\Delta\left(\frac{u(t-1)}{y(t-1)}\right)\right].$$

But letting $t = a+1$ in $Ly(t) + \lambda r(t)y(t) = 0$, we obtain

$$p(a+1)\Delta y(a+1) = [p(a) - q(a+1) - \lambda r(a+1)]y(a+1).$$

This leads to

$$\lambda \sum_{t=a+1}^{b+1} r(t)u^2(t) = \sum_{t=a+1}^{b+2} p(t-1)[\Delta u(t-1)]^2 - \sum_{t=a+1}^{b+1} q(t)u^2(t)$$
$$- \sum_{t=a+2}^{b+1} p(t-1)y(t)y(t-1)\left[\Delta\left(\frac{u(t-1)}{y(t-1)}\right)\right]^2.$$

Since $y(t)$ is of one sign on $[a+1, b+1]$,

$$\lambda \sum_{t=a+1}^{b+1} r(t)u^2(t) \leq \sum_{t=a+1}^{b+2} p(t-1)[\Delta u(t-1)]^2 - \sum_{t=a+1}^{b+1} q(t)u^2(t),$$

where equality holds if and only if

$$\Delta\left(\frac{u(t-1)}{y(t-1)}\right) = 0$$

for $a+2 \leq t \leq b+1$. It follows that equality holds in the last inequality if and only if $u(t) = Cy(t)$, t is in $[a, b+2]$, $C \neq 0$, that is, if and only if $u(t)$ is an eigenfunction corresponding to λ. This gives us the conclusion of this theorem. ∎

Note by the proof of this last theorem that if we want to find a good upper bound for λ_1 we should use our intuition and take $u(t)$ as close to an eigenfunction corresponding to λ_1 as possible.

Example 7.10. Find an upper bound for the smallest eigenvalue for the Sturm-Liouville problem

$$\Delta^2 y(t-1) + \lambda y(t) = 0,$$
$$y(0) = 0, \quad y(4) = 0.$$

Define $u(t)$ on $[0,4]$ by $u(0) = 0$, $u(1) = 0.6$, $u(2) = 1$, $u(3) = 0.6$, $u(4) = 0$. Then $\sum_{t=1}^{4} p(t-1)[\Delta u(t-1)]^2 = 1.04$ and $\sum_{t=1}^{3} r(t)u^2(t) = 1.72$. Hence by Rayleigh's inequality

$$\lambda_1 \leq 0.604.$$

By Example 7.1 the actual value of λ_1 is 0.586 to three decimal places.

Exercises

Section 7.1

7.1 Find eigenpairs for the Sturm-Liouville problem

$$\Delta^2 y(t-1) + \lambda y(t) = 0,$$
$$y(0) = 0, \quad y(6) = 0.$$

7.2 Find a boundary value problem of the form (7.1)–(7.3) that is self-adjoint and not a Sturm-Liouville problem or a periodic Sturm-Liouville problem.

Section 7.2

7.3 Show that the inner product $< \cdot, \cdot >_r$ with respect to the weight function $r(t)$ satisfies properties (a)-(d) following Definition 7.4.

7.4 Prove that the periodic Sturm-Liouville problem (7.1), (7.5), (7.6) is self-adjoint.

7.5 Show that all eigenvalues of the boundary value problem

$$\Delta[t\Delta y(t-1)] + (\lambda t - \sin^2 \frac{\pi}{3} t) y(t) = 0,$$
$$y(2) - 3\Delta y(2) = 0, \quad \Delta y(50) = 0$$

are positive.

7.6 (a) Find eigenpairs for the Sturm-Liouville problem

$$\Delta^2 y(t-1) + \lambda y(t) = 0,$$
$$y(0) = 0, \quad y(3) = 0.$$

(b) Show directly that eigenfunctions corresponding to distinct eigenvalues are orthogonal as guaranteed by Theorem 7.2.

7.7 Show that corresponding to an eigenvalue of a self-adjoint boundary value problem (7.1)–(7.3) we can always pick a real eigenfunction.

7.8 Use matrix methods to find eigenpairs for the boundary value problem in Exercise 7.6.

7.9 Use matrix methods to find eigenpairs for the boundary value problem in Exercise 7.1.

7.10 How many eigenvalues does the boundary value problem

$$\Delta^2 y(t-1) + [\lambda + \sin t] y(t) = 0,$$
$$y(4) = 0, \quad y(73) = 0$$

have? How many linearly independent eigenfunctions does this boundary value problem have?

7.11 Find the Fourier series of $w(t) = t^2 - 5t + 4$ with respect to the eigenfunctions of the boundary value problem in Example 7.1.

7.12 Find the Fourier series of $w(t) = (t-3)^2$ with respect to the Sturm-Liouville problem in Exercise 7.1.

7.13 Find the Fourier series of $w(t) = t^2 - 4t + 3$ in terms of the eigenfunctions in Exercise 7.6.

Section 7.3

7.14 Show that if $y_1(t)$, $y_2(t)$ are linearly independent solutions of $Ly(t) = 0$, then

$$y_p(t) = -cy_1(t) \sum_{s=1}^{t} y_2(s) f(s) - cy_2(t) \sum_{s=t+1}^{b+1} y_1(s) f(s),$$

where $c^{-1} = p(a) W[y_1, y_2](a)$, is a particular solution of $Ly(t) = f(t)$.

7.15 Find a necessary and sufficient condition for the nonhomogeneous boundary value problem

$$\Delta^2 y(t-1) + y(t) = f(t),$$
$$y(0) = 0, \quad y(3) = 0$$

to have a solution (see Exercise 7.6).

7.16 What do you get if you cross a hurricane with the Kentucky Derby?

7.17 Show that if $(\lambda_i, y_i(t))$ is an eigenpair for (7.1), $y(a) = 0$, $y(b+2) = 0$,

then

$$\lambda_i = \frac{\sum_{t=a+1}^{b+2} p(t-1)[\Delta y_i(t-1)]^2 - \sum_{t=a+1}^{b+1} q(t)y_i^2(t)}{\sum_{t=a+1}^{b+1} r(t)y_i^2(t)}.$$

[Hint: multiply both sides of $Ly_i(t) + \lambda_i r(t) y_i(t) = 0$ by $y_i(t)$ and sum both sides from $a+1$ to $b+1$.]

7.18 Show that if $q(t) \leq 0$ on $[a+1, b+1]$, then all eigenvalues of (7.1), $y(a) = 0 = y(b+2)$, are positive.

7.19 Use the Rayleigh Inequality with the test function $u(t)$ defined on $[0,6]$ by $u(0) = 0$, $u(1) = 0.4$, $u(2) = 0.7$, $u(3) = 1$, $u(4) = 0.7$, $u(5) = 0.4$, $u(6) = 0$ to find an upper bound for the smallest eigenvalue of the Sturm-Liouville problem in Exercise 7.1.

7.20 Use the Rayleigh Inequality with the test function $u(t)$ defined on $[0,3]$ by $u(0) = 0$, $u(1) = 0.9$, $u(2) = 0.9$, $u(3) = 0$ to find an upper bound for the smallest eigenvalue of the Sturm-Liouville problem in Exercise 7.6. Compare your answer to the actual value and explain why you got what you did.

7.21 Use the Rayleigh Inequality with the test function $u(t)$ defined on $[0,6]$ by $u(0) = 0$, $u(1) = 2$, $u(2) = 3$, $u(3) = 4$, $u(4) = 3$, $u(5) = 2$, $u(6) = 0$ to find an upper bound for the smallest eigenvalue of the Sturm-Liouville problem

$$\Delta^2 y(t-1) + \lambda \frac{2t^2+7}{t^2+10} y(t) = 0,$$
$$y(0) = 0 = y(6).$$

Chapter 8
Discrete Calculus of Variations

8.1 Introduction

We first consider a very simple example. Let x_1, x_2, A, and B be numbers with $x_1 < x_2$. We would like to find the shortest polygonal path joining the points (x_1, A) and (x_2, B). The horizontal lengths of the line segments of such a path will be given by $d(t)$ for $t = 1, 2, \cdots, b+2$, where $b+2$ is the number of segments. Then for each function $y(t)$ defined on the integer interval $[0, b+2]$ with $y(0) = A$, $y(b+2) = B$, we obtain a polygonal path (see Fig. 8.1.)

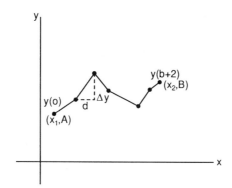

Fig. 8.1 A polygonal path

Mathematically speaking, we would like to minimize

$$\sum_{t=1}^{b+2} \sqrt{[d(t)]^2 + [\Delta y(t-1)]^2}$$

over all functions $y(t)$ defined on $[0, b+2]$ with

$$y(0) = A, \quad y(b+2) = B.$$

Note that the solution to this problem must be a straight line! However, it will serve later in this chapter as a nice illustration of the theory of the optimization of sums, called the discrete calculus of variations.

Let us begin by describing one of the main problems of interest. Assume $f(t, u, v)$ for each fixed t in $[a, b+2]$ is a class C^2 function of (u, v). Let $\mathcal{D}$ be the set of all real valued functions y defined on $[a, b+2]$ with $y(a) = A$, $y(b+2) = B$. The "simplest variational problem" is to extremize (maximize or minimize)

$$J[y] = \sum_{t=a+1}^{b+2} f(t, y(t), \Delta y(t-1))$$

subject to y belonging to the set $\mathcal{D}$.

We say that y_0 in $\mathcal{D}$ minimizes the simplest variation problem if

$$J[y] \geq J[y_0]$$

for all y in $\mathcal{D}$. We say J has a local minimum at y_0 provided there is a $\delta > 0$ such that

$$J[y] \geq J[y_0]$$

for all y in $\mathcal{D}$ with $|y(t) - y_0(t)| < \delta$, $a \leq t \leq b+2$. If in addition $J[y] > J[y_0]$ for $y \neq y_0$ in $\mathcal{D}$ with $|y(t) - y_0(t)| < \delta$, $a \leq t \leq b+2$, then we say J has a proper local minimum at y_0.

8.2 Necessary Conditions

In this section we develop necessary conditions for the simplest variational problem and closely related problems to have a local extremum.

Let $\mathcal{Q}$ be the set of all real valued functions η defined on $[a, b+2]$ such that $\eta(a) = 0 = \eta(b+2)$. Assume the simplest variational problem

8.2 Necessary Conditions

has a local extremum at y_0. Then define

$$\varphi(\epsilon) = J[y_0(t) + \epsilon \eta(t)],$$

where $-\infty < \epsilon < \infty$ and η is a fixed element of $\mathcal{Q}$.

Note that $y_0 + \epsilon \eta$ belongs to $\mathcal{D}$ for all real numbers ϵ. Since φ has a local extremum at $\epsilon = 0$ we have that

$$\varphi'(0) = 0, \qquad (8.1)$$
$$\varphi''(0) \geq 0 \ \{\leq 0\} \qquad (8.2)$$

in the local minimum {maximum} case.

Consider

$$\varphi(\epsilon) = \sum_{t=a+1}^{b+2} f\left(t, y_0(t) + \epsilon \eta(t), \Delta y_0(t-1) + \epsilon \Delta \eta(t-1)\right).$$

Differentiating with respect to ϵ, we have

$$\varphi'(\epsilon) = \sum_{t=a+1}^{b+2} \{\, f_u\left(t, y_0(t) + \epsilon \eta(t), \Delta y_0(t-1) + \epsilon \Delta \eta(t-1)\right) \eta(t)$$
$$+ f_v\left(t, y_0(t) + \epsilon \eta(t), \Delta y_0(t-1) + \epsilon \Delta \eta(t-1)\right) \Delta \eta(t-1)\}.$$

From Eq. (8.1) and the fact that η was arbitrary,

$$J_1[\eta] = 0,$$

for all η in $\mathcal{Q}$, where J_1 is defined on $\mathcal{Q}$ by

$$J_1[\eta] = \sum_{t=a+1}^{b+2} [f_u\left(t, y_0(t), \Delta y_0(t-1)\right) \eta(t)$$
$$+ f_v\left(t, y_0(t), \Delta y_0(t-1)\right) \Delta \eta(t-1)]. \qquad (8.3)$$

Furthermore,

$$\varphi''(\epsilon) = \sum_{t=a+1}^{b+2} \{ f_{uu}\left(t, y_0(t) + \epsilon\eta(t), \Delta y_0(t-1) + \epsilon\Delta\eta(t-1)\right) \eta^2(t)$$
$$+ 2f_{uv}\left(t, y_0(t) + \epsilon\eta(t), \Delta y_0(t-1) + \epsilon\Delta\eta(t-1)\right) \eta(t)\Delta\eta(t-1)$$
$$+ f_{vv}\left(t, y_0(t) + \epsilon\eta(t), \Delta y_0(t-1) + \epsilon\Delta\eta(t-1)\right) [\Delta\eta(t-1)]^2 \}.$$

Using Eq. (8.2) we can conclude that

$$J_2[\eta] \geq 0 \ \{\leq 0\}$$

in the local minimum {maximum} case, where J_2 is defined on Q by

$$J_2[\eta] = \sum_{t=a+1}^{b+2} \{ P(t)\eta^2(t) + 2Q(t)\eta(t)\Delta\eta(t-1) + R(t)[\Delta\eta(t-1)]^2 \} \tag{8.4}$$

and

$$P(t) = f_{uu}\left(t, y_0(t), \Delta y_0(t-1)\right), \tag{8.5}$$
$$Q(t) = f_{uv}\left(t, y_0(t), \Delta y_0(t-1)\right), \tag{8.6}$$
$$R(t) = f_{vv}\left(t, y_0(t), \Delta y_0(t-1)\right). \tag{8.7}$$

We call J_1 the first variation and J_2 the second variation of J corresponding to $y_0(t)$.

We have proved the following theorem.

Theorem 8.1. If the simplest variational problem has a local extremum at $y_0(t)$, then
    a) $J_1[\eta] = 0$ for all η in Q,
    b) $J_2[\eta] \geq 0 \ \{\leq 0\}$ for all η in Q in the minimum {maximum} case,
where J_1 is given by Eq. (8.3) and J_2 is given by Eq. (8.4).

The following lemma will be very useful in the proofs of Theorems 8.2–8.5.

Lemma 8.1. Assume y, η are arbitrary functions defined on $[a, b+2]$ and $J_1[\eta]$ is given by Eq. (8.3) with $y_0(t)$ replaced by $y(t)$. Then

8.2 Necessary Conditions

$$J_1[\eta] = \sum_{t=a+1}^{b+1} \{ f_u(t, y(t), \Delta y(t-1)) - \Delta f_v(t, y(t), \Delta y(t-1)) \} \eta(t)$$
$$- f_v(a+1, y(a+1), \Delta y(a)) \eta(a)$$
$$+ \{ f_u(b+2, y(b+2), \Delta y(b+1))$$
$$+ f_v(b+2, y(b+2), \Delta y(b+1)) \} \eta(b+2).$$

Proof.

$$J_1[\eta] = \sum_{t=a+1}^{b+1} \{ f_u(t, y(t), \Delta y(t-1)) \eta(t)$$
$$+ f_v(t, y(t), \Delta y(t-1)) \Delta \eta(t-1) \}$$
$$+ f_u(b+2, y(b+2), \Delta y(b+1)) \eta(b+2)$$
$$+ f_v(b+2, y(b+2), \Delta y(b+1)) \Delta \eta(b+1).$$

Using summation by parts we have

$$J_1[\eta] = \sum_{t=a+1}^{b+1} f_u(t, y(t), \Delta y(t-1)) \eta(t)$$
$$+ \{ f_v(t, y(t), \Delta y(t-1)) \eta(t-1) \}_{a+1}^{b+2}$$
$$- \sum_{t=a+1}^{b+1} \Delta [f_v(t, y(t), \Delta y(t-1))] \eta(t)$$
$$+ f_u(b+2, y(b+2), \Delta y(b+1)) \eta(b+2)$$
$$+ f_v(b+2, y(b+2), \Delta y(b+1)) [\eta(b+2) - \eta(b+1)],$$

which implies the final result. ∎

Theorem 8.2. *If the simplest variational problem has a local extremum at $y_0(t)$, then $y_0(t)$ satisfies the Euler-Lagrange equation*

$$f_u(t, y(t), \Delta y(t-1)) - \Delta f_v(t, y(t), \Delta y(t-1)) = 0 \qquad (8.8)$$

for t in $[a+1, b+1]$. Since y_0 belongs to $\mathcal{D}$, $y_0(a) = A$, $y_0(b+2) = B$.

Proof. From Theorem 8.1(a), $J_1[\eta] = 0$ for all η in $\mathcal{Q}$, where $J_1[\eta]$ is given by Eq. (8.3). By Lemma 8.1 with $y(t) = y_0(t)$ we have, using $\eta(a) = 0 = \eta(b+2)$,

$$\sum_{t=a+1}^{b+1} [f_u(t, y_0(t), \Delta y_0(t-1)) - \Delta f_v(t, y_0(t), \Delta y_0(t-1))] \eta(t) = 0$$

for all η in $\mathcal{Q}$. Fix s in $[a+1, b+1]$ and let $\eta(t) = 0$, $t \neq s$ and $\eta(s) = 1$. Then

$$f_u(s, y_0(s), \Delta y_0(s-1)) - \Delta f_v(s, y_0(s), \Delta y_0(s-1)) = 0.$$

Since s in $[a+1, b+1]$ is arbitrary we get the desired result. ∎

Example 8.1. Find the Euler–Lagrange equation for

$$J[y] = \sum_{t=1}^{100} \left\{ 5^t y^2(t) - 6^{t-1}[\Delta y(t-1)]^2 \right\}.$$

Here

$$f(t, u, v) = 5^t u^2 - 6^{t-1} v^2.$$

Hence $f_u(t, y(t), \Delta y(t-1)) = 5^t 2y(t)$, $f_v(t, y(t), \Delta y(t-1)) = -2 \cdot 6^{t-1} \Delta y(t-1)$. It follows that the Euler–Lagrange equation is

$$\Delta[6^{t-1} \Delta y(t-1)] + 5^t y(t) = 0.$$

Example 8.2. Find the Euler-Lagrange equation for the problem: minimize

$$J[y] = \sum_{t=1}^{b+2} \sqrt{[d(t)]^2 + [\Delta y(t-1)]^2}$$

subject to y belonging to $\mathcal{D}$ (see the introductory example).

8.2 Necessary Conditions

Here

$$f(t, u, v) = ([d(t)]^2 + v^2)^{\frac{1}{2}},$$

so

$$f_v(t, u, v) = \frac{v}{([d(t)]^2 + v^2)^{\frac{1}{2}}}.$$

The Euler–Lagrange equation is

$$\Delta \left\{ \frac{\Delta y(t-1)}{([d(t)]^2 + [\Delta y(t-1)]^2)^{\frac{1}{2}}} \right\} = 0.$$

It follows that

$$\left\{ \frac{\Delta y(t-1)}{([d(t)]^2 + [\Delta y(t-1)]^2)^{\frac{1}{2}}} \right\} = C$$

or

$$\frac{d(t)}{\Delta y(t-1)} = D;$$

that is, the slope of all line segments of the polygonal path must be equal to a single constant. Thus the polygonal path of shortest length is a straight line, as expected.

In the next two theorems we will eliminate one of the boundary conditions in the simplest variational problem. In Theorem 8.3 we eliminate the boundary condition at $b+2$.

Theorem 8.3. If

$$J[y] = \sum_{t=a+1}^{b+2} f(t, y(t), \Delta y(t-1)),$$

subject to y belonging to $\mathcal{D}_1 \equiv \{$real valued functions y defined on $[a, b+2]$ such that $y(a) = A\}$, has a local extremum at $y_0(t)$, then $y_0(t)$ satisfies the Euler–Lagrange equation (8.8), for t in $[a+1, b+1]$, $y_0(a) = A$, and $y_0(t)$

satisfies the transversality condition

$$f_u(b+2, y(b+2), \Delta y(b+1)) + f_v(b+2, y(b+2), \Delta y(b+1)) = 0. \quad (8.9)$$

Proof. As in the proof of Theorem 8.1(a), $J_1[\eta] = 0$ for all η belonging to $\mathcal{Q}_1 \equiv \{$ real valued functions η defined on $[a, b+2]$ such that $\eta(a) = 0\}$. By Lemma 8.1 with $y(t) = y_0(t)$ and using $\eta(a) = 0$,

$$\sum_{t=a+1}^{b+1} \{f_u(t, y_0(t), \Delta y_0(t-1)) - \Delta f_v(t, y_0(t), \Delta y_0(t-1))\} \eta(t)$$
$$+ \{f_u(b+2, y_0(b+2), \Delta y_0(b+1))$$
$$+ f_v(b+2, y_0(b+2), \Delta y_0(b+1))\} \eta(b+2)$$
$$= 0$$

for all η in $\mathcal{Q}_1$. The conclusions of the theorem follow easily. ∎

Next we eliminate the boundary condition at a.

Theorem 8.4. If

$$J[y] = \sum_{t=a+1}^{b+2} f(t, y(t), \Delta y(t-1)),$$

subject to y belonging to $\mathcal{D}_2 \equiv \{$real valued functions defined on $[a, b+2]$ such that $y(b+2) = B\}$, has a local extremum at y_0, then y_0 satisfies the Euler–Lagrange equation (8.8) for t in $[a+1, b+1]$, $y_0(b+2) = B$, and y_0 satisfies the transversality condition

$$f_v(a+1, y(a+1), \Delta y(a)) = 0. \quad (8.10)$$

Proof. The proof is similar to that of Theorem 8.2. From Lemma 8.1,

$$\sum_{t=a+1}^{b+1} [f_u(t, y_0(t), \Delta y_0(t-1)) - \Delta f_v(t, y_0(t), \Delta y_0(t-1))] \eta(t)$$
$$- f_v(a+1, y_0(a+1), \Delta y_0(a)) \eta(a) = 0$$

8.2 Necessary Conditions

for all η in $\mathcal{Q}_2$, where $\mathcal{Q}_2 \equiv$ {real valued functions on $[a, b+2]$ such that $\eta(b+2) = 0$}. The conclusions now follow. ∎

In a similar way, we have

Theorem 8.5. If $y_0(t)$ is a local extremum for $J[y]$ subject to y in $\mathcal{D}_3 \equiv$ {real valued functions defined on $[a, b+2]$}, then $y_0(t)$ satisfies the Euler–Lagrange equation (8.8) for t in $[a+1, b+1]$ and the transversality conditions of Eqs. (8.9), (8.10).

Example 8.3. Assume $J[y] = \sum_{t=1}^{100} \{2\left(\frac{1}{6}\right)^t y^2(t) - \left(\frac{1}{6}\right)^{t-1} [\Delta y(t-1)]^2\}$, subject to y defined on $[0, 100]$ and $y(100) = 3^{101} - 3 \cdot 2^{101}$, has a maximum at $y_0(t)$. Find $y_0(t)$. Here

$$f(t, u, v) = 2\left(\frac{1}{6}\right)^t u^2 - \left(\frac{1}{6}\right)^{t-1} v^2,$$

so $f_u = 4\left(\frac{1}{6}\right)^t u$, $f_v = -2\left(\frac{1}{6}\right)^{t-1} v$. It follows that the Euler–Lagrange equation is

$$\Delta\left[\left(\frac{1}{6}\right)^{t-1} \Delta y(t-1)\right] + 2\left(\frac{1}{6}\right)^t y(t) = 0.$$

This equation is the same as

$$y(t+1) - 5y(t) + 6y(t-1) = 0.$$

Thus,

$$y_0(t) = A 2^t + B 3^t.$$

The transversality condition (8.10) gives the boundary condition

$$\Delta y(0) = 0.$$

Hence $A 2^0 + 2B 3^0 = 0$ implies $A = -2B$, and

$$y_0(t) = B\left(3^t - 2^{t+1}\right).$$

Finally the boundary condition $y_0(100) = 3^{101} - 3 \cdot 2^{101}$ gives us that

$$B\left(3^{100} - 2^{101}\right) = 3^{101} - 3 \cdot 2^{101},$$

so $B = 3$ and

$$y_0(t) = 3\left(3^t - 2^{t+1}\right).$$

Example 8.4. Assume

$$J[y] = \sum_{t=1}^{100} \left\{3\left[\Delta y(t-1)\right]^2 + 4y^2(t)\right\},$$

subject to $y(t)$ defined on $[0, 100]$ with $y(0) = 3$, has a minimum at $y_0(t)$. Find $y_0(t)$. In this example

$$f(t, u, v) = 3v^2 + 4u^2,$$

so $f_u = 8u$ and $f_v = 6v$. It follows that the Euler–Lagrange equation is

$$3\Delta^2 y(t-1) - 4y(t) = 0$$

or

$$3y(t+1) - 10y(t) + 3y(t-1) = 0.$$

Hence

$$y_0(t) = A\left(\frac{1}{3}\right)^t + B3^t$$

$$y_0(0) = 3 = A + B,$$

8.2 Necessary Conditions

and

$$y_0(t) = A\left(\frac{1}{3}\right)^t + 3^{t+1} - A3^t.$$

The tranversality condition (8.9) leads to the boundary condition

$$4y(100) + 3\Delta y(99) = 0$$

or

$$7y(100) - 3y(99) = 0,$$

so

$$7A\left(\frac{1}{3}\right)^{100} + 7\cdot 3^{101} - 7A\cdot 3^{100} - 3A\left(\frac{1}{3}\right)^{99} - 3\cdot 3^{100} + 3A3^{99} = 0,$$

$$-2A\left(\frac{1}{3}\right)^{100} - 6A\cdot 3^{100} = -6\cdot 3^{101},$$

$$A = \frac{-6\cdot 3^{101}}{-2\left(\frac{1}{3}\right)^{100} - 6\cdot 3^{100}}$$
$$= \frac{3^{202}}{1 + 3^{201}}.$$

Finally,

$$y_0(t) = \frac{3^{202}}{1 + 3^{201}}\left[\left(\frac{1}{3}\right)^t - 3^t\right] + 3^{t+1}.$$

The following theorem gives another reason why the self-adjoint second order difference equation is important.

Theorem 8.6. The Euler–Lagrange equation for the second variation J_2 is a self-adjoint second order difference equation.

Proof. By Eq. (8.4), for η in $\mathcal{Q}$

$$J_2[\eta] = \sum_{t=a+1}^{b+2} \left\{ P(t)\eta^2(t) + 2Q(t)\eta(t)\Delta\eta(t-1) + R(t)[\Delta\eta(t-1)]^2 \right\}.$$

Note that

$$\begin{aligned} 2Q(t)\eta(t)\Delta\eta(t-1) &= Q(t)\eta(t)[\eta(t) - \eta(t-1)] \\ &\quad + Q(t)[\eta(t-1) + \Delta\eta(t-1)]\Delta\eta(t-1) \\ &= Q(t)\eta^2(t) - Q(t)\eta(t)\eta(t-1) \\ &\quad + Q(t)\eta(t-1)\Delta\eta(t-1) + Q(t)[\Delta\eta(t-1)]^2 \\ &= Q(t)\eta^2(t) + Q(t)[\Delta\eta(t-1)]^2 - Q(t)\eta^2(t-1). \end{aligned}$$

Hence

$$\begin{aligned} J_2[\eta] &= \sum_{t=a+1}^{b+2} \left\{ [P(t) + Q(t)]\eta^2(t) - Q(t)\eta^2(t-1) \right. \\ &\qquad \left. + [R(t) + Q(t)][\Delta\eta(t-1)]^2 \right\} \\ &= \sum_{t=a+1}^{b+2} \left\{ [P(t) + Q(t) - Q(t+1)]\eta^2(t) + [R(t) + Q(t)][\Delta\eta(t-1)]^2 \right\} \end{aligned}$$

because $\eta(a) = \eta(b+2) = 0$. Therefore

$$J_2[\eta] = \sum_{t=a+1}^{b+2} \left\{ p(t-1)[\Delta\eta(t-1)]^2 - q(t)\eta^2(t) \right\}, \tag{8.11}$$

where

$$p(t-1) = R(t) + Q(t), \tag{8.12}$$
$$q(t) = \Delta Q(t) - P(t). \tag{8.13}$$

8.2 Necessary Conditions

It follows from Eq. (8.11) that the Euler–Lagrange equation for J_2 is

$$\Delta\left[p(t-1)\Delta y(t-1)\right] + q(t)y(t) = 0. \qquad (8.14)\blacksquare$$

The self-adjoint equation (8.14) where $p(t-1)$ is given by Eq. (8.12) and $q(t)$ is given by Eq. (8.13) is called the Jacobi equation for J.

We will next write J_2 as a quadratic form. By Eq. (8.11)

$$\begin{aligned}J_2[\eta] &= \sum_{t=a+1}^{b+2}\left\{p(t-1)[\Delta\eta(t-1)]^2 - q(t)\eta^2(t)\right\} \\ &= \sum_{t=a+1}^{b+2}\left\{p(t-1)\eta^2(t) - 2p(t-1)\eta(t)\eta(t-1)\right. \\ &\qquad\left. + p(t-1)\eta^2(t-1) - q(t)\eta^2(t)\right\}.\end{aligned}$$

Since $\eta(a) = \eta(b+2) = 0$,

$$J_2[\eta] = -\sum_{t=a+1}^{b+1}\left[c(t)\eta^2(t) + 2p(t-1)\eta(t)\eta(t-1)\right],$$

where $c(t)$ is given by Eq. (6.4). It follows that

$$J_2[\eta] = -u^T S u, \qquad (8.15)$$

where $u = [\eta(a+1)\eta(a+2),\cdots\eta(b+1)]^T$ and S is as in Chapter 7:

$$S = \begin{bmatrix} c(a+1) & p(a+1) & 0 & \cdots & 0 \\ p(a+1) & c(a+2) & p(a+2) & \cdots & 0 \\ 0 & p(a+2) & c(a+3) & \cdots & 0 \\ \vdots & \ddots & \ddots & \ddots & \vdots \\ 0 & \cdots & 0 & p(b) & c(b+1) \end{bmatrix}.$$

Theorem 8.7. If the simplest variation problem has a local minimum {maximum} at $y_0(t)$ then the Legendre necessary condition

$$c(t) \leq 0 \ \{c(t) \geq 0\}, \quad (a+1 \leq t \leq b+1) \tag{8.16}$$

is satisfied.

Proof. By Theorem 8.1(b) $J_2[\eta] \geq 0 \ \{\leq 0\}$ in the local minimum {maximum} case. Take $u = e_{t_0-a}$, where e_{t_0-a} is the unit column vector in $\Re^N$, $N = b - a + 1$, $a + 1 \leq t_0 \leq b + 1$, in the $t_0 - a$ direction; then by Eq. (8.15)

$$J_2[\eta] = -c(t_0) \geq 0 \ \{\leq 0\}.$$

Hence $c(t_0) \leq 0 \ \{\geq 0\}$ in the local minimum {maximum} case. ∎

Lemma 8.2. Assume $z(t)$ is a solution of the Jacobi equation (8.14) where $p(t)$ and $q(t)$ are given by Eqs. (8.12), (8.13), and $a \leq \alpha \leq \beta \leq b+1$. If

$$\eta(t) = \begin{cases} z(t), & (\alpha + 1 \leq t \leq \beta) \\ 0, & (\text{otherwise}), \end{cases}$$

it follows that

$$J_2[\eta] = z(\alpha)p(\alpha)z(\alpha+1) + z(\beta)p(\beta)z(\beta+1).$$

Proof. By Eq. (8.15)

$$J_2[\eta] = -u^T S u,$$

where in this case $u = [0, \cdots, 0, z(\alpha+1), \cdots, z(\beta), 0, \cdots, 0]^T$.

8.2 Necessary Conditions

Hence

$$Su = \begin{bmatrix} 0 \\ \vdots \\ 0 \\ p(\alpha)z(\alpha+1) \\ c(\alpha+1)z(\alpha+1) + p(\alpha+1)z(\alpha+2) \\ p(\alpha+1)z(\alpha+1) + c(\alpha+2)z(\alpha+2) + p(\alpha+2)z(\alpha+3) \\ \vdots \\ p(\beta-2)z(\beta-2) + c(\beta-1)z(\beta-1) + p(\beta-1)z(\beta) \\ p(\beta-1)z(\beta-1) + c(\beta)z(\beta) \\ p(\beta)z(\beta) \\ 0 \\ \vdots \\ 0 \end{bmatrix}.$$

Since $z(t)$ is a solution of Eq. (8.14), which can be written in the form of Eq. (6.6),

$$p(t)z(t+1) + c(t)z(t) + p(t-1)z(t-1) = 0,$$

we have that Su is equal to

$$[0, \cdots, 0, p(\alpha)z(\alpha+1), -p(\alpha)z(\alpha), 0, \cdots,$$
$$\cdots, 0, -p(\beta)z(\beta+1), p(\beta)z(\beta), 0, \cdots, 0]^T.$$

It follows that

$$\begin{aligned} J_2[\eta] &= -u^T Su \\ &= z(\alpha+1)p(\alpha)z(\alpha) + z(\beta)p(\beta)z(\beta+1). \end{aligned} \qquad \blacksquare$$

Theorem 8.8. If the simplest variation problem has a local minimum at $y_0(t)$, and if $z(t)$ is a solution of Eq. (8.14) with

$$z(a)p(a)z(a+1) \le 0, \quad (z(a+1) \ne 0),$$

then
$$z(t)p(t)z(t+1) > 0, \quad (a+1 \leq t \leq b)$$
$$\geq 0, \quad (t = b+1).$$

Proof. Let $z(t)$ be a solution of Eq. (8.14) with $z(a)p(a)z(a+1) \leq 0$, $z(a+1) \neq 0$. By Lemma 8.2 and Theorem 8.1 (b) we have, for

$$\eta(t) = \begin{cases} z(t), & a+1 \leq t \leq \beta \\ 0, & \text{otherwise} \end{cases}$$

where $a+1 \leq \beta \leq b+1$, that

$$J_2[\eta] = z(a)p(a)z(a+1) + z(\beta)p(\beta)z(\beta+1) \geq 0.$$

Hence

$$z(\beta)p(\beta)z(\beta+1) \geq -z(a)p(a)z(a+1) \geq 0.$$

Since β in $[a+1, b+1]$ is arbitrary,

$$z(t)p(t)z(t+1) \geq 0, \ a+1 \leq t \leq b+1.$$

It follows from Lemma 6.1 that $z(t)p(t)z(t+1) > 0$ on $[a+1, b]$. ∎

It follows from Theorem 8.8 and the Sturm Separation Theorem that if the simplest variational problem has a local minimum at $y_0(t)$, then the Jacobi equation (8.14) corresponding to $y_0(t)$ is disconjugate on $[a, b+1]$ (possibly not on $[a, b+2]$). In the next section, we will see the consequences of the case where we can find a $y_0(t)$ such that Eq. (8.14) is disconjugate on the larger interval $[a, b+2]$.

8.3 Sufficient Conditions and Disconjugacy

In Section 8.2 we obtained several necessary conditions for various variational problems to have a local extremum. This section contains some sufficient conditions. In the discrete calculus of variations the sufficient conditions are much easier to come by than in the continuous calculus of

8.3 Sufficient Conditions and Disconjugacy

variations. One reason for this is that in the continuous case it is easy to have $\max_{c \le x \le d} |y(x) - z(x)|$ small but $\max_{c \le x \le d} |y'(x) - z'(x)|$ large. In the discrete case, if $\max_{[a,b+2]} |y(t) - z(t)|$ is small, then it follows that $\max_{[a,b+1]} |\Delta y(t) - \Delta z(t)|$ is small, too.

We will see that disconjugacy is closely related to these sufficient conditions. In fact, the disconjugacy of Eq. (8.14) is equivalent to a certain quadratic functional being positive definite. This result will enable us to prove several interesting results concerning the self-adjoint difference equation (6.1). In particular we will prove some comparison theorems, one of which is a more general Sturm comparison theorem than the one given in Chapter 6, as well as some interesting theorems concerning necessary conditions for disconjugacy.

Theorem 8.9. If y_0 in $\mathcal{D}$ satisfies the Euler-Lagrange equation (8.8) and the corresponding second variation J_2 is positive {negative} definite on $\mathcal{Q}$ then the simplest variational problem has a proper local minimum {maximum} at $y_0(t)$. If in addition f_{uu}, f_{uv}, and f_{vv} are functions of t only, then the simplest variational problem has a proper (global) minimum {maximum} at $y_0(t)$.

Proof. Assume $y_0(t)$ satisfies the Euler-Lagrange equation (8.8) and that the corresponding second variation J_2 is positive definite on $\mathcal{Q}$. We will show that J has a proper local minimum at $y_0(t)$ and in particular that there is $\delta > 0$ such that if y belongs to $\mathcal{D}$, $y \ne y_0$ with $|y(t) - y_0(t)| < \delta$, t in $[a+1, b+1]$, then $J[y_0] < J[y]$.

Let y belong to $\mathcal{D}$, set $\eta = y - y_0$, and consider

$$\varphi(\epsilon) \equiv J[y_0 + \epsilon y],$$

$-\infty < \epsilon < \infty$. By Taylor's theorem

$$\varphi(1) = \varphi(0) + \frac{\varphi'(0)}{1!} + \frac{\varphi''(\xi)}{2!}, \qquad (8.17)$$

where ξ belongs to $(0, 1)$. Now

$$\varphi'(0) = J_1[\eta] = \sum_{t=a+1}^{b+2} \{f_u(t, y_0(t), \Delta y_0(t-1)) \eta(t) \\ + f_v(t, y_0(t), \Delta y_0(t-1)) \Delta \eta(t-1)\}.$$

Using Lemma 8.1 and $\eta(a) = \eta(b+2) = 0$, we have that

$$\varphi'(0) = \sum_{t=a+1}^{b+1} \left[f_u\left(t, y_0(t), \Delta y_0(t-1)\right) - \Delta f_v\left(t, y_0(t), \Delta y_0(t-1)\right) \right] \eta(t).$$

But since $y_0(t)$ satisfies the Euler-Lagrange equation (8.8) for $a+1 \le t \le b+1$, $\varphi'(0) = 0$. Since $\varphi(0) = J[y_0]$ and $\varphi(1) = J[y]$, it follows from Eq. (8.17) that

$$J[y] - J[y_0] = \frac{1}{2} \varphi''(\xi). \tag{8.18}$$

Next, we calculate

$$\varphi''(\xi) = \sum_{t=a+1}^{b+2} \bigl\{ f_{uu}\left(t, y_0(t) + \xi\eta(t), \Delta y_0(t-1) + \xi\Delta\eta(t-1)\right) \eta^2(t)$$
$$+ 2 f_{uv}\left(t, y_0(t) + \xi\eta(t), \Delta y_0(t-1) + \xi\Delta\eta(t-1)\right) \eta(t) \Delta\eta(t-1)$$
$$+ f_{vv}\left(t, y_0(t) + \xi\eta(t), \Delta y_0(t-1) + \xi\Delta\eta(t-1)\right) [\Delta\eta(t-1)]^2 \bigr\}$$
$$= J_2[\eta; y_0 + \xi\eta].$$

Here

$$J_2[\eta; y] \equiv \sum_{t=a+1}^{b+2} \bigl\{ P(t;y) \eta^2(t) + 2Q(t;y) \eta(t) \Delta\eta(t-1)$$
$$+ R(t;y) [\Delta\eta(t-1)]^2 \bigr\},$$

where

$$P(t;y) = f_{uu}\left(t, y(t), \Delta y(t-1)\right),$$
$$Q(t;y) = f_{uv}\left(t, y(t), \Delta y(t-1)\right),$$
$$R(t;y) = f_{vv}\left(t, y(t), \Delta y(t-1)\right).$$

8.3 Sufficient Conditions and Disconjugacy

By a derivation similar to that of Eq. (8.11) (see proof of Theorem 8.6), we have

$$J_2[\eta; y] = \sum_{t=a+1}^{b+2} \left\{ p(t-1; y)[\Delta \eta(t-1)]^2 - q(t; y)\eta^2(t) \right\},$$

where

$$p(t-1; y) = R(t; y) + Q(t; y)$$
$$q(t; y) = \Delta Q(t; y) - P(t; y).$$

Hence $J_2[\eta; y]$ is a quadratic form in $u = [\eta(a+1), \cdots, \eta(b+1)]^T$, and there is a symmetric matrix $S(y)$ (see how S was obtained from $J_2[\eta] = J_2[\eta; y_0]$) such that

$$J_2[\eta; y] = -u^T S(y) u.$$

Since $J_2[\eta; y_0] = -u^T S(y_0) u$ is positive definite, the maximum eigenvalue of $S(y_0)$ satisfies

$$\lambda_{max} S(y_0) < 0.$$

Now $\lambda_{max} S(y)$ is a continuous function of y, so there are constants $\mu > 0$ and $\delta > 0$ such that

$$\lambda_{\max} S(y) \leq \mu < 0, \qquad (8.19)$$

whenever $|y(t) - y_0(t)| < \delta$, $a+1 \leq t \leq b+1$. From Eq. (8.18)

$$J[y] - J[y_0] = \frac{1}{2} J_2[\eta; y_0 + \xi \eta], \qquad (8.20)$$

where ξ is in $(0, 1)$. Assume $|y(t) - y_0(t)| < \delta$ for $a+1 \leq t \leq b+1$; then

$$|(y_0(t) + \xi \eta(t)) - y_0(t)| = |\xi|\,|\eta(t)|$$
$$= |\xi|\,|y(t) - y_0(t)| < \delta,$$

for $a+1 \leq t \leq b+1$. From Eq. (8.19)

$$J_2[\eta; y_0 + \xi\eta] = -u^T S(y_0 + \xi\eta)u > 0$$

(strict inequality follows since $y \neq y_0$ implies $u \neq 0$), so by Eq. (8.20)

$$J[y] > J[y_0]$$

when $|y(t) - y_0(t)| < \delta$, $y \neq y_0$.

The last statement of the theorem follows from the preceding discussion and the fact that the matrix S in this case is independent of y. ∎

Assume as in Chapter 6 that $p(t) > 0$ on $[a, b+1]$ and $q(t)$ is real valued on $[a+1, b+1]$. Define the quadratic functional Q on $\mathcal{Q}$ by

$$Q[\eta] = \sum_{t=a+1}^{b+2} \left\{ p(t-1)[\Delta\eta(t-1)]^2 - q(t)\eta^2(t) \right\}. \tag{8.21}$$

Note that $\eta(b+2) = 0$, so the value of q at $b+2$ is arbitrary. We have seen that the second variation J_2 is of this form for appropriate $p(t)$ and $q(t)$.

Theorem 8.10. The self-adjoint equation $Ly(t) = 0$ is disconjugate on $[a, b+2]$ if and only if Q is positive definite on $\mathcal{Q}$.

Proof. Assume $Ly(t) = 0$ is disconjugate on $[a, b+2]$. By Theorem 6.10 $Ly(t) = 0$ has a positive solution $y(t)$ on $[a, b+2]$. Make the Riccati substitution

$$z(t) = \frac{p(t-1)\Delta y(t-1)}{y(t-1)},$$

t in $[a+1, b+2]$; then by Theorem 6.15

$$z(t) + p(t-1) > 0$$

on $[a+1, b+2]$ and $z(t)$ satisfies the Riccati equation

$$\Delta z(t) + q(t) + \frac{z^2(t)}{z(t) + p(t-1)} = 0$$

8.3 Sufficient Conditions and Disconjugacy

on $[a+1, b+1]$. Let η belong to $\mathcal{Q}$ and use Eq. (6.32) with $u(t) = \eta(t)$ to obtain

$$Q[\eta] = z(t)\eta^2(t-1)\Big]_{a+1}^{b+3}$$
$$+ \sum_{t=a+1}^{b+2} \left\{ \frac{z(t)\eta(t)}{\sqrt{z(t)+p(t-1)}} - \sqrt{z(t)+p(t-1)}\Delta\eta(t-1) \right\}^2.$$

(Here one has to check that Eq. (6.32) is correct for $t = b+2$.) Since $\eta(a) = \eta(b+2) = 0$,

$$Q[\eta] = \sum_{t=a+1}^{b+2} \left\{ \frac{z(t)\eta(t)}{\sqrt{z(t)+p(t-1)}} - \sqrt{z(t)+p(t-1)}\Delta\eta(t-1) \right\}^2$$
$$\geq 0.$$

Note that $Q[\eta] = 0$ only if

$$\sqrt{z(t)+p(t-1)}\Delta\eta(t-1) = \frac{z(t)\eta(t)}{\sqrt{z(t)+p(t-1)}}$$

for t in $[a+1, b+2]$. Hence $Q[\eta] = 0$ only if

$$\eta(t) = \frac{z(t)+p(t-1)}{p(t-1)}\eta(t-1).$$

Since $\eta(a) = 0$ it follows that $\eta(t) \equiv 0$ on $[a, b+2]$. Hence Q is positive definite on $\mathcal{Q}$.

Conversely assume Q is positive definite on $\mathcal{Q}$. Let $y(t)$ be the solution of $Ly(t) = 0$ such that $y(a) = 0$, $y(a+1) = 1$. Define η in $\mathcal{Q}$ by

$$\eta(t) = \begin{cases} y(t), & (a+1 \leq t \leq \beta) \\ 0, & \text{(otherwise)} \end{cases}$$

for β in $[a+1, b+1]$.

Chapter 8. Discrete Calculus of Variations

By Lemma 8.2 with $J_2 = Q$,

$$\begin{aligned} Q[\eta] &= y(a)p(a)y(a+1) + y(\beta)p(\beta)y(\beta+1) \\ &= y(\beta)p(\beta)y(\beta+1) > 0, \end{aligned}$$

as Q is positive definite and $\eta \neq 0$ because $\eta(a+1) = y(a+1) = 1$. Since β in $[a+1, b+1]$ is arbitrary,

$$y(t) > 0$$

on $[a+1, b+2]$. But then $Ly(t) = 0$ must be disconjugate on $[a, b+2]$. ∎

Corollary 8.1. *If y_0 in $\mathcal{D}$ satisfies the Euler-Lagrange equation (8.8) and the corresponding Jacobi equation (8.14) is disconjugate on $[a, b+2]$, then the simplest variational problem has a local minimum at y_0. If, in addition, f_{uu}, f_{uv}, and f_{vv} depend only on t, then the simplest variational problem has a proper global minimum at y_0.*

Example 8.5. Show that

$$J[y] = \sum_{t=1}^{400} \left(\frac{1}{5}\right)^{t-1} [\Delta y(t-1)]^2,$$

subject to y defined on $[0, 400]$, $y(0) = 50$, $y(400) = 5^{401} + 45$, has a proper global minimum at some y_0. Find y_0.

Here

$$f(t, u, v) = \left(\frac{1}{5}\right)^{t-1} v^2,$$

so

$$f_u = 0, \quad f_v = 2\left(\frac{1}{5}\right)^{t-1} v, \quad f_{uu} = f_{uv} = 0, \quad f_{vv} = 2\left(\frac{1}{5}\right)^{t-1}.$$

8.3 Sufficient Conditions and Disconjugacy

Note that f_{uu}, f_{uv}, and f_{vv} depend only on t. From Eqs. (8.5)–(8.7)

$$P(t) = 0, \quad Q(t) = 0, \quad R(t) = 2\left(\frac{1}{5}\right)^{t-1}.$$

Hence by Eqs. (8.12) and (8.13),

$$p(t-1) = 2\left(\frac{1}{5}\right)^{t-1}, \quad q(t) = 0.$$

Since $q(t) \leq 0$ on $[1, 399]$, we have by Corollary 6.7 that the Jacobi equation

$$\Delta\left[2\left(\frac{1}{5}\right)^{t-1} \Delta\eta(t-1)\right] = 0$$

is disconjugate on $[0, 400]$. It follows from Corollary 8.1 that this problem has a proper global minimum at some y_0. To find y_0, note that the Euler–Lagrange equation is

$$0 - \Delta\left[2\left(\frac{1}{5}\right)^{t-1} \Delta y(t-1)\right] = 0$$

or

$$\Delta\left[\left(\frac{1}{5}\right)^{t-1} \Delta y(t-1)\right] = 0.$$

Hence

$$\Delta y_0(t) = A5^t.$$

It follows that

$$\begin{aligned} y_0(t) &= B5^t + C, \\ y_0(0) &= 50 = B + C, \\ y_0(t) &= B5^t - B + 50, \\ y_0(400) &= B5^{400} - B + 50 = 5^{401} + 45, \\ B &= 5. \end{aligned}$$

Finally,

$$y_0(t) = 5^{t+1} + 45.$$

Example 8.6. Show that the answer we got in Example 8.2 is a proper local minimum. In Example 8.2 (introductory example)

$$f(t, u, v) = ([d(t)]^2 + v^2)^{\frac{1}{2}}.$$

Then

$$f_{uu} = 0, \quad f_{uv} = 0, \quad f_{vv} = \frac{1}{([d(t)]^2 + v^2)^{\frac{3}{2}}}.$$

The Jacobi equation is

$$\Delta \left\{ \frac{1}{\{[d(t)]^2 + \Delta y_0(t-1)]^2\}^{\frac{3}{2}}} \Delta \eta(t-1) \right\} = 0.$$

Since this equation is disconjugate on $[a, b+2]$, by Corollary 8.1 there is a proper local minimum at $y_0(t)$.

One can actually show that there is a proper global minimum at $y_0(t)$ because in this case for any y in $\mathcal{D}$

$$J_2[\eta; y] = \sum_{t=a+1}^{b+2} \frac{1}{\{[d(t)]^2 + \Delta y(t-1)]^2\}^{\frac{3}{2}}} [\Delta \eta(t-1)]^2 \geq 0$$

8.3 Sufficient Conditions and Disconjugacy

for all η, and $J_2[\eta; y] = 0$ only if $\eta = 0$. By the proof of Theorem 8.9 we get the desired result.

Theorem 8.11. If $Ly(t) = 0$ is disconjugate on $[a, b+1]$, then

$$\sum_{t=a+1}^{b+2} \{p(t-1)\chi_S(t)\chi_{S^\sim}(t-1) + p(t)\chi_S(t)\chi_{S^\sim}(t+1) - q(t)\chi_S(t)\} > 0,$$

where S is a nonempty subset of $[a+1, b+1]$, $S^\sim = [a, b+2]\setminus S$, and χ_S is the characteristic function on S, that is, $\chi_S(t) = 1$, t in S, $\chi_S(t) = 0$, and t in $S^\sim$.

Proof. Assume $Ly(t) = 0$ is disconjugate on $[a, b+2]$. Then by Theorem 8.10, Q defined by Eq. (8.21) is positive definite on $\mathcal{Q}$. Define η on $[a, b+2]$ by

$$\eta(t) = \chi_S(t).$$

Since $S \subset [a+1, b+1]$, $\eta(a) = \eta(b+2) = 0$, η belongs to $\mathcal{Q}$. Consequently $Q[\eta] > 0$, that is,

$$\sum_{t=a+1}^{b+2} \left\{ p(t-1)\left[\chi_S(t) - \chi_S(t-1)\right]^2 - q(t)\chi_S^2(t) \right\} > 0.$$

Then

$$\sum_{t=a+1}^{b+1} p(t-1)\chi_S(t)\left[\chi_S(t) - \chi_S(t-1)\right]$$
$$+ \sum_{t=a+2}^{b+2} p(t-1)\chi_S(t-1)\left[\chi_S(t-1) - \chi_S(t)\right] - \sum_{t=a+1}^{b+2} q(t)\chi_S(t) > 0,$$

so

$$\sum_{t=a+1}^{b+1} p(t-1)\chi_S(t)\chi_{S^\sim}(t-1) + \sum_{t=a+2}^{b+2} p(t-1)\chi_S(t-1)\chi_{S^\sim}(t)$$
$$- \sum_{t=a+1}^{b+1} q(t)\chi_S(t) > 0.$$

The conclusion of the theorem follows. ■

Corollary 8.2. If $Ly(t) = 0$ is disconjugate on $[a, b+2]$, then

$$\sum_{t=a+1}^{b+1} q(t) < p(a) + p(b+1).$$

Proof. Let $S = [a+1, b+1]$ in Theorem 8.11. Then $S^\sim = \{a, b+2\}$, and

$$\sum_{t=a+1}^{b+1} \{p(t-1)\chi_S(t)\chi_{S^\sim}(t-1) + p(t)\chi_S(t)\chi_{S^\sim}(t+1) - q(t)\chi_S(t)\}$$
$$= p(a) + p(b+1) - \sum_{t=a+1}^{b+1} q(t) > 0$$

by Theorem 8.11. ■

Corollary 8.3. If $Ly(t) = 0$ is disconjugate on $[a, b+2]$, then

$$q(t) < p(t) + p(t-1)$$

(same as $c(t) < 0$) for t in $[a+1, b+1]$.

Proof. Let $S = \{t_0\}$ for some t_0 in $[a+1, b+1]$ in Theorem 8.11. Then

$$\sum_{t=a+1}^{b+1} \{p(t-1)\chi_S(t)\chi_{S^\sim}(t-1) + p(t)\chi_S(t)\chi_{S^\sim}(t+1) - q(t)\chi_S(t)\}$$
$$= p(t_0-1) + p(t_0) - q(t_0) > 0$$

8.3 Sufficient Conditions and Disconjugacy

by Theorem 8.11. Since t_0 in $[a+1, b+1]$ is arbitrary the result follows. ∎

We are now going to prove comparison theorems for the two equations

$$L_i y(t) = \Delta[p_i(t-1)\Delta y(t-1)] + q_i(t)y(t) = 0,$$

$i = 1, 2$, where $p_i(t) > 0$ in $[a, b+1]$, $i = 1, 2$, and $q_i(t)$ is defined on $[a+1, b+1]$, $i = 1, 2$.

Theorem 8.12. (Sturm Comparison Theorem) Assume $q_1(t) \geq q_2(t)$ on $[a+1, b+1]$ and $p_2(t) \geq p_1(t) > 0$ on $[a, b+1]$. If $L_1 y(t) = 0$ is disconjugate on $[a, b+2]$, then $L_2 y(t) = 0$ is disconjugate on $[a, b+2]$.

Proof. Assume $L_1 y(t) = 0$ is disconjugate on $[a, b+2]$. Then by Theorem 8.10, Q_1 is positive definite on $\mathcal{Q}$ where

$$Q_1[\eta] \equiv \sum_{t=a+1}^{b+2} \left\{ p_1(t-1)[\Delta \eta(t-1)]^2 - q_1(t)\eta^2(t) \right\}.$$

From our assumptions on $p_i(t)$ and $q_i(t)$, $i = 1, 2$, we have for η in $\mathcal{Q}$

$$\begin{aligned}
Q_2[\eta] &\equiv \sum_{t=a+1}^{b+2} \left\{ p_2(t-1)[\Delta \eta(t-1)]^2 - q_2(t)\eta^2(t) \right\} \\
&\geq \sum_{t=a+1}^{b+2} \left\{ p_1(t-1)[\Delta \eta(t-1)]^2 - q_1(t)\eta^2(t) \right\} \\
&= Q_1[\eta].
\end{aligned}$$

It follows that Q_2 is positive definite on $\mathcal{Q}$. Hence by Theorem 8.10, $L_2 y(t) = 0$ is disconjugate on $[a, b+2]$. ∎

Theorem 8.13. If $L_i y(t) = 0$ is disconjugate on $[a, b+2]$ for $i = 1, 2$, and if

$$\begin{aligned}
p(t) &= \lambda_1 p_1(t) + \lambda_2 p_2(t), \\
q(t) &= \lambda_1 q_1(t) + \lambda_2 q_2(t),
\end{aligned}$$

where $\lambda_1 > 0$, $\lambda_2 > 0$, then $Ly(t) = 0$ is disconjugate on $[a, b+2]$.

Proof. Since $L_i y(t) = 0$ is disconjugate on $[a, b+2]$, $i = 1, 2$, the quadratic functionals Q_i, $i = 1, 2$ (defined in the previous proof) are positive definite on $\mathcal{Q}$. Hence for η in $\mathcal{Q}$

$$\begin{aligned} Q[\eta] &= \sum_{t=a+1}^{b+2} \left\{ p(t-1)[\Delta \eta(t-1)]^2 - q(t)\eta^2(t) \right\} \\ &= \sum_{t=a+1}^{b+2} \left\{ (\lambda_1 p_1(t-1) + \lambda_2 p_2(t-1)) [\Delta \eta(t-1)]^2 \right. \\ &\quad \left. - (\lambda_1 q_1(t) + \lambda_2 q_2(t)) \eta^2(t) \right\} \\ &= \lambda_1 Q_1[\eta] + \lambda_2 Q_2[\eta]. \end{aligned}$$

It follows that Q is positive definite on $\mathcal{Q}$, and so by Theorem 8.10, $Ly(t) = 0$ is disconjugate on $[a, b+2]$. ∎

Theorem 8.14. (Weierstrass Summation Formula) If $y(t)$ is a solution of the boundary value problem $Ly(t) = 0$, $y(a) = A$, $y(b+2) = B$, η is in $\mathcal{Q}$, and $z = y + \eta$, then

$$Q[z] = Q[y] + Q[\eta].$$

Proof. Define

$$\varphi(\epsilon) = Q[y + \epsilon \eta].$$

By Taylor's Theorem

$$\varphi(1) = \varphi(0) + \frac{\varphi'(0)}{1!} + \frac{\varphi''(\xi)}{2!} \tag{8.22}$$

for some ξ in $(0, 1)$. Since

$$\varphi(\epsilon) = \sum_{t=a+1}^{b+2} \left\{ p(t-1)[\ \Delta y(t-1) + \epsilon \Delta \eta(t-1)]^2 \right. \\ \left. - q(t)[y(t) + \epsilon \eta(t)]^2 \right\},$$

8.3 Sufficient Conditions and Disconjugacy

it follows that

$$\varphi'(0) = 2 \sum_{t=a+1}^{b+2} \{p(t-1)\Delta y(t-1)\Delta \eta(t-1) - q(t)y(t)\eta(t)\}.$$

Applying summation by parts,

$$\varphi'(0) = 2p(t-1)\Delta y(t-1)\eta(t-1) \,]_{a+1}^{b+3}$$
$$- 2 \sum_{t=a+1}^{b+2} \Delta[p(t-1)\Delta y(t-1)]\eta(t)$$
$$- 2 \sum_{t=a+1}^{b+2} q(t)y(t)\eta(t)$$
$$= 0.$$

The second derivative is

$$\varphi''(\epsilon) = 2 \sum_{t=a+1}^{b+2} \left\{ p(t-1)\left[\Delta \eta(t-1)\right]^2 - q(t)\eta^2(t) \right\}$$
$$= 2Q[\eta].$$

Since $\varphi(1) = Q[z]$, $\varphi(0) = Q[y]$, (8.22) implies that

$$Q[z] = Q[y] + Q[\eta] \qquad \blacksquare$$

Corollary 8.4. If $y(t)$ is a solution of the boundary value problem $Ly(t) = 0$, $y(a) = A$, $y(b+2) = B$, z in $\mathcal{D}$ and Q is positive definite on $\mathcal{Q}$, then

$$Q[z] \geq Q[y],$$

where equality holds only if $z = y$.

Proof. Let $\eta = z - y$; then η is in $\mathcal{Q}$ and $z = y + \eta$. By Theorem 8.14

$$Q[z] = Q[y] + Q[\eta].$$

Since Q is positive definite on $\mathcal{Q}$

$$Q[z] \geq Q[y],$$

with equality holding if and only if $\eta(t) \equiv 0$. Now $y = z$ if and only if $\eta(t) \equiv 0$, and the result follows. ∎

Using Corollary 8.4 and Theorem 6.7, we obtain a final result.

Corollary 8.5. If $Ly(t) = 0$ is disconjugate on $[a, b+2]$, then the problem of minimizing

$$Q[y] = \sum_{t=a+1}^{b+2} \left\{ p(t-1)[\Delta y(t-1)]^2 - q(t) y^2(t) \right\},$$

subject to y in $\mathcal{D}$, has a proper global minimum at $y_0(t)$, where $y_0(t)$ is the solution of the boundary value problem $Ly(t) = 0$, $y(a) = A$, $y(b+2) = B$.

Many of the results in this chapter are due to Ahlbrandt and Hooker [10]. For generalizations of some of these results to the matrix case, see Ahlbrandt and Hooker [9],[11],[12],[13], Peil and Peterson [205], and Peterson and Ridenhour [212],[214].

Exercises

Section 8.1

8.1 Find the $J[y]$ to be minimized in order to minimize the surface area obtained by rotating the curve in Fig. 8.1 ($y(a) = A > 0$, $y(b+2) = B > 0$) about the x-axis.

Section 8.2

8.2 Find the Euler-Lagrange equation for each of the following:
  (a) $J[y] = \sum_{t=0}^{99}\{4y^2(t) + 3[\Delta y(t-1)]^2\}$,
  (b) $J[y] = \sum_{t=0}^{49}\{y^2(t) + 2y(t)\Delta y(t-1) + 6[\Delta y(t-1)]^2\}$.

8.3 In each of the following problems assume there is a local extremum $y_0(t)$. Find $y_0(t)$.
  (a) $J[y] = \sum_{t=1}^{100}[\Delta y(t-1)]^2$, $y(0) = 2$, $y(100) = 200$.
  (b) same J, $y(0) = 2$.
  (c) same J, $y(100) = 200$.
  (d) same J, $y(0)$ free, $y(100)$ free.

8.4 By considering $\sum_{t=a+1}^{b+2} f(t, y(t), \Delta y(t-1))$, $y(a) = A$, $y(b+2) = B$ as a function of the $b - a + 1$ variables $y(a+1), \cdots, y(b+1)$ show that if the values $y_0(a+1), \cdots, y_0(b+1)$ render this function an extremum, then $y_0(t)$ satisfies the Euler-Lagrange equation for $a+1 \leq t \leq b+1$.

8.5 Assume

$$J[y] = \sum_{t=1}^{400}\left\{\left(\frac{1}{8}\right)^{t-1}[\Delta y(t-1)]^2 - 3\left(\frac{1}{8}\right)^t y^2(t)\right\}$$

subject to y defined on $[0, 400]$ and $y(0) = 0$, $y(400) = 2^{402} - 4^{401}$, has a minimum at $y_0(t)$. Find $y_0(t)$.

8.6 Assume

$$J[y] = \sum_{t=1}^{100} \left\{ y^2(t) + 2[\Delta y(t-1)]^2 \right\},$$

(a) subject to y defined on $[0, 100]$, has a minimum at $y_0(t)$. Find $y_0(t)$. Calculate $J[y_0(t)]$ and explain whether your answer makes sense.

(b) subject to y defined on $[0, 100]$ and $y(0) = 1$, has a minimum $y_0(t)$. Find $y_0(t)$.

8.7 Prove Lemma 8.2 using Eq. (8.11).

Section 8.3

8.8 Use Theorem 8.10 to show that if $q(t) \leq 0$ in $[a+1, b+1]$, then $Ly(t) = 0$ is disconjugate on $[a, b+2]$.

8.9 Show that

$$J[y] = \sum_{t=1}^{500} \left\{ \left(\frac{1}{6}\right)^{t-1} [\Delta y(t-1)]^2 - 2\left(\frac{1}{6}\right)^t y^2(t) \right\},$$

subject to y defined on $[0, 500]$ with $y(0) = 5$, $y(500) = 10$, has a proper global minimum at some y_0. Find y_0.

Chapter 9
Boundary Value Problems for Nonlinear Equations

9.1 Introduction

In this chapter we will consider boundary value problems for nonlinear equations, specifically

$$\Delta^2 y(t-1) = f(t, y(t)), \quad (t \text{ in } [a, b+2]), \tag{9.1}$$
$$y(a) = A, \quad y(b+2) = B. \tag{9.2}$$

Here $f(t, y)$ is a function defined for all t in $[a, b+2]$ and all real numbers y. There are fundamental questions that arise for Eqs. (9.1), (9.2). Does a solution exist, is it unique, and how can solutions be approximated?

Suppose that $y(t)$ is a solution of Eqs. (9.1), (9.2). From Corollary 6.4

$$y(t) = \sum_{s=a+1}^{b+1} G(t,s) f(s, y(s)) + w(t), \tag{9.3}$$

where $G(t, s)$ is the Green's function for

$$\Delta^2 y(t-1) = 0,$$
$$y(a) = 0, \quad y(b+2) = 0,$$

and

$$w(t) = A + \frac{B-A}{b+2-a}(t-a).$$

Let $\mathcal{B} = \{$real valued functions defined on $[a, b+2]\}$ and define $T: \mathcal{B} \to \mathcal{B}$ by

$$Ty(t) = \sum_{s=a+1}^{b+1} G(t,s) f(s, y(s)) + w(t)$$

for t in $[a, b+2]$. Then $Ty(t) = y(t)$, i.e., y is a "fixed point" of T. Thus solutions of Eqs. (9.1), (9.2) are necessarily fixed points of the operator T. Since the steps of the analysis are reversible, it is also true that all fixed points of T are solutions of (9.1), (9.2).

In order to establish a theorem on the existence of fixed points of T, we need the concept of the "norm" of a vector. Let R^n denote the set of ordered n-tuples of real numbers. A norm on R^n is a function $\|\cdot\|: R^n \to R$ having the following properties:
  a) $\|x\| \geq 0$ for all x in R^n;
  b) $\|x\| = 0$ if and only if $x = (0, 0, \ldots, 0)$;
  c) $\|cx\| = |c| \|x\|$ for all c in R and x in R^n;
  d) $\|x + y\| \leq \|x\| + \|y\|$ for all x, y in R^n.
Here are some examples of norms:
  a) $\|(x_1, \ldots, x_n)\| = (x_1^2 + x_2^2 + \cdots + x_n^2)^{\frac{1}{2}}$;
  (Euclidean norm)
  b) $\|(x_1, \ldots, x_n)\| = \max_{1 \leq i \leq n}\{|x_i|\}$;
  (Maximum norm)
  c) $\|(x_1, \ldots, x_n)\| = |x_1| + \cdots + |x_n|$.
  (Traffic norm)

Let $\|\cdot\|$ be any norm on R^n. A sequence $\{x_k\}$ in R^n converges to x in R^n if $\lim_{k \to \infty} \|x_k - x\| = 0$. A Cauchy sequence $\{x_k\}$ in R^n satisfies the following property: given $\epsilon > 0$, there is an M so that $\|x_{k+l} - x_k\| < \epsilon$ for all $l \geq 0$ whenever $k \geq M$. Actually, it can be shown that these concepts are norm-independent since if a sequence is convergent (Cauchy) with respect to one norm, then it is convergent (Cauchy) with respect to every norm. In the following theorem, we will make use of the fact that every Cauchy sequence is convergent (see [21]).

Theorem 9.1. (Contraction Mapping Theorem) Let $\|\cdot\|$ be a norm for R^n and S a closed subset of R^n. Assume $T: S \to S$ is a contraction mapping: there is an α, $0 \leq \alpha < 1$, such that $\|Tx - Ty\| \leq \alpha \|x - y\|$ for all x, y in S.

9.1 Introduction

Then T has a unique fixed point z in S. Furthermore, if y_0 is in S and we set $y_k = Ty_{k-1}$ for $k \geq 1$ (the "Picard iterates"), then

$$\|y_k - z\| \leq \frac{\alpha^k}{1-\alpha}\|y_1 - y_0\| \quad (k \geq 0). \tag{9.4}$$

Proof. We will show that $\{y_k\}$ is a Cauchy sequence. First note that

$$\begin{aligned}
\|y_{k+1} - y_k\| &= \|Ty_k - Ty_{k-1}\| \\
&\leq \alpha\|y_k - y_{k-1}\| \\
&= \alpha\|Ty_{k-1} - Ty_{k-2}\| \\
&\leq \alpha^2\|y_{k-1} - y_{k-2}\| \\
&\vdots \\
&\leq \alpha^k\|y_1 - y_0\|.
\end{aligned}$$

For $l = 1, 2, \cdots,$

$$\begin{aligned}
\|y_{k+l} - y_k\| &\leq \|y_{k+l} - y_{k+l-1}\| + \cdots + \|y_{k+1} - y_k\| \\
&\leq \alpha^{k+l-1}\|y_1 - y_0\| + \cdots + \alpha^k\|y_1 - y_0\| \\
&\leq \alpha^k(1 + \alpha + \alpha^2 + \cdots)\|y_1 - y_0\|,
\end{aligned}$$

so

$$\|y_{k+l} - y_k\| \leq \frac{\alpha^k}{1-\alpha}\|y_1 - y_0\|. \tag{9.5}$$

Now Eq. (9.5) implies that $\{y_k\}$ is a Cauchy sequence. Then $\{y_k\}$ converges to some z in R^n, and z is in S since S is closed.

Since T is a contraction, it is continuous on S (see Exercise 9.2).

Hence

$$Tz = T\left(\lim_{k\to\infty} y_k\right)$$
$$= \lim_{k\to\infty} T(y_k)$$
$$= \lim_{k\to\infty} y_{k+1}$$
$$= z,$$

so z is a fixed point for T. If w in S is also a fixed point for T, then

$$\|z - w\| = \|Tz - Tw\| \leq \alpha\|z - w\|.$$

Since $\alpha < 1$, $\|z - w\| = 0$, so $z = w$. We conclude that z is the only fixed point of T in S.

Finally, Eq. (9.4) is obtained by letting $l \to \infty$ in Eq. (9.5). ∎

9.2 The Lipschitz Case

We begin our study of Eqs. (9.1), (9.2) in this section by showing that there is a unique solution if $f(t, y)$ satisfies a growth condition with respect to y, known as a "Lipschitz condition."

Definition 9.1. Suppose that there is a constant $K \geq 0$ so that

$$|f(t, y) - f(t, x)| \leq K|y - x|$$

for all integers t in $[a, b + 2]$ and all x, y in R. Then we say f satisfies a "Lipschitz condition" with respect to y on $[a, b + 2] \times R$.

The constant K in Definition 9.1 is called a Lipschitz constant for f. The contraction mapping theorem can now be used to obtain a unique solution for the boundary value problem (9.1), (9.2).

Theorem 9.2. Assume $f(t, y)$ satisfies a Lipschitz condition with respect to y on $[a, b + 2] \times R$ with Lipschitz constant K. If $b + 2 - a < \sqrt{\frac{8}{K}}$, then (9.1), (9.2) has a unique solution.

Proof. As in Section 9.1, $\mathcal{B} = \{$real-valued functions on $[a, b + 2]\}$ and

9.2 The Lipschitz Case

$T : \mathcal{B} \to \mathcal{B}$ is given by

$$Ty(t) = \sum_{s=a+1}^{b+1} G(t,s)f(s,y(s)) + w(t).$$

Note that $\mathcal{B}$ is equivalent to R^{b-a+3}. We will use the maximum norm on $\mathcal{B}$:

$$\|y\| = \max\{|y(t)| : t \text{ is in } [a, b+2]\}.$$

Let us show that T is a contraction mapping on $\mathcal{B}$. Consider

$$\begin{aligned}
|Ty(t) - Tx(t)| &= \left| \sum_{s=a+1}^{b+1} G(t,s)\left[f(s,y(s)) - f(s,x(s))\right] \right| \\
&\leq \sum_{s=a+1}^{b+1} |G(t,s)| K |y(s) - x(s)| \\
&\leq K \sum_{s=a+1}^{b+1} |G(t,s)| \|y - x\| \\
&\leq K \frac{(b+s-a)^2}{8} \|y - x\|
\end{aligned}$$

for t in $[a, b+2]$ by Exercise 6.20.
Hence

$$\|Ty - Tz\| \leq \alpha \|y - z\|,$$

where $\alpha = \frac{K(b+2-a)^2}{8} < 1$, so T is a contraction mapping and has a unique fixed point by Theorem 9.1. It follows from the discussion in Section 9.1 that (9.1), (9.2) has a unique solution. ∎

Example 9.1. Show that the problem

$$\begin{aligned}
\Delta^2 y(t-1) &= -0.1 \cos y(t), \\
y(0) &= 0 = y(8)
\end{aligned}$$

has a unique solution and find an approximation of the solution.

For $f(y) = -0.1\cos y$, $f'(y) = 0.1\sin y$, and by the mean value theorem

$$|f(x) - f(y)| \leq K|x-y| = 0.1|x-y|$$

for all x, y. Then $\sqrt{\frac{8}{K}} = \sqrt{80} > 8$, and Theorem 9.2 is applicable.

To obtain an approximation of the solution of the boundary value problem, we start with the initial guess $y_0(t) \equiv 0$ and compute the next Picard iterate. Now

$$y_1(t) = Ty_0(t) = \sum_{s=1}^{7} G(t,s)[-0.1\cos y_0(s)]$$

$$= -\sum_{s=1}^{7} 0.1 G(t,s),$$

where

$$G(t,s) = \begin{cases} -\frac{(8-s)t}{8}, & 0 \leq t \leq s \leq 7, \\ -\frac{(8-t)s}{8}, & 1 \leq s \leq t \leq 8. \end{cases}$$

Then

$$\begin{aligned} y_1(t) &= 0.1 \sum_{s=1}^{t-1} \frac{(8-t)s}{8} - 0.1 \sum_{s=t}^{7} \frac{(s-8)t}{7} \\ &= 0.1 \frac{8-t}{8} \left[\frac{s^{(2)}}{2}\right]_{s=1}^{t} - 0.1 \frac{t}{8} \left[\frac{(s-8)^{(2)}}{2}\right]_{s=t}^{8} \\ &= 0.05 t(8-t), \quad (0 \leq t \leq 8). \end{aligned}$$

By using a slightly more complicated norm than the maximum norm, we get the following generalization of Theorem 9.2.

Theorem 9.3. Assume there is a $k(t) \geq 0$ on $[a, b+2]$ such that

$$|f(t,y) - f(t,x)| \leq k(t)|y-x|$$

9.2 The Lipschitz Case

for all t in $[a, b+2]$ and x, y in R. If $\Delta^2 y(t-1) + k(t)y(t) = 0$ is disconjugate on $[a, b+2]$, then the boundary value problem (9.1), (9.2) has a unique solution.

Proof. Since $\Delta^2 y(t-1) + k(t)y(t) = 0$ is disconjugate on $[a, b+2]$, we know from Theorem 6.10 that there is a positive solution $y(t)$ on $[a, b+2]$. Consequently, for α less than one and sufficiently near one, the equation $\Delta^2 u(t-1) + \frac{k(t)}{\alpha} u(t) = 0$ has a positive solution $u(t)$ on $[a, b+2]$ (see Exercise 9.7).

Let $p(t)$ be the unique solution of

$$\Delta^2 p(t-1) = 0,$$
$$p(a) = u(a), \quad p(b+2) = u(b+2).$$

Then

$$u(t) = \sum_{s=a+1}^{b+1} G(t,s) \left[-\frac{1}{\alpha} k(s) u(s) \right] + p(t),$$

where $G(t,s)$ is the Green's function for $\Delta^2 v(t-1) = 0$, $v(a) = v(b+2) = 0$. Since $p(t) > 0$ on $[a, b+2]$, we have

$$u(t) > \sum_{s=a+1}^{b+1} G(t,s) \left[-\frac{1}{\alpha} k(s) u(s) \right]$$

on $[a, b+2]$. Rearranging the last inequality,

$$\alpha > \frac{1}{u(t)} \sum_{s=a+1}^{b+1} |G(t,s)| k(s) u(s),$$

so

$$\alpha > \max_{[a,b+2]} \frac{1}{u(t)} \sum_{s=a+1}^{b+1} |G(t,s)| k(s) u(s).$$

Now define a norm on $\mathcal{B}$ by

$$\|x\| = \max_{[a,b+2]} \left\{ \frac{|x(t)|}{u(t)}, \ a \le t \le b+2 \right\}.$$

Let $T: \mathcal{B} \to \mathcal{B}$ be given by

$$Tx(t) = \sum_{s=a+1}^{b+1} G(t,s) f(s, x(s)) + w(s),$$

where $\Delta^2 w(t-1) = 0$, $w(a) = A$, $w(b+2) = B$. Then

$$|Tx(t) - Tz(t)| \le \sum_{s=a+1}^{b+1} |G(t,s)| k(s) |x(s) - z(s)|,$$

so

$$\frac{|Tx(t) - Tz(t)|}{u(t)} \le \frac{1}{u(t)} \sum_{s=a+1}^{b+1} |G(t,s)| k(s) u(s) \frac{|x(s) - z(s)|}{u(s)}$$

$$\le \frac{1}{u(t)} \sum_{s=a+1}^{b+1} |G(t,s)| k(s) u(s) \|x - z\|$$

$$\le \alpha \|x - z\|$$

for $a \le t \le b+2$. Hence $\|Tx - Tz\| \le \alpha \|x - z\|$, and the proof is completed by applying the contraction mapping theorem. ∎

Corollary 9.1. Assume there is a K in $(0, 4)$ such that $|f(t,y) - f(t,z)| \le K|y - z|$ for all t in $[a, b+2]$, y, z in R. If $b+2-a < \frac{\pi}{\arccos \frac{2-K}{2}}$, then (9.1), (9.2) has a unique solution.

Proof. According to Theorem 9.3, it suffices to show that $\Delta^2 u(t-1) + K u(t) = 0$ is disconjugate on $[a, b+2]$. The characteristic equation for this difference equation is $m^2 + (K-2)m + 1 = 0$, so

$$m = \frac{2 - K \pm \sqrt{(K-2)^2 - 4}}{2}.$$

9.2 The Lipschitz Case

Since $0 < K < 4$, we can choose θ in $(0, \pi)$ so that $2 - K = 2\cos\theta$. Then $m = \pm e^{i\theta}$ with $\theta = \arccos\frac{2-K}{2}$, and $u(t) = \sin\theta(t-a)$ is a nontrivial solution with $u(a) = 0$. Consequently, $\Delta^2 u(t-1) + Ku(t) = 0$ is disconjugate on $[a, b+2]$ if $b + 2 - a < \frac{\pi}{\theta}$. ∎

Example 9.2. Show that the boundary value problem

$$\Delta^2 y(t-a) = -0.1\cos y(t),$$
$$y(0) = A, \quad y(9) = B$$

has a unique solution.

As in Example 9.1, $K = 0.1$ is a Lipschitz constant for $f(y) = -0.1\cos y$. However, Theorem 9.2 does not apply since $\sqrt{\frac{8}{K}} = \sqrt{80} < 9$. Corollary 9.1 does provide a unique solution because

$$b + 2 - a = 9 < \frac{\pi}{\arccos 0.95} \simeq 9.89.$$

The assumption that f in Eq. (9.1) satisfies a Lipschitz condition on $[a, b+2] \times R$ is quite strong and will not be satisfied in most cases.

Example 9.3.

$$\Delta^2 y(t-1) = -0.01 e^y,$$
$$y(a) = 0, \quad y(b+2) = 0.$$

Here $f(t, y) = -0.01 e^y$ does not satisfy a Lipschitz condition for y in R. To see this, simply note that

$$\frac{f(y) - f(0)}{y - 0} = \frac{-0.01 e^y + 0.01}{y} \to -\infty, \quad (y \to \infty),$$

by L'Hospital's rule.

The next theorem utilizes the contraction mapping theorem to obtain a unique solution in some cases where f does not satisfy a Lipschitz condition on R.

Theorem 9.4. Assume there are positive constants N and K so that $|f(t,y) - f(t,z)| \le K|y - z|$ for t in $[a+1, b+1]$ and y, z in $[-N, N]$. Set

$m = \max\{|f(t,0)| : a+1 \le t \le b+1\}$ and $M = \max\{|f(t,y)| : a+1 \le t \le b+1, |y| \le N\}$. If $\alpha \equiv K\frac{(b+2-a)^2}{8} < 1$ and either $\frac{m(b+2-a)^2}{8} \le N(1-\alpha)$ or $\frac{M(b+2-a)^2}{8} \le N$, then Eq. (9.1) with homogeneous boundary conditions $y(a) = y(b+2) = 0$ has a unique solution $y(t)$ with $|y(t)| \le N$ for $a \le t \le b+2$.

Proof. Let $\mathcal{C} = \{$functions y on $[a,b+2]$ such that $y(a) = y(b+2) = 0$ and $|y(t)| \le N$, $a \le t \le b+2\}$. Then $\mathcal{C}$ is a closed subset of R^{b-a+3}. Let $\|\cdot\|$ be the maximum norm. Define T on $\mathcal{C}$ by

$$Ty(t) = \sum_{s=a+1}^{b+1} G(t,s) f(s, y(s)),$$

where $G(t,s)$ is the Green's function for the boundary value problem $\Delta^2 u(t-1) = 0$, $u(a) = u(b+2) = 0$.

For y and z in $\mathcal{C}$ consider

$$|Ty(t) - Tz(t)| = \left| \sum_{s=a+1}^{b+1} G(t,s) [f(s,y(s)) - f(s,z(s))] \right|$$

$$\le \sum_{s=a+1}^{b+1} |G(t,s)| K |y(s) - z(s)|$$

$$\le K \frac{(b+2-a)^2}{8} \|y - z\|.$$

Then $\|Ty - Tz\| \le \alpha \|y - z\|$ for y, z in $\mathcal{C}$.

It remains to show $T : \mathcal{C} \to \mathcal{C}$. First assume $\frac{m(b+2-a)^2}{8} \le N(1-\alpha)$. Then

$$|T(0)(t)| = \left| \sum_{s=a+1}^{b+1} G(t,s) f(s,0) \right|$$

$$\le \frac{m(b+2-a)^2}{8}$$

$$\le N(1-\alpha)$$

9.2 The Lipschitz Case

for t in $[a, b+2]$, so $\|T(0)\| \leq N(1-\alpha)$. For y in $\mathcal{C}$,

$$\begin{aligned} \|Ty\| &\leq \|T(y) - T(0)\| + \|T(0)\| \\ &\leq \alpha\|y - 0\| + N(1-\alpha) \\ &\leq N, \end{aligned}$$

and we have $T : \mathcal{C} \to \mathcal{C}$ in this case.

Finally assume $\frac{M(b+2-a)^2}{8} \leq N$. For y in $\mathcal{C}$

$$\begin{aligned} |Ty(t)| &\leq \sum_{s=a+1}^{b+1} |G(t,s)| \, |f(s, y(s))| \\ &\leq \frac{M(b+2-a)^2}{8} \\ &\leq N, \end{aligned}$$

and $\|Ty\| \leq N$, so $T : \mathcal{C} \to \mathcal{C}$ in this case also.

The contraction mapping theorem can be applied to obtain a unique solution of Eq. (9.1) in C, and the proof is complete. ∎

Example 9.3. (continued) With $f = -0.01e^y$, we have

$$\begin{aligned} |f(y) - f(z)| &= 0.01e^c|y - z| \\ &\leq 0.01e^N|y - z| \end{aligned}$$

if $|y|, |z| \leq N$, by the Mean Value Theorem.

We need

$$\alpha \equiv 0.01 e^N \frac{(b+2-a)^2}{8} < 1.$$

Note that

$$\frac{M(b+2-a)^2}{8} = 0.01 e^N \frac{(b+2-a)^2}{8} \leq N$$

will then be true for $N = 1$. It follows from Theorem 9.4 that

$$\Delta^2 y(t-1) = -0.01e^y,$$
$$y(a) = y(b+2) = 0$$

has a unique solution y with $|y(t)| \leq 1$ for $a \leq t \leq b+2$ if $(b+2-a)^2 < \frac{800}{e}$.

9.3 Existence of Solutions

Since boundary value problems for difference equations often have multiple solutions, it is useful to have a collection of results that yield existence of solutions without the implication that the solutions must be unique. The existence theorems of this type to be presented in this section will be based on the following version of the Brouwer Fixed Point Theorem.

Theorem 9.5. (Brouwer Fixed Point Theorem) Let $K = \{(x_1, \ldots, x_n) : c_i \leq x_i \leq d_i, \ i = 1, \ldots, n\}$ and suppose $T : K \to K$ is continuous. Then T has a fixed point in K.

A proof of Theorem 9.5 can be found in [62]. See Exercise 9.10 for the case $n = 1$. Here is our basic existence theorem.

Theorem 9.6. Assume for each t in $[a+1, b+1]$ that $f(t, y)$ is a continuous function of y. If $M \geq \max\{|A|, |B|\}$ and $b + 2 - a \leq \sqrt{\frac{8M}{Q}}$, where $Q = \max\{|f(t,y)| : a+1 \leq t \leq b+1, \ |y| \leq 2M\}$, then (9.1), (9.2) has a solution.

Proof. Let $K = \{y : |y(t)| \leq 2M, \ a \leq t \leq b+2\}$. Note that K is the type of subset of R^{b-a+3} to which the Brouwer fixed point theorem is applicable. Define T on K by

$$Ty(t) = \sum_{s=a+1}^{b+1} G(t,s) f(s, y(s)) + w(t),$$

$a \leq t \leq b+2$, where $G(t,s)$ is the Green's function for $\Delta^2 y(t-1) = 0$, $y(a) = y(b+2) = 0$ and w is the solution of $\Delta^2 w(t-1) = 0$, $w(a) = A$, $w(b+2) = B$. It is easily checked that T is continuous on K.

9.3 Existence of Solutions

We now show that $T : K \to K$. Let y belong to K and consider

$$\begin{aligned}
|Ty(t)| &= \left| \sum_{s=a+1}^{b+1} G(t,s) f(s, y(s)) + w(t) \right| \\
&\leq Q \sum_{s=a+1}^{b+1} |G(t,s)| + M \\
&\leq \frac{(b+2-a)^2}{8} Q + M \\
&\leq 2M,
\end{aligned}$$

$a \leq t \leq b+2$. Hence Ty is in K, and the conclusion follows from the Brouwer fixed point theorem. ∎

Corollary 9.2. If $f(t, y)$ is continuous in y for each t in $[a, b+2]$ and is bounded on $[a+1, b+2] \times R$, then (9.1), (9.2) has a solution.

Proof. Choose $P > \sup\{|f(t, y)| : a+1 \leq t \leq b+1,\ y \text{ in } R\}$. Pick M large enough so that $b+2-a < \sqrt{\frac{8M}{P}}$ and $|A|, |B| \leq M$. For the Q defined in Theorem 9.6, $Q \leq P$, so

$$b + 2 - a < \sqrt{\frac{8M}{Q}},$$

and by Theorem 9.6, (9.1), (9.2) has a solution. ∎

A very powerful technique for establishing existence of one or more solutions of a nonlinear boundary value problem is the construction of comparison functions that satisfy a relation like Eq. (9.1) with equality replaced by inequality.

Definition 9.2. A real-valued function $\alpha(t)$ on $[a, b+2]$ is a "lower solution" for (9.1), (9.2), if

$$\Delta^2 \alpha(t-1) \geq f(t, \alpha(t))$$

for t in $[a+1, b+1]$, $\alpha(a) \leq A$ and $\alpha(b+2) \leq B$. Similarly, $\beta(t)$ is an "upper solution" for (9.1), (9.2) if

$$\Delta^2 \beta(t-1) \leq f(t, \beta(t))$$

for t in $[a+1, b+1]$, $\beta(a) \geq A$ and $\beta(b+2) \geq B$.

Theorem 9.7. Assume $f(t, y)$ is continuous in y for each t in $[a+1, b+1]$, $\alpha(t)$ and $\beta(t)$ are lower and upper solutions, respectively, for (9.1), (9.2) and $\alpha(t) \leq \beta(t)$ on $[a, b+2]$. Then (9.1), (9.2) has a solution $y(t)$ with $\alpha(t) \leq y(t) \leq \beta(t)$ for t in $[a, b+2]$.

Proof. Define $F(t, y)$ for $a+1 \leq t \leq b+1$, y in R, by

$$F(t, y) = \begin{cases} f(t, \beta(t)) + \frac{y - \beta(t)}{1 + |y|} & \text{if } y \geq \beta(t), \\ f(t, y) & \text{if } \alpha(t) \leq y \leq \beta(t), \\ f(t, \alpha(t)) + \frac{y - \alpha(t)}{1 + |y|} & \text{if } y \leq \alpha(t). \end{cases}$$

Note that $F(t, y)$ is continuous as a function of y for each t. Furthermore, F is bounded and agrees with f when $\alpha(t) \leq y \leq \beta(t)$. By Corollary 9.2, the boundary value problem

$$\Delta^2 y(t-1) = F(t, y(t)),$$
$$y(a) = A, \quad y(b+2) = B$$

has a solution $y(t)$.

We claim that $y(t) \leq \beta(t)$ for t in $[a, b+2]$. If not, then $y(t) - \beta(t)$ has a positive maximum at some t_0 in $[a+1, b+2]$. Consequently, we must have $\Delta^2(y - \beta)(t_0 - 1) \leq 0$. On the other hand,

$$\begin{aligned}
\Delta^2(y - \beta)(t_0 - 1) &\geq F(t_0, y(t_0)) - f(t_0, \beta(t_0)) \\
&= f(t_0, \beta(t_0)) + \frac{y(t_0) - \beta(t_0)}{1 + |y(t_0)|} - f(t_0, \beta(t_0)) \\
&= \frac{y(t_0) - \beta(t_0)}{1 + |y(t_0)|} > 0,
\end{aligned}$$

which is a contradiction. It follows that $y(t) \leq \beta(t)$ on $[a, b+2]$.

Similarly, $\alpha(t) \leq y(t)$ on $[a, b+2]$ (see Exercise 9.11). Thus $y(t)$ is a solution of (9.1), (9.2). ∎

9.3 Existence of Solutions

Example 9.4. Consider the boundary value problem

$$\Delta^2 y(t-1) = -0.1\cos y(t),$$
$$y(0) = 0 = y(b+2).$$

First, note that $\alpha(t) = 0$ is a lower solution for this problem since it satisfies the boundary conditions and

$$\Delta^2(0) = 0 > -0.1\cos(0).$$

Next, let $\beta(t) = 0.05t(b+2-t)$. Then $\beta(0) = \beta(b+2) = 0$ and

$$\begin{aligned}\Delta^2\beta(t-1) &= 0.05\Delta^2[(t-1)(b+3-t)] \\ &= 0.05(-2) \\ &= -0.1 \\ &\leq -0.1\cos\beta(t),\end{aligned}$$

so $\beta(t)$ is a lower solution. We can conclude that there is a solution $y(t)$ with

$$0 \leq y(t) \leq 0.05t(b+2-t), \quad (0 \leq t \leq b+2).$$

Compare this example with Example 9.1.

Corollary 9.3. Assume for each t in $[a+1, b+1]$, $f(t, y)$ is continuous and nondecreasing in y, $-\infty < y < \infty$. Then (9.1), (9.2) has a solution $y(t)$. Furthermore, if $f(t, y)$ is strictly increasing in y, then the solution is unique.

Proof. Choose $M \geq \max\{|f(t, 0)| : a \leq t \leq b+2\}$. Let $u(t)$ be the solution of

$$\Delta^2 u(t-1) = M,$$
$$u(a) = 0 = u(b+2).$$

Note that $u(t) \leq 0$ on $[a, b+2]$. Pick $K \geq \max\{|A|, |B|\}$ and let $\alpha(t) = u(t) - K$. Then

$$\begin{aligned}
\Delta^2 \alpha(t-1) &= \Delta^2 u(t-1) \\
&= M \\
&\geq f(t, 0) \\
&\geq f(t, \alpha(t))
\end{aligned}$$

since f is nondecreasing in y and $\alpha(t) \leq 0$. Also, $\alpha(a) \leq A$, $\alpha(b+2) \leq B$, so $\alpha(t)$ is a lower solution for (9.1), (9.2).

An upper solution can be similarly constructed. Hence (9.1), (9.2) has a solution $y(t)$.

Suppose $f(t, y)$ is strictly increasing in y. Let $x(t)$ be a second solution of (9.1), (9.2). Define $z(t) = x(t) - y(t)$. If $x(t)$ is larger than $y(t)$ for some t, then $z(t)$ has a positive maximum at some t_0, so $\Delta^2 z(t_0 - 1) \leq 0$. However,

$$\begin{aligned}
\Delta^2 z(t_0 - 1) &= f(t_0, x(t_0)) - f(t_0, y(t_0)) \\
&> 0
\end{aligned}$$

since $x(t_0) > y(t_0)$, and we have a contradiction. Likewise we can show that $y(t)$ is nowhere larger than $x(t)$. As a result, $x(t) \equiv y(t)$, and solutions are unique in this case. ∎

The following is an immediate consequence of Corollary 9.3.

Example 9.5. The boundary value problem

$$\begin{aligned}
\Delta^2 y(t-1) &= c(t)y + d(t)y^3 + e(t), \\
y(a) &= A, \quad y(b+2) = B,
\end{aligned}$$

where $c(t) \geq 0$, $d(t) \geq 0$ on $[a+1, b+1]$, has a solution. The solution is unique if $c^2(t) + d^2(t) > 0$ for t in $[a+1, b+1]$.

The next theorem is a generalization of the uniqueness of solutions of initial value problems for Eq. (9.1) (see Exercise 9.1).

Theorem 9.8. Assume $f(t, y)$ is continuous in y for each t, $\Delta^2 \alpha(t-1) - f(t, \alpha(t)) \geq 0 \geq \Delta^2 \beta(t-1) - f(t, \beta(t))$, $\alpha(t) \leq \beta(t)$ for t in $[a, b+2]$, and there is a t_0, $a+1 \leq t_0 \leq b+1$, where $\alpha(t_0) = \beta(t_0)$, $\Delta\alpha(t_0) = \Delta\beta(t_0)$. Then $\alpha(t) \equiv \beta(t)$ on $[a, b+2]$ (so it is a solution of Eq. (9.1)).

9.3 Existence of Solutions

Proof. Assume $\alpha(t) \not\equiv \beta(t)$ on $[a, b+2]$. Then there is an integer t_1 where $\beta(t_1) < \alpha(t_1)$ and either $t_0 + 1 < t_1 \leq b+2$ or $a \leq t_1 < t_0$. We consider only the first case (see Exercise 9.12 for the other case).

By Theorem 9.7, there are solutions $y_1(t)$, $y_2(t)$ of Eq. (9.1) so that $y_i(t_0) = \alpha(t_0)$ and $\alpha(t) \leq y_i(t) \leq \beta(t)$ for $t_0 \leq t \leq t_1$, $i = 1, 2$, and $y_1(t_1) = \alpha(t_1)$, $y_2(t_1) = \beta(t_1)$. It follows that $y_1(t_0) = y_2(t_0)$, $y_1(t_0+1) = y_2(t_0+1)$, but $y_1(t_1) \neq y_2(t_1)$, which violates the uniqueness of solutions of initial value problems for Eq. (9.1). This contradiction implies that $\alpha(t) \equiv \beta(t)$ on $[a, b+2]$. ∎

The final existence theorem of this section involves the case where $f(t, y)$ satisfies a one-sided Lipschitz condition with respect to y.

Theorem 9.9. Assume $f(t, y)$ is continuous in y for each t and there is a function $k(t)$ defined on $[a+1, b+1]$ such that

$$f(t, u) - f(t, v) \geq k(t)(u - v)$$

for $u \geq v$, t in $[a+1, b+1]$, and $\Delta^2 y(t-1) = k(t)y(t)$ is disconjugate on $[a, b+2]$. Then (9.1), (9.2) has a unique solution.

Proof. Let $y(t, m)$ be the solution of

$$\Delta^2 y(t-1) = f(t, y(t)),$$
$$y(a) = A, \quad y(a+1) = m.$$

Define $S = \{y(b+2, m) : m \text{ is in } R\}$. By continuity of solutions with respect to initial values, S is an interval. We will complete the (existence) proof by showing that S is bounded neither above nor below.

Fix $m_1 > m_2$ and let $w(t) = y(t, m_1) - y(t, m_2)$. We show by induction that $w(t) > 0$ on $[a+1, b+2]$. First, note that $w(a+1) = m_1 - m_2 > 0$. Let $t_0 > a+1$ and assume $w(t) > 0$ on $[a+1, t_0-1]$. For t in $[a+1, t_0-1]$,

$$\Delta^2 w(t-1) = f(t, y(t, m_1)) - f(t, y(t, m_2))$$
$$\geq k(t)[y(t, m_1) - y(t, m_2)]$$
$$= k(t)w(t).$$

By Theorem 6.6,

$$w(t) \geq (m_1 - m_2)u(t)$$

on $[a, t_0]$, where $u(t)$ is the solution of

$$\Delta^2 u(t-1) = k(t)u(t),$$
$$u(a) = 0, \quad u(a+1) = 1.$$

Now the disconjugacy of $\Delta^2 u(t-1) = k(t)u(t)$ on $[a, b+2]$ implies $u(t) > 0$ on $[a+1, b+2]$, and we have $w(t_0) \geq (m_1 - m_2)u(t_0) > 0$. By induction $w(t) > 0$ on $[a+1, b+2]$. In particular

$$w(b+2) \geq (m_1 - m_2)u(b+2).$$

Keeping m_2 fixed and letting $m_1 \to \infty$, we find that S is not bounded above. Fixing m_1 and letting $m_2 \to -\infty$, we have that S is not bounded below, so $S = R$ and, as a result, (9.1), (9.2) has a solution.

Uniqueness follows immediately from the fact that $m_1 > m_2$ implies $y(b+2, m_1) > y(b+2, m_2)$. ∎

9.4 Boundary Value Problems for Differential Equations

Mathematical modeling of problems arising in the physical sciences relies heavily on the use of differential equations together with initial or boundary conditions. We consider briefly in this section the relationship between the boundary value problems for difference equations of the last two sections and boundary value problems for differential equations of the type

$$y'' = f(x, y), \quad (0 \leq x \leq p) \tag{9.6}$$
$$y(0) = y(p) = 0. \tag{9.7}$$

The function f is assumed to be continuous in x and y. More general boundary value problems are considered by Gaines [82].

The following lemma will tell us which difference equations to use to approximate solutions of Eq. (9.6).

9.4 Boundary Value Problems for Differential Equations

Lemma 9.1. Assume $y(x)$ has a continuous second derivative on $[0, p]$. Let $\epsilon > 0$. For n sufficiently large and $1 \leq t \leq n - 1$,

$$\left| \frac{n^2}{p^2} \left[y\left(\frac{p}{n}(t+1)\right) - 2y\left(\frac{p}{n}t\right) + y\left(\frac{p}{n}(t-1)\right) \right] - y''\left(\frac{p}{n}t\right) \right| < \epsilon.$$

Proof. By Taylor's Theorem,

$$y\left(\frac{p}{n}(t+1)\right) = y\left(\frac{p}{n}t\right) + \frac{p}{n}y'\left(\frac{p}{n}t\right) + \frac{p^2}{2n^2}y''(c_1),$$

$$y\left(\frac{p}{n}(t-1)\right) = y\left(\frac{p}{n}t\right) - \frac{p}{n}y'\left(\frac{p}{n}t\right) + \frac{p^2}{2n^2}y''(c_2),$$

where $\frac{p}{n}t < c_1 < \frac{p}{n}(t+1)$ and $\frac{p}{n}(t-1) < c_2 < \frac{p}{n}t$. Adding these two equations, we have

$$y\left(\frac{p}{n}(t+1)\right) + y\left(\frac{p}{n}(t-1)\right) = 2y\left(\frac{p}{n}t\right) + \frac{p^2}{n^2}\left(\frac{y''(c_1) + y''(c_2)}{2}\right).$$

Finally,

$$\left| \frac{n^2}{p^2} \left[y\left(\frac{p}{n}(t+1)\right) - 2y\left(\frac{p}{n}t\right) + y\left(\frac{p}{n}(t-1)\right) \right] - y''\left(\frac{p}{n}t\right) \right|$$

$$= \left| \frac{y''(c_1) + y''(c_2)}{2} - y''\left(\frac{p}{n}t\right) \right|$$

$$\leq \frac{1}{2}\left| y''(c_1) - y''\left(\frac{p}{n}t\right) \right| + \frac{1}{2}\left| y''(c_2) - y''\left(\frac{p}{n}t\right) \right|$$

$$< \epsilon$$

for n sufficiently large and $1 \leq t \leq n - 1$ since y'' is uniformly continuous on $[0, p]$. ∎

Lemma 9.1 indicates that the second derivative term y'' in Eq. (9.6) can be approximated by $\frac{n^2}{p^2}[y(\frac{p}{n}(t+1)) - 2y(\frac{p}{n}t) + y(\frac{p}{n}(t-1))]$ when n is large. To obtain a difference equation of familiar form, we write $z(t)$

for $y(\frac{p}{n}t)$. The corresponding difference equation is then

$$\Delta^2 z(t-1) = \frac{p^2}{n^2} f\left(\frac{p}{n}t, z(t)\right) \quad (t=1,\cdots,n-1) \qquad (9.8)$$

with boundary conditions

$$z(0) = z(n) = 0. \qquad (9.9)$$

We now state a lemma that gives the fundamental relationship between solutions of (9.8), (9.9) and (9.6), (9.7). The proof is omitted since it would involve a detour into the theory of differential equations (see Gaines [82]).

Lemma 9.2. Assume

a) there is an n_0 so that (9.8), (9.9) has a solution $z_n(t)$ for $n \geq n_0$;

b) there are positive constants N and Q so that

$$|z_n(t)| \leq N, \quad n|\Delta z_n(t-1)| \leq Q$$

for $1 \leq t \leq n$ and $n \geq n_0$.

There is a subsequence $\{z_{n_k}(t)\}$ and a solution $y(x)$ of (9.6), (9.7) so that

$$\lim_{k\to\infty} \max_{0 \leq t \leq n_k} \left| z_{n_k}(t) - y\left(\frac{pt}{n_k}\right) \right| = 0.$$

If it is known that (9.6), (9.7) has at most one solution, then the original sequence $\{z_n(t)\}$ will converge to y in the sense of Lemma 9.2 (see Exercise 9.16).

The use of Lemma 9.2 is simplified by the next lemma.

Lemma 9.3. Let $\{z_n(t)\}$ be a sequence of solutions of (9.8), (9.9), and assume there is a positive constant N so that $|z_n(t)| \leq N$ for $1 \leq t \leq n-1$ and all n. Then there is a positive constant Q so that $n|\Delta z_n(t-1)| \leq Q$ for $1 \leq t \leq n$ and all n.

9.4 Boundary Value Problems for Differential Equations

Proof. Since f is continuous, there is a constant Q so that $|f(x,y)| \leq \frac{Q}{p^2}$ for $0 \leq x \leq p$ and $|y| \leq N$. By Eq. (9.3)

$$z_n(t) = \sum_{s=1}^{n-1} G(t,s) \frac{p^2}{n^2} f\left(\frac{p}{n}s, z_n(s)\right),$$

where by Example 6.12

$$G(t,s) = \begin{cases} -\frac{(n-s)t}{n} & (t \leq s) \\ -\frac{(n-t)s}{n} & (s \leq t). \end{cases}$$

Then

$$|\Delta z_n(t-1)| = \left| \sum_{s=1}^{n-1} \Delta_t G(t-1,s) \frac{p^2}{n^2} f\left(\frac{p}{n}s, z_n(s)\right) \right|$$

$$\leq \sum_{s=1}^{t-1} \frac{sp^2}{n^3} \left| f\left(\frac{ps}{n}, z_n(s)\right) \right| + \sum_{s=t}^{n-1} \frac{p^2(n-s)}{n^3} \left| f\left(\frac{p}{n}s, z_n(s)\right) \right|$$

$$\leq \frac{Q}{n^3} \left[\frac{t(t-1)}{2} + \frac{(n-t+1)(n-t)}{2} \right]$$

$$\leq \frac{Q}{n^3} \left[\frac{n(n-1)}{2} + \frac{n(n-1)}{2} \right]$$

$$= Q \frac{n-1}{n^2}.$$

We have

$$n|\Delta z_n(t-1)| \leq Q \frac{n-1}{n} \leq Q$$

for $1 \leq t \leq n$ and all n. ∎

As a first application of these lemmas, we give a version of Theorem 9.4 for differential equations.

Theorem 9.10. Assume there are constants N and K so that $|f(x,y_1) - f(x,y_2)| \leq K|y_1 - y_2|$ for $0 \leq x \leq p$ and y_1, y_2 in $[-N, N]$. Set $M = \max\{|f(x,y)| : 0 \leq x \leq p, |y| \leq N\}$. If $\frac{Kp^2}{8} < 1$ and $\frac{Mp^2}{8} \leq N$, then (9.6), (9.7) has a solution $y(x)$ that is the limit of a sequence of solutions of (9.8), (9.9) in the sense of Lemma 9.2.

Proof. Note that

$$\left|\frac{p^2}{n^2}f\left(\frac{p}{n}t, y_1\right) - \frac{p^2}{n^2}f\left(\frac{p}{n}t, y_2\right)\right| \leq K\frac{p^2}{n^2}|y_1 - y_2|$$

and

$$\max\left\{\frac{p^2}{n^2}\left|f\left(\frac{p}{n}t, y\right)\right| : 0 \leq t \leq n, |y| \leq N\right\} = \frac{p^2}{n^2}M.$$

Theorem 9.4 can now be applied to (9.8), (9.9) since $\alpha = K\frac{p^2}{n^2}\frac{n^2}{8} < 1$ and $\frac{p^2 M}{n^2}\frac{n^2}{8} \leq N$. For each n, (9.8), (9.9) has a unique solution $z_n(t)$ so that $|z_n(t)| \leq N$ for $0 \leq t \leq n$. By Lemmas 9.2 and 9.3, there is a subsequence $\{z_{n_k}\}$ and solution $y(x)$ of (9.6), (9.7) so that

$$\lim_{k \to \infty} \max_{0 \leq t \leq n_k} \left|z_{n_k}(t) - y\left(\frac{pt}{n_k}\right)\right| = 0. \qquad \blacksquare$$

Remark. It can be shown that the solution $y(x)$ given by Theorem 9.10 is unique (see [113]), so in fact the original sequence $\{z_n\}$ converges to y.

Our final theorem shows that the existence of continuous lower and upper solutions for (9.8), (9.9) implies that (9.6), (9.7) has a solution.

Theorem 9.11. Suppose $\alpha(x)$ and $\beta(x)$ are continuous functions on $[0, p]$ so that $\alpha(x) \leq \beta(x)$, $0 \leq x \leq p$, and $\alpha(\frac{p}{n}t)$ and $\beta(\frac{p}{n}t)$ are lower and upper solutions, respectively, for (9.8), (9.9) if n is sufficiently large. Then (9.6), (9.7) has a solution $y(x)$ with $\alpha(x) \leq y(x) \leq \beta(x)$, $0 \leq x \leq p$, and $y(x)$ is the limit of a sequence of solutions of (9.8), (9.9) in the sense of Lemma 9.2.

9.4 Boundary Value Problems for Differential Equations

Proof. For n sufficiently large, Theorem 9.7 implies that (9.8), (9.9) has a solution $z_n(t)$ with $\alpha(\frac{pt}{n}) \leq z_n(t) \leq \beta(\frac{pt}{n})$ for $0 \leq t \leq n$. Since α and β are continuous, there is an N so that $|z_n(t)| \leq N$ for $0 \leq t \leq n$. Using Lemmas 9.2 and 9.3, we obtain a solution $y(x)$ of (9.8), (9.9) so that some subsequence $\{z_{n_k}\}$ satisfies

$$\lim_{k \to \infty} \max_{0 \leq t \leq n_k} \left| z_{n_k}(t) - y\left(\frac{pt}{n_k}\right) \right| = 0.$$

It follows that $\alpha(x) \leq y(x) \leq \beta(x)$ for $0 \leq x \leq p$. ■

Alternatively, we can ask that α and β be lower and upper solutions for (9.6), (9.7) as defined in the following corollary.

Corollary 9.4. Assume α and β have continuous second derivatives on $[0,p]$, $\alpha(x) \leq \beta(x)$ for $0 \leq x \leq p$, $\alpha(0) \leq 0$, $\alpha(p) \leq 0$, $\beta(0) \geq 0$, $\beta(p) \geq 0$, and

$$\alpha''(x) - f(x, \alpha(x)) > 0,$$
$$\beta''(x) - f(x, \beta(x)) < 0$$

for $0 \leq x \leq p$. Then the conclusion of Theorem 9.11 holds.

Proof. Let $\epsilon = -\max_{0 \leq x \leq p} \{\beta''(x) - f(x, \beta(x))\} > 0$. By Lemma 9.1,

$$\frac{n^2}{p^2} \Delta^2 \beta\left(\frac{p}{n}(t-1)\right) - \beta''\left(\frac{p}{n}t\right) < \epsilon, \quad (1 \leq t \leq n-1),$$

if n is sufficiently large. Then

$$\frac{n^2}{p^2} \Delta^2 \beta\left(\frac{p}{n}(t-1)\right) - f\left(\frac{p}{n}t, \beta(\frac{p}{n}t)\right)$$
$$= \frac{n^2}{p^2} \Delta^2 \beta\left(\frac{p}{n}(t-1)\right) - \beta''\left(\frac{p}{n}t\right) + \beta''\left(\frac{p}{n}t\right) - f\left(\frac{p}{n}t, \beta(\frac{p}{n}t)\right)$$
$$< \epsilon - \epsilon = 0,$$

394 Chapter 9. BVPs for Nonlinear Equations

so $\beta(\frac{p}{n}t)$ is an upper solution for (9.8), (9.9) if n is sufficiently large. Similarly, $\alpha(\frac{p}{n}t)$ is a lower solution for (9.8), (9.9) if n is sufficiently large, and the conclusion of Theorem 9.11 holds. ∎

Example 9.6. The boundary value problem

$$(9.10) \qquad y'' + ry\left(1 - \frac{y}{K}\right) = 0,$$
$$(9.11) \qquad y(0) = y(p) = 0,$$

where r, K and p are positive constants, arises in the study of patches of plankton at the surface of the ocean (see [23]).

Note that $y(x) \equiv 0$ is a solution of the problem, but only positive solutions are of interest since $y(x)$ represents the density of plankton at position x inside the patch. First, let $\beta(x) \equiv K$. It is easily checked that $\beta(\frac{p}{n}t)$ is an upper solution for the discrete boundary value problem corresponding to (9.10), (9.11).

Now let $\alpha(x) = a\sin(\frac{\pi x}{p})$, where a is a positive constant. We want to show (under certain conditions) that $\alpha(\frac{p}{n}t) = a\sin(\frac{\pi}{n}t)$ is a lower solution for the discrete problem when a is small. Clearly, $\alpha(0) = \alpha(\frac{p}{n}n) = 0$. Using Theorem 2.2, we have

$$\Delta^2 a\sin\left(\frac{\pi}{n}(t-1)\right) + \frac{p^2}{n^2}ra\sin\left(\frac{\pi}{n}t\right)\left(1 - \frac{a\sin(\frac{\pi}{n}t)}{K}\right) \qquad (9.12)$$

$$= -4a\sin^2\left(\frac{\pi}{2n}\right)\sin\left(\frac{\pi}{n}t\right) + \frac{p^2}{n^2}ra\sin\left(\frac{\pi}{n}t\right)\left(1 - \frac{a\sin(\frac{\pi}{n}t)}{K}\right)$$

$$= a\sin\left(\frac{\pi}{n}t\right)\left[-4\sin^2\left(\frac{\pi}{2n}\right) + \frac{p^2}{n^2}r\left(1 - \frac{a\sin(\frac{\pi}{n}t)}{K}\right)\right].$$

Since $\theta > \sin\theta$ for $\theta > 0$, $\frac{\pi^2}{n^2} > 4\sin^2(\frac{\pi}{2n})$. Consequently, the expression in Eq. (9.12) is at least

$$a\sin\left(\frac{\pi}{n}t\right)\left[-\frac{\pi^2}{n^2} + \frac{p^2}{n^2}r\left(1 - \frac{a\sin(\frac{\pi}{n}t)}{K}\right)\right] > 0$$

if $rp^2 > \pi^2$ and a is sufficiently small. Thus $\alpha(\frac{p}{n}t)$ is a lower solution for all n. Of course, $\alpha(x) < \beta(x)$ is also true for small a. Theorem 9.11 yields

9.4 Boundary Value Problems for Differential Equations

a solution $y(x)$ of (9.10), (9.11) so that

$$a \sin\left(\frac{\pi x}{p}\right) \leq y(x) \leq K$$

for $0 \leq x \leq p$. The condition $rp^2 > \pi^2$ is interpreted to mean that the patch must have a width p of more than $\frac{\pi}{\sqrt{r}}$ in order that the colony of plankton be viable.

Exercises

Section 9.1

9.1 Show that solutions of initial value problems for Eq. (9.1) are unique and exist on $[a, b+2]$.

9.2 Show that if T is a contraction mapping on S, then T is continuous on S.

9.3 Show that if the condition $0 \leq \alpha < 1$ in Theorem 9.1 is replaced by $0 \leq \alpha \leq 1$, then T need not have a fixed point.

9.4 Give an example to show that the condition $\|Tx - Ty\| < \|x - y\|$ for all $x \neq y$ in R^n does not imply the existence of a fixed point for T.

9.5 Show that Theorem 9.1 is still true with a suitable change in Eq. (9.4) if we assume only that T^m is a contraction mapping on S for some integer $m \geq 1$.

Section 9.2

9.6 Verify that the maximum norm defined in the proof of Theorem 9.2 is a norm on R^{b-a+3}.

9.7 Prove that if $\Delta^2 y(t-1) + k(t)y(t) = 0$ is disconjugate on $[a, b+2]$, then for α near one and less than one the equation $\Delta^2 u(t-1) + \frac{k(t)}{\alpha} u(t) = 0$ has a positive solution on $[a, b+2]$.

9.8 For what values of b does the boundary value problem

$$\Delta^2 y(t-1) = \frac{0.2}{1 + y^2(t)},$$
$$y(0) = 0 = y(b+2)$$

have a unique solution?

9.9 Use Theorem 9.4 to show that

$$\Delta^2 y(t-1) = 0.1 y^2(t) + 1,$$
$$y(0) = 0 = y(3)$$

has a unique solution $y(t)$ with $|y(t)| \leq 2$ for $0 \leq t \leq 3$.

Exercises

Section 9.3

9.10 (a) Prove Theorem 9.5 in the case $n = 1$.
(b) Show that the fixed point in Theorem 9.5 need not be unique.

9.11 Show that $\alpha(t) \leq y(t)$ on $[a, b+2]$ in the proof of Theorem 9.7.

9.12 Complete the proof of Theorem 9.8 by treating the case $a \leq t_1 < t_0$.

9.13 Show that the boundary value problem $\Delta^2 y(t-1) = e^y$, $y(a) = A$, $y(b+2) = B$, has a unique solution for all A, B.

9.14 Show that the boundary value problem

$$\Delta^2 y(t-1) = y^2(t),$$
$$y(0) = 0, \quad y(10) = 10$$

has a solution $y(t)$ so that $0 \leq y(t) \leq t$ for $0 \leq t \leq 10$.

9.15 Consider the following special case of Example 9.4:

$$\Delta^2 y(t-1) = -0.1 \cos y(t),$$
$$y(0) = 0 = y(14).$$

Show there is a solution $y(t)$ so that $0.02t(14-t) \leq y(t) \leq 0.05t(14-t)$ for $0 \leq t \leq 14$.

Section 9.4

9.16 Show that if (9.6), (9.7) has at most one solution, then the original sequence $\{z_n\}$ in Lemma 9.2 converges to the solution of (9.6), (9.7).

9.17 Assume $f(x, y)$ is bounded and continuous on $[0, p] \times R$. Show that (9.6), (9.7) has a solution $y(x)$.

9.18 Use Corollary 9.4 to show that the boundary value problem

$$y'' + y - y^2 + 1.1 = 0,$$
$$y(0) = y(\pi) = 0$$

has a solution $y(x)$ such that $\sin x \leq y(x) \leq 1.7$ for $0 \leq x \leq \pi$.

Chapter 10
Partial Difference Equations

10.1 Discretization of Partial Differential Equations

Partial difference equations are difference equations that involve functions of two or more independent variables. We encountered several examples earlier in the book (see Example 1.5, Exercises 1.10 and 1.11, and Exercise 2.34). Such equations occur frequently in combinatorics and in the approximation of solutions of partial differential equations by finite difference methods. We will find in the examples that follow that the type of initial and boundary conditions needed to produce a unique solution of a partial difference equation depends on the form of the equation and on the domain in which the equation is to be solved.

Let us begin by considering the approximation of solutions of the heat equation

$$\frac{\partial u}{\partial t} = \frac{\partial^2 u}{\partial x^2}. \qquad (10.1)$$

With $u(x,t)$ denoting the temperature at position x and time t, the heat equation models the flow of thermal energy in one space dimension. Actually, there are numerous problems in the physical and biological sciences that involve diffusion and for which the heat equation gives a useful mathematical description.

In order to obtain an appropriate difference equation for Eq. (10.1), let h and k be small positive step sizes and define the grid points

$$x_i = ih, \quad t_j = jk,$$

for certain integral values of i and j.

By Taylor's formula,

$$u(x_i, t_j + k) = u(x_i, t_j) + \frac{\partial u}{\partial t}(x_i, t_j)k$$
$$+ \frac{\partial^2 u}{\partial t^2}(x_i, c_{ij})\frac{k^2}{2}$$

for some c_{ij} between t_j and $t_j + k$, so

$$\frac{\partial u}{\partial t}(x_i, t_j) = \frac{u(x_i, t_j + k) - u(x_i, t_j)}{k} + \mathcal{O}(k), \qquad (10.2)$$

provided that $\frac{\partial^2 u}{\partial t^2}$ exists and is bounded. Also,

$$u(x_i + h, t_j) = u(x_i, t_j) + \frac{\partial u}{\partial x}(x_i, t_j)h$$
$$+ \frac{\partial^2 u}{\partial x^2}(x_i, t_j)\frac{h^2}{2} + \frac{\partial^3 u}{\partial x^3}(x_i, t_j)\frac{h^3}{3!}$$
$$+ \frac{\partial^4 u}{\partial x^4}(d_{ij}, t_j)\frac{h^4}{4!}$$

for some d_{ij} between x_i and $x_i + h$ if $\frac{\partial^4 u}{\partial x^4}$ exists and is bounded.

Now add the preceding expression for $u(x_i + h, t_j)$ to the analogous expression for $u(x_i - h, t_j)$ and rearrange to obtain

$$\frac{\partial^2 u}{\partial x^2}(x_i, t_j) = \frac{u(x_i + h, t_j) - 2u(x_i, t_j) + u(x_i - h, t_j)}{h^2} + \mathcal{O}(h^2). \quad (10.3)$$

Then from Eqs. (10.2), (10.3),

$$\frac{\partial u}{\partial t}(x_i, t_j) - \frac{\partial^2 u}{\partial x^2}(x_i, t_j) = \frac{u(x_i, t_j + k) - u(x_i, t_j)}{k}$$
$$- \frac{u(x_i + h, t_j) - 2u(x_i, t_j) + u(x_i - h, t_j)}{h^2}$$
$$+ \mathcal{O}(k) + \mathcal{O}(h^2).$$

10.1 Discretization of Partial Differential Equations

Let $y(i,j) = u(x_i, t_j)$. The approximating difference equation for Eq. (10.1) is

$$y(i, j+1) = \left(1 - \frac{2k}{h^2}\right) y(i,j) + \frac{k}{h^2}\left(y(i+1,j) + y(i-1),j)\right). \quad (10.4)$$

We see that Eq. (10.4) permits us to compute the value of y at $(i, j+1)$ if the values of y at (i,j), $(i+1,j)$, and $(i-1,j)$ are known. Any such group of four points in the grid is called a computational "molecule" (see Fig. 10.1).

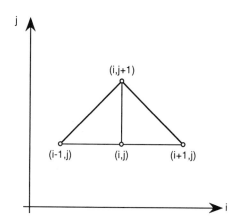

Fig. 10.1 A molecule for the heat equation

For example, suppose we wish to solve Eq. (10.1) together with an initial condition

$$u(x, 0) = g(x), \quad (-\infty < x < \infty).$$

Let $f(i) = g(x_i)$, where i ranges over the set of all integers. Then for Eq. (10.4) we have the initial condition

$$y(i, 0) = f(i), \quad (i = 0, \pm 1, \pm 2, \cdots).$$

Using the molecules with lower base on the i axis, we can compute from Eq. (10.4) and the initial values $f(i)$ all the values $y(i,1)$. Similarly, these values can be used to compute $y(i,2)$ for all i. Thus $y(i,j)$ is uniquely determined for all i and $j = 0, 1, 2, \cdots$ in this manner.

Next, consider the problem of obtaining a unique solution of Eq. (10.1) in the first quadrant $\{(x,t) : x \geq 0,\ t \geq 0\}$. If we know $u(x,0) = g(x)$ for $x \geq 0$, then we have the initial values $y(i,0) = f(i)$ for $i = 0, 1, \cdots$. Observe that the molecules determine $y(i,j)$ only for $i \geq j$ (see Fig. 10.2). In order to find $y(i,j)$ for $j > i$ we need additional information such as the values of $y(0,j)$ for $j \geq 1$. Thus a unique solution of Eq. (10.4) is obtained by iteration if we have the additional condition that $u(0,t)$ is prescribed for all $t \geq 0$. In the heat flow problem, this condition corresponds to knowing the temperature at $x = 0$ for all $t \geq 0$.

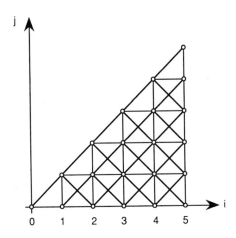

Fig. 10.2 Grid points reached from the initial axis

A related problem involves a finite space domain, say $0 \leq x \leq l$. We take $h = \frac{l}{N}$ for some positive integer N and have $x_i = ih$, $i = 0, 1, 2, \cdots, N$. Now the initial values $y(i,0) = f(i)$ ($i = 0, \cdots, N$) determine $y(i,j)$ in a very limited region (see Fig. 10.3). In this case we need the values $y(0,j)$ and $y(N,j)$ for $j = 1, 2, \cdots$ in order to find unique values for $y(i,j)$, $i = 0, \cdots, N$, $j = 0, 1, \cdots$.

10.1 Discretization of Partial Differential Equations

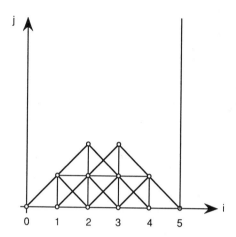

Fig. 10.3 Grid points reached without boundary values

Specifically, we consider the problem

$$\begin{aligned}
\frac{\partial u}{\partial t} &= \frac{\partial^2 u}{\partial x^2}, & (0 < x < l,\ t > 0) \\
u(x, 0) &= g(x), & (0 \leq x \leq l) \\
u(0, t) &= u(l, t) = 0, & (t > 0),
\end{aligned} \qquad (10.5)$$

where $g(0) = g(l) = 0$. The corresponding discrete problem is

$$\begin{aligned}
y(i, j+1) &= (1 - 2\alpha) y(i, j) + \alpha \left(y(i+1, j) + y(i-1), j) \right), \\
& \qquad (i = 1, \cdots, N-1;\ j = 0, 1, \cdots) \\
y(i, 0) &= f(i), \qquad (i = 0, \cdots, N) \\
y(0, j) &= y(N, j) = 0, \qquad (j = 1, 2, \cdots),
\end{aligned} \qquad (10.6)$$

where $\alpha = \frac{k}{h^2}$, $h = \frac{l}{N}$, and $f(i) = g(ih)$. Now Eq. (10.6) can be written in matrix form by defining the $N-1$ by $N-1$ matrix

$$A = \begin{bmatrix} 1-2\alpha & \alpha & 0 & \cdots & 0 \\ \alpha & 1-2\alpha & \alpha & \cdots & 0 \\ 0 & \alpha & 1-2\alpha & \ddots & \vdots \\ \vdots & \ddots & \ddots & \ddots & \alpha \\ 0 & \cdots & 0 & \alpha & 1-2\alpha \end{bmatrix},$$

and the vectors

$$v(j) = \begin{bmatrix} y(1,j) \\ y(2,j) \\ \vdots \\ y(N-1,j) \end{bmatrix}, \quad v^0 = \begin{bmatrix} f(1) \\ f(2) \\ \vdots \\ f(N-1) \end{bmatrix}.$$

Then Eq. (10.6) is equivalent to

$$\begin{aligned} v(j+1) &= Av(j), \quad (j=0,1,\cdots) \\ v(0) &= v^0. \end{aligned} \quad (10.7)$$

We have a system of ordinary difference equations with constant coefficients like those studied in Chapter 4. Recall that the solution of the initial value problem (10.7) is $v(j) = A^j v^0$.

In order to analyze Eq. (10.7) further, we will compute the eigenvalues of the matrix A. Note that A is symmetric, so it has real eigenvalues and $N-1$ independent eigenvectors. From Section 7.2, we have that the eigenvalues of A are the same as the eigenvalues of the Sturm-Liouville problem

$$\begin{aligned} \alpha w(t+1) + (1-2\alpha)w(t) + \alpha w(t-1) &= \lambda w(t), \\ w(0) = 0, \quad w(N) &= 0. \end{aligned}$$

A rearrangement of the difference equation yields

$$w(t+1) + (\mu - 2)w(t) + w(t-1) = 0,$$

10.1 Discretization of Partial Differential Equations

where $\mu = \frac{1-\lambda}{\alpha}$. The eigenvalues can be found as in Example 7.1:

$$\mu_n = 2 - 2\cos\frac{n\pi}{N}$$
$$= 4\sin^2\frac{n\pi}{2N}, \quad (n = 1, \cdots, N-1).$$

Thus,

$$\lambda_n = 1 - 4\alpha\sin^2\frac{n\pi}{2N}, \quad (n = 1, \cdots, N-1),$$

are the eigenvalues of A.

Moreover, the discussion in Section 7.2 allows us to compute $N-1$ independent eigenvectors of A using the eigenfunctions $\sin\frac{n\pi}{N}t$, $(n = 1, \cdots, N-1)$, of the Sturm-Liouville problem. The resulting matrix of eigenvectors is

$$M = \begin{bmatrix} \sin\frac{\pi}{N} & \sin 2\frac{\pi}{N} & \cdots & \sin(N-1)\frac{\pi}{N} \\ \sin 2\frac{\pi}{N} & \sin 2 \cdot 2\frac{\pi}{N} & \cdots & \sin 2(N-1)\frac{\pi}{N} \\ \vdots & \vdots & \ddots & \vdots \\ \sin(N-1)\frac{\pi}{N} & \sin 2(N-1)\frac{\pi}{N} & \cdots & \sin(N-1)^2\frac{\pi}{N} \end{bmatrix}.$$

Let

$$\begin{bmatrix} b_1 \\ \vdots \\ b_{N-1} \end{bmatrix} = M^{-1}v^0.$$

From Eq. (4.6), the solution of Eq. (10.7) is given by

$$v(j) = M \begin{bmatrix} b_1 \lambda_1^j \\ \cdot \\ \cdot \\ \cdot \\ b_{N-1}\lambda_{N-1}^j \end{bmatrix}$$

for $j = 0, 1, 2, \cdots$.

Does the computed solution $v(j)$ serve as a good approximation to the solution of Eq. (10.5)? Note that $|\lambda_n| < 1$ for $n = 1, \cdots, N-1$ if and only if

$$\frac{k}{h^2} \sin^2 \frac{N-1}{N} \frac{\pi}{2} < \frac{1}{2}. \tag{10.8}$$

Thus if Eq. (10.8) holds, then the solution $v(j)$ goes to zero as $j \to \infty$. It is easily shown (for example by Fourier analysis) that the solution u of Eq. (10.5) also has the property $\lim_{t \to \infty} u(x,t) = 0$ for each x in $[0, l]$. On the other hand, if Eq. (10.8) is violated, then in most cases $v(j)$ will not converge to zero as $j \to \infty$ and thus will be a very poor approximation of $u(x,t)$ as t increases. For this reason the present method is said to be "conditionally stable." It can be shown that if $\frac{k}{h^2} \leq \frac{1}{2}$ and the initial function g is sufficiently smooth, then the $v(j)$ are $\mathcal{O}(k + h^2)$ approximations to the solution of Eq. (10.5) (see [136]).

An alternate approach for discretizing the heat equation is to approximate $\frac{\partial u}{\partial t}$ as follows:

$$u(x_i, t_j - k) = u(x_i, t_j) - \frac{\partial u}{\partial t}(x_i, t_j)k$$
$$+ \frac{\partial^2 u}{\partial t^2}(x_i, c_{ij})\frac{k^2}{2},$$

where c_{ij} is between t_j and $t_j - k$, so

$$\frac{\partial u}{\partial t}(x_i, t_j) = \frac{u(x_i, t_j) - u(x_i, t_j - k)}{k} + \mathcal{O}(k),$$

a "backwards difference quotient." Using this formula together with Eq. (10.3) in Eq. (10.1), we arrive at the difference equation

$$y(i, j-1) = \left(1 + \frac{2k}{h^2}\right) y(i,j) - \frac{k}{h^2} \left(y(i+1, j) + y(i-1, j)\right), \tag{10.9}$$

where $y(i,j) = u(x_i, t_j)$ (see Exercise 10.2). Equation (10.9) has molecules of the type shown in Fig. 10.4. Thus with this approach the solution of Eq. (10.5) cannot be approximated by explicit iteration of Eq. (10.9), and we have what is known as an implicit method.

10.1 Discretization of Partial Differential Equations

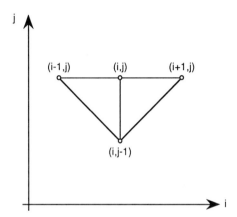

Fig. 10.4 A molecule for the implicit method

However, if we define α, $v(j)$ and $v(0)$ as before and

$$B = \begin{bmatrix} 1+2\alpha & -\alpha & 0 & \cdots & 0 \\ -\alpha & 1+2\alpha & -\alpha & \cdots & 0 \\ 0 & -\alpha & 1+2\alpha & \ddots & \vdots \\ \vdots & \ddots & \ddots & \ddots & -\alpha \\ 0 & \cdots & 0 & -\alpha & 1+2\alpha \end{bmatrix},$$

then Eq. (10.9) plus the auxiliary conditions can be written

$$Bv(j) = v(j-1), \quad (j = 1, 2, \cdots)$$
$$v(0) = v^0.$$

Now the eigenvalues of B are

$$1 + 4\alpha \sin^2 \frac{n\pi}{2N}, \quad (n = 1, \cdots, N-1),$$

so B is invertible, and we have

$$v(j) = B^{-1}v(j-1), \quad (j=1,2,\cdots) \tag{10.10}$$
$$v(0) = v^0.$$

The system (10.10) can be solved explicitly as in the earlier method (see Exercise 10.3). Also, note that the eigenvalues of B^{-1} are

$$0 < \frac{1}{1+4\alpha\sin^2\frac{n\pi}{2N}} < 1, \quad (n=1,\cdots,N-1).$$

Consequently, this implicit method is unconditionally stable and can be used to approximate the solution of Eq. (10.5) for all small values of h and k.

Difference equations that approximate solutions of other partial differential equations can be obtained in much the same way using Taylor's formula. For example, for Laplace's equation

$$\frac{\partial^2 u}{\partial x^2} + \frac{\partial^2 u}{\partial y^2} = 0,$$

we can derive the difference equation

$$2\left[\left(\frac{h}{k}\right)^2 + 1\right]z(i,j) = z(i+1,j) + z(i-1,j)$$
$$+ \left(\frac{h}{k}\right)^2 (z(i,j+1) + z(i,j-1)), \tag{10.11}$$

where $z(i,j) = u(x_i, y_i)$, $x_i = ih$ and $y_j = jk$ (see Exercise 10.4). Since Eq. (10.11) has diamond-shaped molecules such as the one shown in Figure 10.5, it turns out that the values of $z(i,j)$ on the boundary of a closed region such as the rectangle in Fig. 10.5 must be known in order to compute the values of $z(i,j)$ throughout the region.

10.2 Solutions of Partial Difference Equations

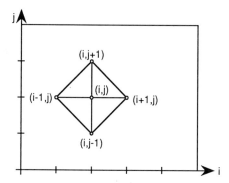

Fig. 10.5 A molecule for Laplace's equation

10.2 Solutions of Partial Difference Equations

In Section 10.1, we were able to give an explicit solution for a partial difference equation with certain auxiliary conditions by writing it as an initial value problem for a system of ordinary difference equations. This approach was possible because of the special nature of the problem: the domain of interest was finite in one direction and semi-infinite in the other, and the lateral boundary conditions were homogeneous.

This section will summarize briefly a number of other special methods for finding explicit solutions of linear partial difference equations. In order to get a general idea of what the nature of solutions may be, consider first the linear equation with two terms

$$y(i,j) = p(i)y(i+a, j+b).$$

For simplicity, we have assumed that the coefficient function p is a function of only one variable and p does not vanish. The shifts a and b are arbitrary but fixed integers with $a \neq 0$. Let f be an arbitrary function and try a solution of the form

$$y(i,j) = z(i)f(aj - bi). \qquad (10.12)$$

We get

$$\begin{aligned}z(i)f(aj - bi) &= p(i)z(i+a)f\left(a(j+b) - b(i+a)\right) \\ &= p(i)z(i+a)f(aj - bi),\end{aligned}$$

so

$$z(i) = p(i)z(i+a).$$

The last equation can be solved by iteration and has $|a|$ independent solutions $z_1, \cdots, z_{|a|}$. From Eq. (10.12) and the fact that the equation is homogeneous, we have that

$$y(i,j) = \sum_{i=1}^{|a|} f_n(aj - bi)z_n(i)$$

is a solution of the two term difference equation for any arbitrary functions $f_1, \cdots, f_{|a|}$.

Example 10.1. Solve the equation

$$y(i,j) = \frac{1}{i}y(i+2, j+1).$$

Letting $y(i,j) = z(i)f(2j - i)$, we find that z satisfies

$$z(i+2) = iz(i).$$

10.2 Solutions of Partial Difference Equations

By iteration, two independent solutions are

$$z_1(i) = \begin{cases} \frac{(2k-1)!}{2^{k-1}(k-1)!} & \text{if } i = 2k+1, \\ 0 & \text{if } i = 2k, \end{cases}$$

$$z_2(i) = \begin{cases} 0 & \text{if } i = 2k+1, \\ (k-1)!2^{k-1} & \text{if } i = 2k, \end{cases}$$

for $i = 1, 2, \cdots$. Thus the original equation has solutions of the form

$$y(i,j) = f_1(2j-i)z_1(i) + f_2(2j-i)z_2(i),$$

where f_1 and f_2 are arbitrary functions.

Equations with three or more terms can be solved via the substitution (10.12) in special cases where the shifts satisfy a certain compatibility condition. Consider

$$y(i,j) = p(i)y(i+a, j+b) + q(i)y(i+c, j+d). \tag{10.13}$$

Again we try $y(i,j) = z(i)f(aj - bi)$:

$$\begin{aligned} z(i)f(aj-bi) &= p(i)z(i+a)f(aj-bi) \\ &\quad + q(i)z(i+c)f(aj-bi+ad-bc). \end{aligned}$$

The f's will cancel to give an ordinary difference equation if the shifts satisfy the condition $ad - bc = 0$. In this case the equation is

$$z(i) = p(i)z(i+a) + q(i)z(i+c).$$

Example 10.2. Two players P and Q play a game where at each stage P wins a chip from Q with probability p and Q wins a chip from P with probability $q = 1 - p$. The game ends when one player is out of chips. Let $y(i,j)$ denote the probability that P wins the game if P starts the game with i chips and Q starts with j chips. Find $y(i,j)$ for $i,j = 0, 1, \cdots$.

Consider the first stage of the game. There are two mutually exclusive possibilities: either P wins a chip (with probability p) and then has probability $y(i+1, j-1)$ of winning the game or P loses a chip and has probability $y(i-1, j+1)$ of winning. Thus y satisfies

$$y(i,j) = py(i+1, j-1) + qy(i-1, j+1),$$

which is of the form of Eq. (10.13) with $a = d = 1$ and $b = c = -1$. Since $ad - bc = 0$, we substitute $y(i,j) = z(i)f(i+j)$ and find

$$z(i) = pz(i+1) + qz(i-1).$$

This second order equation has characteristic roots $\lambda = \frac{q}{p}, 1$, so $z(i) = 1$, $\left(\frac{q}{p}\right)^i$ are linearly independent solutions. Then

$$y(i,j) = f_1(i+j) + f_2(i+j)\left(\frac{q}{p}\right)^i.$$

We need two boundary conditions in order to compute f_1 and f_2. Since P has no chance of winning if $i = 0$ and no change of losing of $j = 0$, we have

$$y(0,j) = 0, \quad y(i,0) = 1$$

for $i, j = 1, 2, \cdots$. Thus,

$$f_1(j) + f_2(j) = 0,$$
$$f_1(i) + f_2(i)\left(\frac{q}{p}\right)^i = 1,$$

so

$$f_1(i+j) = -f_2(i+j) = \frac{1}{1 - \left(\frac{q}{p}\right)^{i+j}},$$

10.2 Solutions of Partial Difference Equations

and finally

$$y(i,j) = \frac{1}{1-\left(\frac{q}{p}\right)^{i+j}}\left[1-\left(\frac{q}{p}\right)^i\right].$$

Check this answer for $i = j = 1!$. Also, compare with Exercise 3.71.

When the shifts in the partial difference equation (10.13) fail to satisfy the compatibility condition, it may still be possible to find an exact solution by the operator method or by the use of the z-transform (or equivalently a generating function). We will use both of these methods to solve the same difference equation in the next example.

Example 10.3. Solve

$$y(i,j) = y(i-1,j) + y(i-1,j-1).$$

In order to formulate an equivalent operator equation, we introduce the shift operators

$$E_1 y(i,j) = y(i+1,j), \quad E_2 y(i,j) = y(i,j+1).$$

The equation now reads

$$y(i,j) = \left(E_1^{-1} + E_1^{-1} E_2^{-1}\right) y(i,j)$$
$$= E_1^{-1}\left(I + E_2^{-1}\right) y(i,j).$$

If we apply E_1 to both sides, we get

$$y(i+1,j) = \left(I + E_2^{-1}\right) y(i,j).$$

We can regard this last equation as a difference equation with variable i in which $(I+E_2^{-1})$ acts like a constant since it has no effect on i. By iteration,

$$y(i,j) = \left(I + E_2^{-1}\right)^i f(j),$$

where f is arbitrary. The Binomial Theorem gives

$$y(i,j) = \sum_{n=0}^{i} \binom{i}{n} E_2^{-n} f(j)$$

$$= \sum_{n=0}^{i} \binom{i}{n} f(j-n).$$

Now we will solve the equation using the z-transform. It will be easier to apply the results of Section 3.7 if the equation is written with positive shifts:

$$y(i+1, j+1) = y(i, j+1) + y(i,j). \tag{10.14}$$

Also, we can simplify the calculation by imposing initial conditions at $i = 0$ and $j = k$, say

$$y(0,j) = \delta_{kj}, \quad y(i,k) = i+1, \tag{10.15}$$

for $j \geq k$, $i \geq 0$.

Let $Y(z,j)$ denote the z-transform of $y(i,j)$ with respect to i:

$$Y(z,j) = \sum_{n=0}^{\infty} \frac{y(n,j)}{z^n}.$$

Applying the z-transform to Eq. (10.14), we have by Theorem 3.12

$$zY(z, j+1) - zy(0, j+1) = Y(z, j+1) + Y(z,j).$$

The first equation in (10.15) implies $y(0, j+1) = 0$ for $j \geq k$, so

$$Y(z, j+1) = \frac{1}{z-1} Y(z,j).$$

10.2 Solutions of Partial Difference Equations

By iteration,

$$Y(z,j) = \frac{1}{(z-1)^{j-k}} Y(z,k).$$

From Eq. (10.15) and Table 3.1,

$$Y(z,k) = z(i+1) = \frac{z}{(z-1)^2} + \frac{z}{z-1},$$

so

$$Y(z,j) = \frac{z}{(z-1)^{j-k+2}} + \frac{z}{(z-1)^{j-k+1}}.$$

By Exercise 3.107,

$$\begin{aligned} y(i,j) &= \binom{i}{j-k+1} + \binom{i}{j-k} \\ &= \binom{i+1}{j-k+1} \end{aligned}$$

is the unique solution of Eqs. (10.14), (10.15).

We can obtain the same answer we got by operator techniques if we note that Eq. (10.14) is homogeneous, and consequently sums of solutions of Eq. (10.14) are also solutions. Let g be an arbitrary function. Then we have the solution

$$\begin{aligned} y(i,j) &= \sum_{k=j-i}^{j+1} \binom{i+1}{j-k+1} g(k) \\ &= \sum_{m=0}^{i+1} \binom{i+1}{m} g(j-m+1), \end{aligned}$$

which is equivalent to the previous result.

In Example 1.5 we showed that the number of ways $W(n,k)$ of obtaining k red marbles in n draws with replacement from a sack with r

red marbles and g green marbles satisfies the equation

$$W(n,k) = rW(n-1, k-1) + gW(n-1, k),$$

which is more general than the equation in Example 10.3. It can also be solved by operator or transform methods (see Exercise 10.9). The exercises also contain a number of nonhomogeneous equations that can be solved by these methods.

The solution of linear partial difference equations with variable coefficients can occasionally be carried out using Stirling numbers of the first or second kind. Stirling numbers of the second kind were introduced in Exercise 2.34. Here we consider only Stirling numbers of the first kind.

We define the Stirling number of the first kind, written $\begin{bmatrix} i \\ j \end{bmatrix}$, to be the solution $y(i,j) = \begin{bmatrix} i \\ j \end{bmatrix}$ of the equation

$$y(i+1, j+1) = iy(i, j+1) + y(i, j) \quad (i, j \geq 0), \tag{10.16}$$

which satisfies the initial conditions

$$y(i, 0) = \delta_{i0}, \quad y(0, j) = \delta_{0j} \quad (i, j \geq 0). \tag{10.17}$$

It is easy to check by sketching the computational molecules that (10.16), (10.17) has a unique solution $y(i, j)$ for $i, j \geq 0$. Knuth [143] contains a number of formulas and tables which provide much useful information about the Stirling numbers.

Let us compute the z-transform of $\begin{bmatrix} i \\ j \end{bmatrix}$ with respect to j by applying the z-transform to both sides of Eq. (10.16):

$$zY(i+1, z) - zy(i+1, 0) = izY(i, z) + Y(i, z).$$

From Eq. (10.17), $y(i+1, 0) = 0$ for $i \geq 0$, so

$$Y(i+1, z) = (i + \frac{1}{z})Y(i, z).$$

10.2 Solutions of Partial Difference Equations

Then

$$Y(i,z) = \prod_{k=0}^{i-1}(k + \frac{1}{z})Y(0,z).$$

Again by Eq. (10.17), $Y(0,z) = Z(\delta_{0j}) = 1$, so the z-transform of $\begin{bmatrix} i \\ j \end{bmatrix}$ with respect to j is

$$Z\left(\begin{bmatrix} i \\ j \end{bmatrix}\right) = \prod_{k=0}^{i-1}(k + \frac{1}{z}).$$

From Definition 2.3,

$$\begin{aligned}\left(-\frac{1}{z}\right)^{(i)} &= -\frac{1}{z}\left(-\frac{1}{z} - 1\right)\left(-\frac{1}{z} - 2\right) \cdot \cdots \cdot \left(-\frac{1}{z} - (i-1)\right) \\ &= (-1)^i \frac{1}{z}\left(\frac{1}{z} + 1\right) \cdot \cdots \cdot \left(\frac{1}{z} + (i-1)\right),\end{aligned}$$

so the preceding formula can be written in terms of the factorial function:

$$Z\left(\begin{bmatrix} i \\ j \end{bmatrix}\right) = (-1)^i \left(-\frac{1}{z}\right)^{(i)}.$$

The following example was suggested by Knuth [143].

Example 10.4. Suppose i distinct numbers are placed in a hat and drawn out one by one at random to find the largest. At each stage the number drawn is compared to the largest number found so far. If the number drawn is smaller we discard it, and if it is larger we replace the previous largest number by it. Let $p(i,j)$ be the probability that exactly j replacements are needed.

To find an equation for $p(i,j)$ consider the case that the last number drawn is the largest. This event occurs with probability $\frac{1}{i}$, and the number of replacements in this case is one more than the number needed for the remaining $i-1$ numbers. If the last number is not the largest (probability

$= \frac{i-1}{i}$), the number of replacements is the same. Thus

$$p(i,j) = \frac{1}{i}p(i-1,j-1) + \frac{i-1}{i}p(i-1,j). \qquad (10.18)$$

The initial conditions are chosen to be

$$p(1,j) = \delta_{j0}, \quad p(i,-1) = 0, \qquad (10.19)$$

for $i \geq 1$, $j \geq 0$. The reader should check that the second equation in (10.19) is consistent with Eq. (10.18) and the known values $p(i,0)$.

Since we are planning to take the z-transform with respect to the second variable, we would like the second condition in Eq. (10.19) to be given at zero. Thus we make the change of variables

$$k = j+1, \quad y(i,k) = p(i,k-1).$$

Now y satisfies

$$y(i+1,k+1) = \frac{1}{i+1}y(i,k) + \frac{i}{i+1}y(i,k+1), \qquad (10.20)$$
$$y(i,0) = 0, \quad y(1,k) = \delta_{0,k-1}, \qquad (10.21)$$

for $i \geq 1$, $k \geq 0$. Now apply the z-transform (with respect to k) to Eq. (10.20):

$$zY(i+1,z) = \frac{i}{i+1}zY(i,z) + \frac{1}{i+1}Y(i,z),$$

where we have used Eq. (10.21). Then

$$Y(i+1,z) = \left(\frac{i+\frac{1}{z}}{i+1}\right)Y(i,z).$$

10.2 Solutions of Partial Difference Equations

Iterating, we have

$$Y(i,z) = \frac{1}{i!} \prod_{n=1}^{i-1} (n + \frac{1}{z}) Y(1,z).$$

From Eq. (10.21), $Y(1,z) = Z(\delta_{0,k-1}) = \frac{1}{z}$, so

$$Y(i,z) = \frac{1}{i!} \prod_{n=0}^{i-1} (n + \frac{1}{z}).$$

It follows that

$$y(i,k) = \frac{1}{i!} \begin{bmatrix} i \\ k \end{bmatrix},$$

and finally

$$p(i,j) = \frac{1}{i!} \begin{bmatrix} i \\ j+1 \end{bmatrix}.$$

Exercises

Section 10.1

10.1 Solve the problem:
$$y(i, j+1) = 1/2\, y(i,j) + 1/4\, (y(i+1,j) + y(i-1,j)),$$
$$y(0, j) = y(4, j) = 0, \quad (j \geq 0)$$
$$y(i, 0) = \sin i\pi/4, \quad (i = 1, 2, 3),$$
for all $j \geq 1$.

10.2 Derive the difference equation (10.9) for the heat equation.

10.3 Show how to obtain an explicit solution for the system (10.10).

10.4 Derive the partial difference equation (10.11) for LaPlace's equation.

10.5 For the wave equation $\frac{\partial^2 u}{\partial t^2} - \frac{\partial^2 u}{\partial x^2} = 0$,
(a) derive a difference equation of the form

$$y(i, j+1) = 2(1-\alpha^2) y(i,j) + \alpha^2 \left(y(i+1, j) + y(i-1, j) \right) - y(i, j-1).$$

(b) Sketch a typical computational molecule for the difference equation in (a).

Section 10.2

10.6 Solve the equation $y(i, j) = 4y(i+2, j+1)$.

10.7 Solve the nonhomogeneous equation $y(i, j) = 2y(i-1, j-1) + 3^i$.

10.8 Use the substitution (10.12) to find solutions of
(a) $y(i, j) = 2y(i-1, j+3) - y(i-2, j+6)$,
(b) $y(i, j) - 5y(i-1, j+1) + 6y(i-2, j+2) = 3i$.

10.9 Use the operator method to solve the problem in Example 1.5:
$$W(n, k) = rW(n-1, k-1) + gW(n-1, k),$$
$$W(n, 0) = g^n.$$

10.10 Show that
$$y(i+1, j) = ay(i, j+1) + by(k, j)$$

(a,b constants) has solution

$$y(i,j) = b^i \sum_{n=0}^{i} \binom{i}{n} \left(\frac{a}{b}\right)^n f(j+n),$$

using the operator method.

10.11 Suppose that in a certain game P needs i points to win while Q needs j points to win. At each stage P wins a point with probability p and Q wins a point with probability $q = 1 - p$. Let $y(i,j)$ be the probability that P wins the game.

(a) Show that $y(i,j)$ satisfies

$$y(i,j) = py(i-1,j) + qy(i,j-1)$$
$$y(0,j) = 1 \quad (j \geq 1), \quad y(i,0) = 0 \quad (i \geq 1).$$

(b) Find $y(i,j)$.

10.12 Solve the equation

$$y(i+1, j+1) + y(i,j) = 2ij$$

by substituting a trial solution $y(i,j) = aij + bi + cj + d$ into the equation, solving for a, b, c, and d, and adding this particular solution to the general solution of the associated homogeneous equation.

10.13 Compute the Stirling numbers $\begin{bmatrix} i \\ j \end{bmatrix}$ for $0 \leq i, j \leq 5$.

10.14 (a) Show that

$$Z\left(2^j \begin{bmatrix} i \\ j \end{bmatrix}\right) = \prod_{k=0}^{i-1}\left(k + \frac{2}{z}\right).$$

(b) For $i \geq 1$, $j \geq 0$, use the z-transform to solve

$$y(i+1, j+1) = (i-1)y(i, j+1) + 2y(i,j),$$
$$y(i,0) = \delta_{i1}, \quad y(1,j) = \delta_{j0},$$

Answers to Selected Problems

Chapter 1

[1.1] $909.70, 11.6 years.

[1.2] 168,750.

[1.3] $(10,000)10^{\frac{t}{2}}$.

[1.4] 1.1 hours.

[1.8] $R(t) = \frac{1}{2}t^2 + \frac{1}{2}t + 1$.

[1.9] $a_{k+2} = \frac{k}{(k+1)(k+2)} a_k$, $k \geq 1$ and odd, a_0 arbitrary, $a_{2k} = 0$, $k \geq 1$.

[1.13] $\Gamma(5/2) = 3/4 \sqrt{\pi}$, $\Gamma(-3/2) = 4/3 \sqrt{\pi}$.

Chapter 2

[2.10] No. For example $t^{(1)}t^{(1)} = t^2 \neq t^{(2)}$.

[2.13] (a) $y(t) = \frac{1}{4}t^{(4)} + \frac{1}{2}3^t + C(t)$.

(b) $y(t) = \binom{t}{7} + tC(t) + D(t)$ where $C(t+1) = C(t)$, $D(t+1) = D(t)$.

[2.18] $y(t+1) = y(t) + t$, $y(t) = \frac{1}{2}t^{(2)}$.

[2.20] $-\frac{t\cos(t-\frac{1}{2})}{2\sin\frac{1}{2}} + \frac{\sin t}{4\sin^2 \frac{1}{2}} + C(t)$.

[2.21] $-\frac{1}{2}t \cdot t^{(2)} - \frac{1}{2}(t+1)^{-1} + C(t)$.

[2.22] (a) $\frac{1}{2}3^t(t^2 - 3t + 3) + C(t)$.

(b) $\binom{t}{2}\binom{t}{8} - t\binom{t+1}{9} + \binom{t+2}{10} + C(t)$.

(c) $\binom{t}{2}\binom{t}{3} - t\binom{t+1}{4} + \binom{t+2}{5} + C(t)$.

[2.23] $\frac{1}{2}n3^n - \frac{3}{4}3^n - \frac{3}{4}$.

[2.33] $\frac{1}{n+1}$.

[2.36] (a) e^x (b) $\cos x$ (c) $\frac{2x}{(1-2x)^2}$, $|x| < 1/2$ (d) $-\ln(1-2x)$, $|x| < 1/2$.

[2.37] $L_0(t) = 1$, $L_1(t) = 1-t$, $L_2(t) = 1 - 2t + \frac{1}{2}t^2$, $L_3(t) = \frac{1}{6}(6 - 14t + 3t^2 - t^3)$.

[2.38] $P_0(t) = 1$, $P_1(t) = t$, $P_2(t) = \frac{3}{2}t^2 - \frac{1}{2}$, $P_3(t) = \frac{5}{2}t^3 - \frac{3}{2}t$.

[2.39] $H_0(t) = 1$, $H_1(t) = 2t$, $H_2(t) = -1 + 2t^2$, $H_3(t) = -2t + \frac{4}{3}t^3$.

[2.48] $\frac{1}{6}(n-1)n(2n-1)$.

Chapter 3

[3.2] (a) $u(t) = \frac{A}{t}$.

(b) $u(t) = \frac{A}{(3t+4)(3t+1)}$.

[3.3] (a) $u(t) = Ce^{\frac{3}{2}t(t-1)}$.

(b) $u(t) = Ce^{\frac{\sin 2(t-\frac{1}{2})}{2\sin 1}}$.

[3.4] (a) $y(t) = A2^t - 5$.

(c) $y(t) = 5^t[\frac{1}{5}t + C]$.

[3.5] (b) $y(t) = \frac{1}{6}t(t+1)(2t+1)$.

[3.6] $y(t) = \frac{2e-7}{3(e-3)}3^{t-1} + \frac{e^t}{e-3}$.

[3.7] $y(t) = \frac{1}{(3t+1)(3t+7)}(\frac{1}{2}t^2 - \frac{1}{2}t + C)$.

[3.11] $t \approx 9$ years, $t \approx 14.27$ years.

[3.13] \$81,211.76.

[3.20] $y_n = 2^n - 1$.

[3.22] First order.

[3.25] $u(t) = -\frac{3}{2}2^t + \frac{4}{9}3^t$.

[3.29] $y(t) = A2^t + B3^t - \frac{1}{2}t2^t$.

[3.30] $y(t) = A2^t + \frac{1}{2}\binom{t}{6}2^t$.

[3.31] $y(t) = A + B6^t - \frac{1}{5}t^2 + \frac{8}{25}t$.

Chapter 3 - Answers

[3.35] $y(t) = \frac{2}{3}3^t - \frac{1}{2}2^t - \frac{1}{2}t \cdot 2^t$.

[3.36] (a) linearly independent.
(b) linearly independent.
(c) linearly independent.

[3.38] (a) $u(t) = C_1 6^t + C_2 t 6^t + C_3 t^2 6^t + C_4 t^3 6^t + C_5 t^4 6^t$.
(b) $u(t) = A(-3 + \sqrt{6})^t + B(-3 - \sqrt{6})^t$.
(d) $u(t) = A 2^t + B t 2^t + C(-2)^t + D t (-2)^t$.

[3.39] (a) $u(t) = A \sin \frac{\pi}{2} t + B \cos \frac{\pi}{2} t$.
(b) $u(t) = (4\sqrt{2})^t (A \cos \frac{\pi}{4} t + B \sin \frac{\pi}{4} t)$.
(c) $u(t) = A \cos \frac{\pi}{2} t + B \sin \frac{\pi}{2} t + C t \cos \frac{\pi}{2} t + D t \sin \frac{\pi}{2} t$.

[3.40] $a_n = 2^n + 1$, $R = 1/2$.

[3.42] (a) $y(t) = A(\frac{1}{4})^t + B(\frac{1}{2})^t + \frac{1}{21} 2^t$.
(b) $y(t) = A + Bt + t^2 + \frac{1}{2} t^3$.

[3.57] $x(t) = -\frac{9}{5}(0.7)^t + \frac{19}{5}(0.2)^t$.
$y(t) = \frac{6}{5}(0.7)^t + \frac{19}{5}(0.2)^t$.

[3.69] (b) $x(10) = 18,061$.

[3.73] $u(t) = A(-1)^t + B(-1)^{t+1} \sum (-1)^t \Gamma(t+1)$.

[3.74] (a) $u_n = A(n-1)! + B(n-1)! \sum \frac{1}{n}$.
(b) $u_n = A + B e^{\frac{1}{2} n(n-1)}$.

[3.75] $u_n = An + B 2^n - \frac{3}{4} n 2^n + \frac{1}{4} n^2 2^n$.

[3.77] (a) $u(t) = A t(t+1)(t+2) + \frac{B}{t-1}$.
(b) $u(t) = A(t+1)^{(2)} + B(t+1)^{(2)} \sum \frac{1}{t+2}$.

[3.80] $u_n = \frac{A}{n! 3^n} + \frac{B}{n! 3^n} \sum (n! 3^{n+1})$.

[3.81] $u_n = A + B \sum_{k=1}^{n-1} \frac{(-3)^k}{k}$.

[3.82] $u_n = 2^{n-1} \sum_{k=1}^{n-1} \frac{1}{(k-1)! 2^k}$.

[3.83] $u_n = \frac{1}{n!}[A 2^n + B]$.

Answers to Selected Problems

[3.84] $u(t) = a_0 \sum_{k=0}^{\infty} (-\frac{1}{3})^k \frac{1}{k!} t^{(-k)}$.

[3.86] (b) $y(t) = \frac{(1+C)\cos\frac{\pi}{4}t - (1-C)\sin\frac{\pi}{4}t}{\cos\frac{\pi}{4}t + C\sin\frac{\pi}{4}t}$

and $y(t) = \cot(\frac{\pi}{4}t) + 1$.

[3.87] $y(t) = \frac{2}{t(t-1)+D}$.

[3.88] (b) $y_n = A^{(-1)^n} B^{2^n}$.

[3.89] $y(t) = \pm \left(\frac{t+c}{t}\right)^{\frac{1}{2}}$.

[3.90] $y_n = \sin(A 2^n)$.

[3.99] (b) $U(z) = \frac{z^2 - 3z\cos 2}{z^2 - 6z\cos 2 + 9}$.

(d) $Y(z) = \frac{z^3 + 4z^2 + z}{(z-1)^4}$.

(f) $V(z) = -\frac{2z^2}{(z^2+1)^2}$.

(h) $U(z) = z\sin(\frac{1}{z})$.

[3.102] (a) $y_k = (-1)^k + 4^k$.

(c) $v_k = 2 + 3k$.

(e) $y_k = \cos(\frac{\pi}{3}k)$.

(g) $w_k = 3\delta_k(2) + 5\delta_k(4)$.

[3.104] $Z(k^2) = \frac{z^2+z}{(z-1)^3}$, $Z(k^3) = \frac{z^3+4z^2+z}{(z-1)^4}$.

[3.106] (b) $Y(z) = \frac{z^2}{z^2-1}$.

[3.108] (a) $y_k = 4^k - 3^k$.

(c) $y_k = k 5^k$.

(e) $y_k = 2(-3)^k + 4(-3)^{k-3} u_k(3)$.

[3.109] (a) $y_k = 3 \cdot 2^k - 2 \cdot 3^k$.

(c) $y_k = k \cdot 4^k$.

[3.110] (b) $u_k = \sin\frac{\pi}{2}k$, $v_k = \cos\frac{\pi}{2}k$.

(d) $u_k = 0$, $v_k = k + 2$.

[3.113] (b) $\frac{1}{2}k^2 + \frac{1}{2}k$.

[3.114] (a) $y_k = 3$.

(c) $y_k = \frac{9}{7}5^k + \frac{12}{7}(-2)^k$.

[3.115] (b) $y_k = \frac{3}{2}4^k + \frac{3}{2}(-2)^k$.

[3.117] (a) $y_k = 10 - \frac{210}{2869}k$.

(c) $y_k = -\frac{1}{119}k$.

Chapter 4

[4.1]

$$u(t+1) = \begin{bmatrix} 0 & 1 & 0 & 0 \\ 3 & 6 & -1 & -4 \\ 0 & 0 & 0 & 1 \\ 0 & -3 & 2 & -1 \end{bmatrix} u(t) + \begin{bmatrix} 0 \\ 0 \\ 0 \\ t3^t \end{bmatrix}.$$

[4.3] (a) $\sigma(A) = \{-1, -2\}$, $r(A) = 2$, $(-2, [1 \ \ -2]^T)$, $(-1, [1 \ \ -1]^T)$ are eigenpairs.

(b) $\sigma(A) = \{-1, 8\}$, $r(A) = 8$, $(-1, [1, -2, 0]^T)$, $(-1, [0, -2, 1]^T)$, $(8, [2, 1, 2]^T)$ are eigenpairs.

(c) $\sigma(A) = \{2 + 3i, 2 - 3i\}$, $r(A) = \sqrt{13}$.

(d) $\sigma(A) = \{-1, -2\}$, $r(A) = 2$.

[4.6] $u(t) = 5(-1)^t - 6(-2)^t$.

[4.7]

(a) $\begin{bmatrix} \frac{1}{5}(-2)^t + \frac{4}{5}3^t & -\frac{2}{5}(-2)^t + \frac{2}{5}3^t \\ -\frac{2}{5}(-2)^t + \frac{2}{5}3^t & \frac{4}{5}(-2)^t + \frac{1}{5}3^t \end{bmatrix}$

(b) $\begin{bmatrix} 1 & 0 & 0 \\ t & 1 & 0 \\ \frac{1}{2}t(t-1) & t & 1 \end{bmatrix}$

Answers to Selected Problems

[4.13] (a) asymptotically stable.

(b) not asymptotically stable.

(c) asymptotically stable.

[4.18] (b) The plane $u_3 = 0$.

[4.22] (a) $u = 0, 3/4$.

(b) $u = \frac{n\pi}{8}$, n an integer.

[4.23] (c) all $u \neq 0$ are periodic with period 8.

[4.24] (a) $u = 0, 1$ are fixed points. If $|u(0)| < 1$ then $\lim_{t \to \infty} u(t) = 0$. If $|u(0)| > 1$, then $\lim_{t \to \infty} y(t) = \infty$.

(b) $u = 0, 1, -1$ are fixed points. If $|u(0)| < 1$, then $\lim_{t \to \infty} u(t) = 0$. If $u(0) > 1$, $\lim_{t \to \infty} u(t) = \infty$. If $u(0) < -1$, $\lim_{t \to \infty} u(t) = -\infty$.

(c) $u = -1$ is a fixed point. If $u(0) > -1$, then $\lim_{t \to \infty} u(t) = \infty$. If $u(0) < -1$, then $\lim_{t \to \infty} u(t) = -\infty$.

(d) $u = \frac{1}{2}$ is a fixed point. Every point is periodic with period 2.

[4.25] (a) $u = 0$ is asymptotically stable, $u = 1$ is unstable.

(b) $u = 0$ is asymptotically stable, $u = \pm 1$ are unstable.

(c) $u = -1$ is unstable.

(d) Theorem 4.8 doesn't apply. $u = \frac{1}{2}$ is stable but not asymptotically stable.

[4.26] $u = 0$ is asymptotically stable.

$u = \pm 1$ are unstable.

[4.29] No.

[4.30] If $|u_1| < \frac{1}{|\alpha|^{\frac{1}{n-1}}}$ and $|u_2| < \frac{1}{|\beta|^{\frac{1}{n-1}}}$ and $u(0) = \begin{bmatrix} u_1 \\ u_2 \end{bmatrix}$, then $\lim_{t \to \infty} u(t) = 0$.

[4.36] $\sin^2 \frac{\pi}{15}$ (and many others).

Chapter 5

[5.6]   (b) 10.05, 100.005.

[5.8]   (a) 0.001016.

[5.9]   (a) 0.000064.

[5.13]

$$\sum_{k=1}^{n-1} \frac{1}{(2k)!} = \cosh(1) - 1 + \mathcal{O}\left(\frac{1}{(2n)!}\right) \quad (n \to \infty).$$

[5.21] $-\frac{139}{51840n^3}$.

[5.22] $u(t) \sim C(\frac{2}{e})^t t^{t-\frac{3}{2}}$ $(t \to \infty)$.

[5.28]   (a) If $|x| > 1$, then the equation has independent solutions u_1, u_2 so that

$$\lim_{t \to \infty} \frac{u_1(t+1)}{u_1(t)} = x + \sqrt{x^2 - 1}, \quad \lim_{t \to \infty} \frac{u_2(t+1)}{u_2(t)} = x - \sqrt{x^2 - 1}.$$

[5.40] $u(t) - 1 \sim C2^{-t}(u(0) - 1)$ $(t \to \infty)$.

[5.44] $|u(t) - 1| \sim 2\sqrt{2}(C|u(0) - 1|)^{3^t}$ $(t \to \infty)$.

[5.49] $u(t) = \sqrt{2t}[1 + \mathcal{O}(\frac{\log t}{t})]$ $(t \to \infty)$.

[5.51] $u(t) = \sqrt{\frac{3}{t}}[1 + \mathcal{O}(\frac{\log t}{t})]$ $(t \to \infty)$.

Chapter 6

[6.1]   (a) $\Delta(3^{t-1}\Delta y(t-1)) + e^t y(t) = 0$.

(b) $\Delta(\cos \frac{t-1}{100} \Delta y(t-1)) + (2^t + \cos \frac{t}{100} + \cos \frac{t-1}{100}) y(t) = 0$.

(c) $\Delta^2 y(t-1) + 4y(t) = 0$.

[6.2]   (a) $\Delta((t-2)!\Delta y(t-1)) = 0$, $y(t) = A \sum_{s=1}^{t} \frac{1}{(s-1)!} + B$.

(b) $\Delta[(t-2)!\Delta y(t-1)] + (t-2)! y(t) = 0$.

(c) $\Delta[(\frac{1}{6})^{t-1}\Delta y(t-1)] + 2(\frac{1}{6})^t y(t) = 0$.

(d) $\Delta^2 y(t-1) = 0$, $y(t) = A + Bt$.

[6.3] $\Delta^2 z(\lambda - 1) + (2 - \frac{2\lambda}{t}) z(\lambda) = 0$.

[6.4] (a) $w[2^t, 3^t] = 6^t$.

(b) $w[1, t] = 1$.

(c) $w[\cos \frac{\pi}{2}t, \sin \frac{\pi}{2}t] = 1$.

[6.6] (a) $\Delta^2 y(t-1) = 0$.

(c) $\Delta^2(\frac{y(t-1)}{2^{t-1}}) = 0$.

[6.7] (a) $y(t, s) = 2^s 3^t - 3^s 2^t$.

[6.8] (a) $y(t) = \frac{1}{6}t^3 - \frac{19}{6}t + 5$.

(c) $y(t) = \frac{1}{2}6^t - 2 \cdot 3^t + \frac{3}{2} \cdot 2^t$.

[6.10] (a) disconjugate on $(-\infty, \infty)$.

[6.19]

$$H(t, s) = \begin{cases} a - t, & t \leq s \\ a - s, & s \leq t \end{cases}$$

[6.22] $y(t) = \frac{1}{6}t^3 - \frac{32}{3}t$.

[6.24] $y(t) = 5t^2 - 44t + 10$.

[6.30] (a) $y_1(t) = 5^t$, $y_2(t) = t5^t$.

[6.31] (b) $z(t) = \frac{D}{1+Dt}$, $z(t) = \frac{1}{t}$.

[6.36] $u(t) = (-3)^t$ is oscillatory.

$v(t) = 2^t$ is nonoscillatory.

Chapter 7

[7.1] $(2 - \sqrt{3}, \sin \frac{\pi}{6}t)$, $(1, \sin \frac{\pi}{3}t)$, $(2, \sin \frac{\pi}{2}t)$

$(3, \sin \frac{2\pi}{3}t)$, $(2 + \sqrt{3}, \sin \frac{5\pi}{6}t)$.

[7.6] (a) $(1, \sin \frac{\pi}{3}t)$, $(3, \sin \frac{2\pi}{3}t)$.

[7.8] $(1, [\frac{\sqrt{3}}{2}, \frac{\sqrt{3}}{2}]^T), (3, [\frac{\sqrt{3}}{2}, -\frac{\sqrt{3}}{2}]^T)$.

[7.10] $N = b - a + 1 = 68$.

[7.11] $w(t) = -\frac{1}{2}(4 + 3\sqrt{2})\sin\frac{\pi}{4}t + \frac{1}{2}(4 - 3\sqrt{2})\sin\frac{3\pi}{4}t$.

[7.12] $w(t) = \frac{1}{3}(4 + \sqrt{3})\sin\frac{\pi}{6}t + \frac{8}{3}\sin\frac{\pi}{2}t + \frac{1}{3}(4 - \sqrt{3})\sin\frac{5\pi}{6}t$.

[7.15] $f(1) = -f(2)$.

[7.16] A Storm-Louisville problem.

[7.19] .296

[7.20] 1, $u(t)$ is an eigenfunction.

[7.21] .222

Chapter 8

[8.2]   (a) $\Delta^2 y(t-1) - \frac{4}{3}y(t) = 0$.

[8.3]   (a) $y(t) = 2 + 1.98t$.

 (c) $y(t) = 100$.

[8.5] $y_0(t) = 2^{t+2} - 4^{t+1}$

[8.6]   (b) $y_0(t) = \frac{1}{1+2^{201}}(2^{201}(\frac{1}{2})^t + 2^t)$.

[8.9] $y_0(t) = \frac{10 - 5 \cdot 3^{500}}{2^{500} - 3^{500}}(2^t - 3^t) + 5 \cdot 3^t$.

Chapter 9

[9.3]   Let $\mathcal{S}$ be the real numbers and define T on $\mathcal{S}$ by $Tx = x + 1$, then $|Tx - Ty| \leq \alpha |x - y|$ for all x, y in $\mathcal{S}$ with $\alpha = 1$ but T has no fixed points.

[9.8] $1 \leq b \leq 6$.

Chapter 10

[10.1]

$$\begin{bmatrix} y(1,j) \\ y(2,j) \\ y(3,j) \end{bmatrix} = \left(\frac{2 + \sqrt{2}}{4}\right)^j \begin{bmatrix} .5\sqrt{2} \\ 1 \\ .5\sqrt{2} \end{bmatrix}.$$

[10.7] $y(i,j) = 2^{-i}f(i-j) + 3^{i+1}$, f arbitrary.

[10.8a] $y(i,j) = f(j+3i) + g(j+3i)i$, f,g arbitrary.

[10.11b]

$$y(i,j) =$$

$$p^i \left[1 + iq + \frac{i(i+1)}{2!}q^2 + \cdots + \frac{i(i+1)\cdots(i+j-2)}{(j-1)!}q^{j-1} \right].$$

[10.14b]

$$y(i,j) = 2^j \begin{bmatrix} i-1 \\ j \end{bmatrix}.$$

References

[1] R. Agarwal, On multipoint boundary value problems for discrete equations, *J. Math. Anal. Appl.* 96 (1983), 520–534.

[2] R. Agarwal, Initial-value methods for discrete boundary value problems, *J. Math. Anal. Appl.* 100 (1984), 513–529.

[3] R. Agarwal, Difference calculus with applications to difference equations, *Proc. Conf. General Inequalities, Oberwolfach Internat. Ser. Numer. Math.* 71 (1984), 95–110.

[4] R. Agarwal, Initial and boundary value problems for n^{th} order difference equations, *Math Solvaca* 36 (1986), 39–47.

[5] R. Agarwal, Computational methods for discrete boundary value problems, *Appl. Math. Comput.* 18 (1986), 15–41.

[6] R. Agarwal, Properties of solutions of higher order nonlinear difference equations, *An Ati Univ "Al. I. Cuza" Iasi* (to appear).

[7] R. Agarwal and E. Thandapani, On some new discrete inequalities, *Appl. Math. Comp.* 7 (1980), 205–224.

[8] C. Ahlbrandt, Continued fraction representations of maximal and minimal solutions of a discrete matrix Riccati equation (preprint).

[9] C. Ahlbrandt and J. Hooker, Riccati transformations and principal solutions of discrete linear systems, *Proc. 1984 Workshop Spectral Theory of Sturm-Liouville Differential Operators*, Edited by H. Kaper and A. Zettl ANL-84-87, Argonne National Lab., Argonne, Illinois, 1984.

[10] C. Ahlbrandt and J. Hooker, A variational view of nonoscillation theory for linear difference equations, *Proceedings of the Thirteenth Midwest Differential Equations Conference*, Edited by J. Henderson (1985), 1–21.

[11] C. Ahlbrandt and J. Hooker, Disconjugacy criteria for second order linear difference equations, *Qualitative Properties of Differential Equations, Proc. of 1984 Edmonton Conference*, University of Alberta, Edmonton (1987), 15–26.

[12] C. Ahlbrandt and J. Hooker, Recessive solutions of symmetric three-term recurrence relations, *Canadian Mathematical Society Conference Proceedings, Oscillation, Bifurcation and Chaos*, Edited by F. Atkinson, W. Langford and A. Mingarelli (1987), 3–42.

[13] C. Ahlbrandt and J. Hooker, Riccati matrix difference equations and disconjugacy of discrete linear systems, *SIAM J. Math. Anal.* 19 (1988), 1183–1197.

[14] E. Angle and R. Kalaba, A one-sweep numerical method for vector-matrix difference equations with two-point boundary conditions, *J. Optimization Theory and Appl.* 6 (1970), 345–355.

[15] R. Aris, *Discrete Dynamic Programming*, Blaisdell, New York, 1964.

[16] F. Arscott, A Riccati type transformation of linear difference equations, *Congressus Numerantium* 30 (1981), 197–202.

[17] F. Arscott, R. LaCroix and W. Shymanski, A three-term recursion and the computations of Mathieu functions, *Proc. Eighth Manitoba Conference in Numerical Math. and Computing* (1978), 107–115.

[18] R. Askey and M. Ismail, Recurrence relations, continued fractions and orthogonal polynomials, *Mem. Amer. Math. Soc.*, 300(1984), 108.

[19] F. Atkinson, *Discrete and Continuous Boundary Problems*, Academic Press, New York, 1964.

[20] K. Atkinson, *An Introduction to Numerical Analysis*, 2nd ed., Wiley, New York, 1989.

[21] R. Bartle, *The Elements of Real Analysis*, 2nd edition, Wiley, New York, 1976.

[22] P. Batchelder, *An Introduction to Linear Difference Equations*, Harvard University Press, Cambridge, 1927.

References

[23] E. Beltrami, *Mathematics for Dynamic Modeling*, Academic Press, New York, 1987.

[24] C. Bender and S. Orszag, *Advanced Mathematical Methods for Scientists and Engineers*, McGraw-Hill, New York, 1978.

[25] W. Beyn, Discrete Green's functions and strong stability properties of the finite difference method, *Applicable Anal.* 14 (1982), 73–98.

[26] G. Birkhoff and W. Trjitzinksy, Analytic theory of singular difference equations, *Acta Math.* 60 (1932), 1–89.

[27] A. Bishop, *Introduction to Discrete Linear Controls*, Academic Press, New York, 1975.

[28] M. Bocher, Boundary problems and Green's functions for linear differential and difference equations, *Ann. Math.* 13 (1911-2), 71–88.

[29] G. Boole, *Calculus of Finite Differences*, 4th ed., Chelsea, New York, 1958.

[30] L. Brand, *Differential and Difference Equations*, John Wiley, New York, 1966.

[31] R. Burden, J. Faires, and A. Reynolds, *Numerical Analysis*, Prindle, Weber and Schmidt, Boston, 1978.

[32] Y. Bykov and L. Zivogladova, On the oscillation of solutions of nonlinear finite difference equations, *Differencialnye Uravnenija* 9 (1973), 2080–2081.

[33] Y. Bykov, L. Zivogladova, and E. Sevcov, Sufficient conditions for oscillations of solutions of nonlinear finite difference equations, *Differencialnye Uravnenija* 9 (1973), 1523–1524.

[34] J. Cadzow, *Discrete-Time Systems*, Prentice-Hall, Englewood Cliffs, NJ, 1966.

[35] J. Cadzow, Discrete calculus of variations, *Int. J. Control* II (1970), 393–407.

[36] J. Cash, *Finite Recurrence Relations*, Academic Press, New York, 1976.

[37] J. Cash, An extension of Olver's method for the numerical solution of linear recurrence relations, *Math. Comp.* 32 (1978), 497–510.

[38] J. Cash, *Stable Recursions*, Academic Press, London, 1979.

[39] J. Cash, A note on the solution of linear recurrence relations, *Num. Math.* 34 (1980), 371–386.

[40] S. Chen and L. Erbe, Oscillation and nonoscillation for systems of self-adjoint second-order difference equations, *SIAM J. Math. Anal.* 20 (1989), 939–949.

[41] S. Chen and L. Erbe, Riccati techniques and discrete oscillations, *J. Math. Anal. Appl.* (to appear).

[42] S. Cheng, Monotone solutions of $\Delta^2 x(k) = Q(k)x(k+1)$, *Chinese J. Math.* 10 (1982), 71–75.

[43] S. Cheng, On a class of fourth order linear recurrence equations, *Internat. J. Math. Sci.* 7 (1984), 131–149.

[44] S. Cheng, Sturmian comparison theorems for three-term recurrence equations, *J. Math. Anal. Appl.* 111 (1985), 465–474.

[45] S. Cheng and A. Cho, Convexity of nodes of discrete Sturm-Liouville functions, *Hokkaido Math J.* 11 (1982), 8–14.

[46] S. Cheng, H. Li, and W. Patula, Bounded and zero convergent solutions of second order difference equations, *J. Math. Anal. Appl.* (to appear).

[47] F. Chorlten, *Differential and Difference Equations*, Van Nostrand, Princeton, 1965.

[48] Q. Chuanxi, S. Kuruklis and G. Ladas, Sufficient conditions for oscillations of systems of difference equations, preprint.

[49] C. Coffman, Asymptotic behavior of solutions of ordinary difference equations, *Trans. Amer. Math. Soc.* 110 (1964), 22–51.

[50] E. Cogan and R. Norman, *Handbook of Calculus, Difference and Differential Equations*, Prentice-Hall, Englewood Cliffs, NJ, 1958.

[51] C. Corduneanu, Almost periodic discrete processes, *Libertas Math.* 2 (1982), 159–169.

[52] M. Cuenod and A. During, *A Discrete-Time Approach for System Analysis*, Academic Press, New York, 1969.

[53] W. Culmer and W. Harris, Convergent solutions of ordinary linear homogeneous difference equations, *Pacific J. Math.* 13 (1963), 1111–1138.

[54] N. DeBruijn, *Asymptotic Methods in Analysis*, 2nd ed., Wiley, New York, 1961.

[55] V. Derr, Criterion for a difference equation to be non-oscillatory (Russian), *Diff. Arav.* 12 (1976), 747-750 or *Differential Equations* 12 (1976), 524–527.

[56] P. Deuflhard, A summation technique for minimal solutions of linear homogeneous difference equations, *Computing* 18 (1977), 1–13.

[57] R. Devaney, *An Introduction to Chaotic Dynamical Systems*, Benjamin/ Cummings, Menlo Park, CA, 1986.

[58] P. Diamond, Finite stability domains for difference equations, *J. Austral. Soc.* 22A (1976), 177–181.

[59] P. Diamond, Discrete Liapunov function with $V > 0$, *J. Austral. Math. Soc.*, 20B(1978), 280–284.

[60] P. Dorato and A. Levis, Optimal linear regulators: the discrete-time case, *IEEE Transactions on Automatic Control* 16 (1971), 613–620.

[61] R. Driver, Note on a paper of Halanay on stability of finite difference equations, *Arch. Rat. Mech. Anal.* 18 (1965), 241–243.

[62] N. Dunford and J. Schwartz, *Linear Operators (Part I)*, Interscience, New York, 1957.

[63] S. Elaydi and A. Peterson, Stability of difference equations, *Proceedings of the International Conference on Theory and Applications of Differential Equations*, Edited by R. Aftabizadeh, Ohio University Press (1988), 417–422.

[64] P. Eloe, Difference equations and multipoint boundary value problems *Proc. Amer. Math. Soc.* 86 (1982), 253–259.

[65] P. Eloe, A boundary value problem for a system of difference equations, *Nonlinear Analysis* 7 (1983), 813–820.

[66] P. Eloe, Criteria for right disfocality of linear difference equations, *J. Math. Anal. Appl.* 120 (1986), 610–621.

[67] P. Eloe, A comparison theorem for linear difference equations, *Proc. Amer. Math. Soc.* 103 (1988), 451–457.

[68] P. Eloe, Eventual disconjugacy and right disfocality of linear difference equations, *Bulletin of Canad. Math.* 31 (1988), 362–373.

[69] P. Eloe and J. Henderson, Analogues of Fekete and DesCartes systems of solutions for difference equations, *J. Approx. Theory* 59 (1989) 38–52.

[70] L. Erbe and B. Zhang, Oscillation of discrete analogues of delay equations, *Differential and Integral Equations* 2 (1989), 300–309.

[71] L. Erbe and B. Zhang, Oscillation of difference equations, *Proceedings of the International Conference on Theory and Applications of Differential Equations*, Edited by S. Elaydi, Ohio University Press (1988).

[72] H. Esser, Stability inequalities for discrete nonlinear two-point boundary value problems, *Applicable Anal.* 10 (1980), 137–162.

[73] M. Evgrafov, The asymptotic behavior of solutions of difference equations, *Dokl. Akad. Nauk. SSSR* 121 (1958), 26–29 (Russian).

[74] M. Feigenbaum, Quantitative universality for a class of nonlinear transformations, *J. Statist. Phys.* 19 (1978), 25–52.

[75] G. Forsythe and W. Wasow, *Finite-Difference Methods for Partial Differential Equations*, John Wiley, New York, 1960.

[76] T. Fort, Oscillatory and non-oscillatory linear difference equations of the second order, *Quar. J. Pure. Appl. Math.* 45 (1914-15), 239–257.

[77] T. Fort, *Finite Differences and Difference Equations in the Real Domain*, Oxford University Press, London, 1948.

[78] T. Fort, The five-point difference equation with periodic coefficients, *Pacific J. Math.* 7 (1957), 1341–1350.

[79] T. Fort, Partial linear difference equations, *Amer. Math. Monthly* 64 (1957), 161–167.

References

[80] H. Freedman, *Deterministic Mathematical Models in Population Ecology*, Marcel Dekker, New York, 1980.

[81] H. Freedman and J. So, Persistence in discrete semidynamical systems *SIAM J. Math. Anal.* 20 (1989), 930–938.

[82] R. Gaines, Difference equations associated with boundary value problems for second order nonlinear ordinary differential equations, *SIAM J. Numer. Anal.* 11 (1974), 411–434.

[83] W. Gautschi, Computational aspects of three-term recurrence relations, *SIAM Review* 9 (1967), 24–82.

[84] W. Gautschi, Zur Numerik Rekurrenten Relationen, *Computing* 9 (1972), 107–126.

[85] W. Gautschi, Minimal solutions of three term recurrence relations and orthogonal polynomials, *Math. Comp.* 36 (1981), 547–554.

[86] A. Gelfond, *Calculus of Finite Differences*, Hindusten, Delhi, India, 1971.

[87] A. Gelfond and I. Kubenskaya, On a theorem of Perron in the theory of difference equations, *Izv. Akad. Nauk. SSSR Ser Mat.* 17 (1953), 83–86.

[88] S. Godunov and V. Ryabenki, *Theory of Difference Schemes*, North-Holland, 1984.

[89] M. Goldberg, Derivation and validation of initial-value methods for boundary-value problems for difference equations, *J. Optimization Theory and Appl.* (1971), 411–419.

[90] S. Goldberg, *Introduction to Difference Equations*, John Wiley, New York, 1958.

[91] S. Gordon, Stability and summability of solutions of difference equations, *Math. Syst. Theory* 5 (1971), 56–75.

[92] D. Greenspan, *Discrete Models*, Addison-Wesley, Reading, MA, 1973.

[93] S. Grossman, *Elementary Linear Algebra*, Wadsworth, Belmont, CA, 1980.

[94] L. Grujic and D. Siljak, Exponential stability of large scale discrete systems, *J. Control* 19 (1976), 481–491.

[95] I. Gyori and G. Ladas, Linearized oscillations for equations with piecewise constant arguments, *Differential and Integral Equations* 2 (1989), 123–131.

[96] I. Gyori and G. Ladas, *Oscillation Theory of Delay Difference Equations*, Oxford University Press (to appear).

[97] I. Gyori, G. Ladas, and L. Patula, Conditions for the oscillation of difference equations with applications to equations with piecewise constant arguments, (preprint).

[98] I. Gyori, G. Ladas, and P. Vlahos, Global attractivity in a delay difference equation, preprint.

[99] A. Halanay, Solution periodiques et presque-periodiques des systems d'equations aux difference finies, *Arch. Rat. Mech.* 12 (1963), 134–149.

[100] A. Halanay, Quelques questions de la theorie de la stabilite pour les systemes aux differences finies, *Arch. Rat. Mech. Anal.* 12(1963), 150–154.

[101] L. Hall and S. Trimble, Asymptotic behavior of solutions of Poincaré difference equations, *Differential Equations and Applications, Proceedings of the International Conference on Theory and Applications of Differential Equations*, edited by A.R. Aftabizadeh, Ohio University (1988), 412–416.

[102] D. Handelman, Stable positivity of polynomials obtained from three-term difference equations, *SIAM J. Math. Anal.* 21(1990), 1051–1065.

[103] D. Hankerson, An existence and uniqueness theorem for difference equations, *SIAM J. Math. Anal.* 20 (1989), 1208–1217.

[104] D. Hankerson and J. Henderson, Comparison of eigenvalues for n-point boundary value problems for difference equations (preprint).

[105] D. Hankerson and A. Peterson, On a theorem of Elias for difference equations, *Lecture Notes in Pure and Applied Mathematics 109, Nonlinear Analysis and Applications*, Marcel Dekker (1987), 229–234.

References

[106] D. Hankerson and A. Peterson, Comparison theorems for eigenvalue problems for n^{th} order differential equations, *Proceedings of the American Mathematical Society* 104 (1988), 1204-1211.

[107] D. Hankerson and A. Peterson, A classification of the solutions of a difference equation according to their behavior at infinity, *J. Math. Anal. Appl.* 136 (1988), 249-266.

[108] D. Hankerson and A. Peterson, Extremal solutions of an nth order difference equation, *Proceedings of the International Conference on Theory and Applications of Differential Equations*, edited by A.R. Aftabizadeh, Ohio University (1988), 417-422.

[109] D. Hankerson and A. Peterson, A positivity result applied to difference equations, *Journal of Approximation Theory* 59 (1989), 76-86.

[110] D. Hankerson and A. Peterson, Comparison of eigenvalues for focal point problems for n^{th} order difference equations, *Differential and Integral Equations* 3 (1990), 363-380.

[111] D. Hankerson and A. Peterson, Positive solutions of a boundary value problem, *Rocky Mountain Journal of Mathematics* (to appear).

[112] B. Harris, Extremal solutions for n^{th} order linear difference equations (preprint).

[113] P. Hartman, *Ordinary Differential Equations*, John Wiley & Sons, Inc., New York, 1964.

[114] P. Hartman, Difference equations: disconjugacy, principal solutions, Green's functions, complete monotonicity, *Trans. Amer. Math. Soc.* 246 (1978), 1-30.

[115] P. Hartman and A. Wintner, On linear difference equations of the second order, *Amer. J. Math.* (1950), 124-128.

[116] P. Hartman and A. Wintner, Linear differential and difference equations with monotone coefficients, *Amer. J. Math.* 75 (1953), 731-743.

[117] J. Henderson, Focal boundary value problems for nonlinear difference equations I, *J. Math. Anal. Appl.* 141 (1989), 559-567.

[118] J. Henderson, Focal boundary value problems for nonlinear difference equations II, *J. Math. Anal. Appl.* 141 (1989), 568-579

[119] J. Henderson, Existence theorems for boundary value problems for n^{th} order nonlinear difference equations, *SIAM J. Math. Anal.* 20 (1989), 468–478.

[120] J. Henderson, Boundary value problems for nonlinear difference equations (submitted).

[121] M. Hénon, A two-dimensional mapping with a strange attractor, *Commun. Math. Phys.* 50 (1976), 69–77.

[122] P. Henrici, *Discrete Variable Methods in Ordinary Differential Equations*, John Wiley, New York, 1962.

[123] P. Henrici, *Error Propagation for Finite Difference Methods*, John Wiley, New York, 1963.

[124] F. Hildebrand, *Finite-Difference Equations and Simulations*, Prentice-Hall, Englewood Cliffs, NJ, 1968.

[125] D. Hinton and R. Lewis, Spectral analysis of second order difference equations, *J. Math. Anal. Appl.* 63 (1978), 421–438.

[126] D. Hinton and R. Lewis, Oscillation theory of generalized second order differential equations, *Rocky Mountain J. Math.* 10 (1980), 751–766.

[127] M. Hirsch and S. Smale, *Differential Equations, Dynamical Systems, and Linear Algebra*, Academic Press, New York, 1974.

[128] J. Hooker, A Hille-Wintner type comparison theorem for second order difference equations, *Int. J. of Math. and Math. Sci.* 6 (1983), 387–394.

[129] J. Hooker, M. Kwong, and W. Patula, Riccati type transformations for second-order linear difference equations II, *J. Math. Anal. Appl.* 107 (1985), 182–196.

[130] J. Hooker, M. Kwong, and W. Patula, Oscillatory second order linear difference equations and Riccati equations, *SIAM J. Math. Anal.* 18 (1987), 54–63.

[131] J. Hooker and W. Patula, Riccati type transformations for second-order linear difference equations, *J. Math. Anal. Appl.* 82 (1981), 451–462.

[132]   J. Hooker and W. Patula, A second order nonlinear difference equation: oscillation and asymptotic behavior, *J. Math. Anal. Appl.* 91 (1983), 9–29.

[133]   J. Hooker and W. Patula, Growth and oscillation properties of solutions of a fourth order linear difference equation, *J. Austral. Math. Soc.*, Ser. B, 26(1985), 310–328.

[134]   J. Hunt, Some stability theorems for ordinary difference equations, *SIAM JNA* 4 (1967), 582–596.

[135]   G. Immink, *Asymptotics of Analytic Difference Equations*, Springer-Verlag, New York, 1981.

[136]   E. Isaacson and H. Keller, *Analysis of Numerical Methods*, Wiley, New York, 1966.

[137]   R. Johnson, *Theory and Applications of Linear Differential and Difference Equations*, Halsted, 1984.

[138]   G. Jones, Fundamental inequalities for discrete and discontinuous functional equations, *J. Soc. Ind. Appl. Math.* 12 (1964), 43–57.

[139]   C. Jordan, *Calculus of Finite Differences* 3rd ed., Chelsea, New York, 1965.

[140]   E. Jury, *Theory and Applications of the z-transform*, John Wiley, New York, 1964.

[141]   R. Kalman, A new approach to linear filtering and prediction problems, *Trans. ASME, Ser. D, J. Basic Energy* 82 (1960), 35–45.

[142]   Z. Kamont and M. Kwapisz, Difference methods for nonlinear parabolic differential functional systems with initial boundary conditions of the Neumann type, *Comm. Math.* (to appear).

[143]   D. Knuth, *The Art of Computer Programming, Vol. 1, Fundamental Algorithms*, Addison-Wesley, Reading, MA, 2nd edition, 1973.

[144]   H. Kocak, *Differential and Difference Equations Through Computer Experiments*, 2nd ed., Springer-Verlag, New York, 1990.

[145]   S. Kuruklis and G. Ladas, Oscillations and global attractivity in a discrete delay logistic model (preprint).

[146] M. Kwapisz, *Elements of the Theory of Recurrence Equations* (in Polish), The Gdansk University Publications, Gdansk, 1983.

[147] M. Kwapisz, On discrete equations, inequalities and interval evaluations of fixed points of isotone-antitone mappings, *Interval Mathematics*, edited by K. Nickel, Academic Press, New York, (1980), 397–406.

[148] M. Kwapisz, Some remarks on comparison theorems for discrete inequalities, *J. Differential Equations* 79 (1989), 258–265.

[149] M. Kwapisz, On the error evaluation of approximate solutions of discrete equations, *J. Nonlinear Anal: TMA* (to appear).

[150] M. Kwapisz, On boundary value problems for difference equations, *Journal of Math. Anal. Appl.* (to appear).

[151] G. Ladas, Oscillations of difference equations with positive and negative coefficients, G. Butler Memorial Conference on Differential Equations and Population Biology, *Rocky Mountain Journal* (to appear).

[152] G. Ladas, Recent developments in the oscillation of delay difference equations, *Differential Equations: International Conference on Differential Equations, Theory and Applications in Stability and Control*, edited by S. Elaydi, Marcel Dekker, 1990.

[153] G. Ladas, Explicit conditions for the oscillation of difference equations, *J. Math. Anal. Appl.* (to appear).

[154] G. Ladas, C. Philos, and Y. Sficas, Sharp conditions for the oscillation of delay difference equations, *Journal of Applied Mathematics and Simulation* 2 (1989), 101–102.

[155] G. Ladas, C. Philos, and Y. Sficas, Necessary and sufficient conditions for the oscillation of difference equations, *Libertas Mathematica* IX (1989).

[156] G. Ladas, C. Qian, P. Vlahos and J. Yan, Stability of solutions of linear nonautonomous difference equations (preprint).

[157] V. Lakshmikantham and D. Trigiante, *Theory of Difference Equations: Numerical Methods and Applications*, Academic Press, New York, 1988.

[158] J. LaSalle, Stability theory for difference equations, MAA Studies in Mathematics 14, *Studies in Ordinary Differential Equations*, Edited by J. Hale (1977), 1–31.

[159] J. LaSalle *The Stability of Dynamical Systems*, Regional Conference Series in Applied Mathematics, *SIAM*, 1979.

[160] J. LaSalle, *The Stability and Control of Discrete Processes*, Springer-Verlag, New York, 1986.

[161] A. Lasota, A discrete boundary value problem, *Annales Polonici Mathematici* 20 (1968), 183–190.

[162] H. Lauwerier, *Mathematical Models of Epidemics*, Math. Centrum, Amsterdam, 1981.

[163] M. Lees, Discrete methods for nonlinear two-point boundary value problems, *Numerical Solution of Partial Differential Equations*, J. Bramble, ed. Academic Press (1966), 59–72.

[164] J. Leon, Limit cycles in populations with separate generations, *J. Theor. Biol.* 49 (1975), 241–244.

[165] S. Levin and R. May, A note on difference-delay equations, *Theor. Popul. Biol.* 9 (1976), 178–187.

[166] H. Levy and F. Lessman, *Finite Difference Equations*, The Macmillan Company, New York, 1961.

[167] Z. Li, The asymptotic estimates of solutions of difference equations, *J. Math. Anal. Appl.* 94 (1983), 181–192.

[168] J. Logan, First integrals in the discrete variational calculus, *Aequationes Mathematical* 9 (1974), 210–220.

[169] J. Logan, A canonical formalism for systems governed by certain difference equations, *Int. J. Control* 17 (1973), 1095–103.

[170] J. Logan, Some invariance identities for discrete systems, *Int. J. Control* 19 (1974), 919–923.

[171] J. Logan, *Applied Mathematics, A Contemporary Approach*, John Wiley, New York, 1987.

[172] E. Lorenz, Deterministic nonperiodic flow, *J. Atmos. Sci.* 20 (1963), 130–141.

[173] S. Maeda, The similarity method for difference equations, *IMA J. Appl. Math.* 38 (1987), 129–134.

[174] L. Maslovskaya, The stability of difference equations, *Diff. Equations* 2 (1966), 608–611.

[175] A Mate and P. Nevai, Sublinear perturbations of the differential equation $y'' = 0$ and the analogous difference equation, *J. Diff. Eq.* (1984), 234–257.

[176] R. Mattheij, Characterizations of dominant and dominated solutions of linear recursions, *Num. Math.* 35 (1980), 421–442.

[177] R. Mattheij, Accurate estimates for the fundamental solutions of discrete boundary value problems, *J. Math. Anal. Appl.* 10 (1984), 444–464.

[178] R. May, Biological populations obeying difference equations: stable points, stable cycles, and chaos, *J. Theor. Biol.* 51 (1975), 511–524.

[179] P. McCarthy, Note on oscillation of solutions of second order linear difference equations, *Port. Math.* 18 (1959), 203–205.

[180] H. Meschkowski, *Differenzengleichungen*, Vandenhoeck and Ruprecht, Göttingen, 1959.

[181] R. Mickens, *Difference Equations*, Van Nostrand Reinhold Company, New York, 1987.

[182] K. Miller, *An Introduction to the Calculus of Finite Differences and Difference Equations*, Hold and Company, New York, 1960.

[183] K. Miller, *Linear Difference Equations*, W.A. Benjamin Inc., New York, 1968.

[184] L. Milne-Thomson, *The Calculus of Finite Differences*, McMillan and Co., London, 1933.

[185] A. Mingarelli, A limit-point criterion for a three-term recurrence relation, *C.R. Math. Rep. Acad. Sci. Canada* 3 (1981), 171–175.

[186] A. Mingarelli, *Volterra-Stieltjes Integral Equations and Generalized Ordinary Differential Expressions*, Lecture Notes in Mathematics 989, Springer-Verlag, New York, 1983.

[187]   E. Moulton, A theorem in difference equations on the alternation of nodes of linearly independent solutions, *Ann. Math.* 13 (1912), 137–139.

[188]   F. Olver, Numerical solutions of second order linear difference equations, *J. Research, NBS*, 71B (1967), 111–129.

[189]   F. Olver, Bounds for the solutions of second order linear difference equations, *J. Res. Nat. Bur. Standards Sect B* 71 (1967), 161–166.

[190]   F. Olver, *Asymptotics and Special Functions*, Academic Press, New York, 1974.

[191]   F. Olver and D. Sookne, Note on backward recurrence algorithms, *Math. Comput.* 26 (1972), 941–947.

[192]   J. Ortega, Stability of difference equations and convergence of iterative processes, *SIAM JNA* 10 (1973), 268–282.

[193]   R. O'Shea, The extension of Zubov method to sampled data control systems described by difference equations, *IEEE Trans. Auto. Conf.* 9 (1964), 62–69.

[194]   B. Pachpatte, Finite difference inequalities and an extension of Lyapunov method, *Mich. Math. J.* 18 (1971), 385–391.

[195]   B. Pachpatte, On the discrete generalization of Gronwall's inequality, *J. Indian Math. Soc.* 37 (1973), 147–156.

[196]   B. Pachpatte, Finite difference inequalities and their applications, *Proc. Nat. Acad. Sci., India*, 43 (1973), 348–356.

[197]   B. Pachpatte, On some discrete inequalities of Bellman-Bihari type, *Indian J. Pure and Applied Math.* 6 (1975), 1479–1487.

[198]   B. Pachpatte, On some fundamental inequalities and its applications in the theory of difference equations, *Ganita* 27 (1976), 1–11.

[199]   T. Pappas, A. Laub, and N. Sandell, On the numerical solution of the discrete-time algebraic Riccati equation, *IEEE Trans. on Automatic Control* Vol. AC-25 No. 4 (1980) 631–641.

[200]   W. Patula, Growth and oscillation properties of second order linear difference equations, *SIAM J. Math. Anal.* 10 (1979), 55–61.

[201] W. Patula, Growth, oscillation and comparison theorems for second order linear difference equations, *SIAM J. Math. Anal.* 10 (1979), 1272–1279.

[202] T. Peil, Disconjugacy of second order linear difference equations, *International Conference on Differential Equations; Theory and Applications in Stability and Control* (1990), edited by S. Elaydi, 411–416.

[203] T. Peil, Disconjugacy for n^{th} order linear difference equations, *J. Math. Anal. Appl.* (to appear).

[204] T. Peil, Criteria for right disfocality of an nth order linear difference equation, *Rocky Mountain J. Math.* (to appear).

[205] T. Peil and A. Peterson, Criteria for C-disfocality of a self-adjoint vector difference equation (submitted).

[206] A. Peterson, Boundary value problems for an nth order linear difference equation, *SIAM J. Math. Anal.* 15 (1984), 124–132.

[207] A. Peterson, Boundary value problems and Green's functions for linear difference equations, *Proceedings of the Twelfth and Thirteenth Midwest Differential Integral Equations*, edited by J. Henderson (1985), 79–100.

[208] A. Peterson, On $(k, n-k)$-disconjugacy for linear difference equations, *Qualitative Properties of Differential Equations, Proceedings of the 1984 Edmonton Conference*(1986), edited by W. Allegretto and G.J. Butler, 329–337.

[209] A. Peterson, Green's functions for $(k, n-k)$-boundary value problems for linear difference equations, *J. Math. Anal. Appl.* 124 (1987), 127–138.

[210] A. Peterson, Existence and uniqueness theorems for nonlinear difference equations, *J. Math. Anal. Appl.* 125 (1987), 185–191.

[211] A. Peterson, A comparison theorem for linear difference equations, *Proceedings of the International Symposium on Nonlinear Analysis and Applications to Biomathematics*, edited by K.N. Murty and J. Gopalakrishna, Section II (1988), 12–18.

[212] A. Peterson and J. Ridenhour, Oscillations of second order matrix difference equations, *Journal of Differential Equations* (to appear).

References

[213] A. Peterson and J. Ridenhour, Atkinson's superlinear oscillation theorem for matrix difference equations, *SIAM J. Math. Anal.* (to appear).

[214] A. Peterson and J. Ridenhour, Disconjugacy for a second order system of difference equations, *International Conference on Differential Equations; Theory and Applications in Stability and Control*(1990), edited by S. Elaydi, 425–431.

[215] A. Peterson and J. Ridenhour, Oscillation theorems for second order scalar difference equations, *International Conference in Differential Equations; Theory and Applications in Stability and Control*(1990), edited by S. Elaydi, 417–424.

[216] A. Peterson and J. Ridenhour, On a disconjugacy criterion of W.T. Reid for difference equations (submitted).

[217] C. Phillips and R. Harbor, *Feedback Control Systems*, Prentice-Hall, Englewood Cliffs, NJ, 1988.

[218] S. Pincherle, Sur la generation de systemes recurrents au moyen d'une equation lineaire differentielle, *Acta Math* 16 (1892), 341–363.

[219] H. Poincaré, Sur les equation lineaires aux differentielles ordinaires et aux differences finies, *Amer. J. Math.* 7 (1885), 203–258.

[220] J. Popenda, Finite difference inequalities, *Fasciculi Math.* 13 (1981), 79–87.

[221] J. Popenda, On the boundedness of the solutions of difference equations, *Fasciculi Math.* 14 (1985), 101–108.

[222] M. Porter, On the roots of functions connected by a linear recurrent relation of the second order, *Ann. Math.* 3 (1901-2), 55–70.

[223] R. Potts, Exact solution of a difference approximation to Duffing's equation, *J. Austral. Math. Soc. (Series B)* 23 (1981), 64–77.

[224] R. Potts, Nonlinear difference equations, *Nonlinear Anal. TMA* 6 (1982), 659–665.

[225] P. Purdom and C. Brown, *The Analysis of Algorithms*, Holt, Rinehart and Winston, New York, 1985.

[226] R. Redheffer and W. Walter, A comparison theorem for difference inequalities, *J. Differential Equations* 44 (1982), 111–117.

[227] W. Reid, A criterion of oscillation for generalized differential equations, *Rocky Mtn. J. Math.* 7 (1977), 799–806.

[228] P. Ribenboim, *The Book of Prime Number Records*, Springer-Verlag, New York, 1988.

[229] C. Richardson, *An Introduction to the Calculus of Finite Differences*, Van Nostrand, Princeton, 1954.

[230] Y. Rodriguez, On nonlinear discrete boundary value problems, *J. Math. Anal. Appl.* 114 (1986), 398–408.

[231] D. Sherbert, Difference equations with applications, *UMAP*, Unit 332 (1979), 1–34.

[232] J. Smith, *Mathematical Ideas in Biology*, Cambridge Press, Cambridge, 1968.

[233] R. Smith, Sufficient conditions for stability of difference equations, *Duke Math. J.* 33 (1966), 725–734.

[234] B. Smith and W. Taylor, Oscillatory and asymptotic behavior of certain fourth order difference equations, *Rocky Mountain J. Math.* 16 (1986), 403–406.

[235] J. Spanier and K. Oldham, *An Atlas of Functions*, Hemisphere, Washington, 1987.

[236] M. Spiegel, *Calculus of Finite Differences and Difference Equations*, Schaum's Outline Series, McGraw-Hill, New York, 1971.

[237] V. Strejc, *State Space Theory of Discrete Linear Control*, Wiley-Interscience, New York, 1981.

[238] S. Sugiyama, Difference inequalities and their applications to stability problems, Lecture Notes in Math. Springer, 243 (1971), 1–15.

[239] S. Sugiyama, On periodic solutions of difference equations, *Bull. Sci. Eng. Res. Lab Wasenda Univ.* 52 (1971), 87–94.

[240] B. Szmanda, Oscillation of solutions of second order difference equations, *Port. Math.* 37 (1978), 251–254.

[241]   B. Szmanda, Oscillation theorems for nonlinear second-order difference equations, *J. Math. Anal. Appl.* 79 (1981), 90–95.

[242]   D. Trigiante and S. Sivasundaram, A new algorithm for unstable three term recurrence relations, *Appl. Math. and Comp.* 22 (1987), 277–289.

[243]   P. Van der Grayssen, A reformulation of Olver's algorithm for the numerical solution of second order difference equations, *Num. Math.* 32 (1979) 159–166.

[244]   D. Vaughan, A nonrecursive algebraic solution for the discrete Riccati equation, *IEEE Trans. Automatic Control* 15 (1970), 597–599.

[245]   N. Watanabe, Note on the Kalman filter with estimated parameters, *J. Time Ser. Anal.* 6 (1985), 269–278.

[246]   F. Weil, Existence theorem for the difference equation $y_{n+1} - 2y_n + y_{n-1} = h^2 f(y_n)$, *Internat. J. Math. Math. Sci.* 3 (1) (1980), 69–77.

[247]   D. Willett and J. Wong, On the discrete analogues of some generalizations of Gronwall's inequality, *Monatsh. Math.* 69 (1965), 362–367.

[248]   J. Wimp, *Computation with Recurence Relations*, Applicable Mathematics Series, Pitman Advanced Publishing Program, Boston, 1984.

[249]   J. Wimp and Y. Luke, An algorithm for generating sequences defined by non-homogeneous difference equations, *Rend. Circ. Mat Palermos* (2) 18 (1969), 251–275.

[250]   A. Wouk, Difference equations and J-matrices, *Duke. Math. J.* (1953), 141–159.

[251]   M. Yamaguti and H. Matano, Euler's finite difference scheme and chaos, *Proc. Japan Acad.* 55A (1979), 78–80.

[252]   M. Yamaguti and S. Ushiki, Discretization and chaos, *C.R. Acad. Sc. Paris* 290 (1980), 637–640.

Index

Abel's summation formula 33
Airy equation 4,37
Annihilator
   method 73,76,84
Anti difference 24
Approximate summation 36-44
Asymptotic
   approximation 210,212
   behavior 207,222,280
   series 211,218
   stability 162,173
   to 208
Attractor
   Hénon 194
   strange 193,195

Baker map 205
Bernoulli
   numbers 38-41
   polynomials 38-41
Bessel functions 246
Big Oh 209
Binomial coefficients 21
   extended 22
Boundary value problem
   for linear difference
   equations 265,269
   for nonlinear difference
   equations 371
Brouwer fixed point theorem 382
Butterfly effect 190

Calculus of variations 354
Casorati matrix 63
Casoratian 63,93
Cauchy
   -Euler equation 95
   function 257
   product 88

Cayley-Hamilton theorem 155
Chaotic behavior 187
Characteristic
   equation 70
   polynomial 70
   roots 70
Companion matrix 152
Comparison theorem 263,273
Continued fractions 59
Contraction
   mapping 372
   theorem 372
Convergence
   linear 235
   quadratic 238
Convolution
   of sequences 122
   theorem 123
   type 87
Crystal lattice 79

Difference equation
   Fibonacci 78
   full history 135
   Emden-Fowler 302
   Euler-Lagrange 343
   partial 399
   self-adjoint 250
Difference operator 15-23
Disconjugate 263,274,289,358
Discrete calculus of variations 340
Disfocal 306
Dominant solution 283

Eigenfunction 312
Eigenvalue 152,312
   multiple 156,320
   simple 153,316

453

Eigenpair 312
Eigenspace
    generalized 167
Emden-Fowler
    difference equation 302
Epidemiology problem 87
Euler-Lagrange difference equation 343
Euler's
    constant 220
    method 9
    summation formula 37,41,216

Factorial
    function 20
    series 60,99,220
Feigenbaum's number 186
Fibonacci
    difference equation 78,262
    sequence 78,79
First variation 342
Fixed point 171,372
    asymptotically stable 173
    stable 173
Fourier
    coefficients 323
    series 323
Fractal 194
Fredholm summation equation 136
Fundamental theorem of
    discrete calculus 32

Gamma function 7,8,58
Generalized
    eigenspace 167
    zero 261
Generating function 37,38,87,97
Golden section 79
Green's
    function 269,371
    theorem 255

Hénon system 193
Hermite polynomials 48
Homogeneous 52,60

Indefinite sum 24
Initial and final value theorem 114

Initial value problem 61
Inner product 314
    with weight function 315

Jacobi's equation 351

Kernel 124,126
    symmetric 130

Ladder network problem 120
Lagrange identity 254
Laguerre polynomials 48
Legendre
    necessary conditions 352
    polynomials 48
Liapunov function 177
    strict 177
Lie's transformation
    group method 102
Linear
    convergence 235
    equations 51,60
    systems 151
Linearly
    dependent 62,69
    independent 62,69
Liouville's formula 255,281
Lipschitz
    condition 374
    constant 374
    one-sided condition 387
Logistic differential equation 13
Lower solution 383,392

Minimum
    global 355
    local 340,355
    proper 340,355
Much smaller than 208

Newton's method 234,238
Nonhomogeneous 60,324
Nonoscillatory 293
Norm
    Euclidean 372
    maximum 372
    traffic 372

Orthogonal 131
    with weight function 83,316
Oscillatory 293

Partial difference equation 399
Period 171
Periodic
    boundary conditions 312
    point 171,192
    Sturm-Liouville problem 312
Perron's theorem 222, 294
Picard iterates 373
Plankton problem 394
Poincaré
    map 195
    theorem 222
    type 222
Polya factorization 255,275
Predator-prey model 81,172,195
Prime number theorem 208
Principal solution 283
Product rule 17

Quadratic
    convergence 238
    form 351
    functional 358
Quotient rule 17

Rayleigh inequality 331
Recessive solution 283
Reduction of order 94
Riccati
    equation 100,284
    inequality 288
    substitution 100,285

Self-adjoint
    difference equation 249,250,350
    boundary value problem 315
Sensitive dependence on
    initial conditions 189
Separated boundary conditions 311
Shift operator 17,192
Shifting theorem 111
Simple eigenvalue 153

Simplest variational problem 340
Spectral radius 153
Spectrum 153
Stable 173
    asymptotically 162,173
    subspace 169
    subspace theorem 168
Staircase method 173
Stirling's formula 218,219
Strange attractor 193,195
Strict Liapunov function 177
Sturm
    comparison theorem 292,365
    separation theorem 261
Sturm-Liouville
    difference equation 311
    periodic 312
    problem 311,404
Sturmian theory 261
Summation by parts 29,33,214
Summation equations
    Fredholm 126
    Volterra 124
Symbolic dynamics 191
Symmetric kernel 130

Tent map 190
Tiling problem 89
Tower of Hanoi problem 3
Transversality condition 346
Trapezoidal rule 49

Unstable 173
Upper solution 384,392

Variation
    first 342
    second 342
Variation of constants
    formula 53,67,159,258
Volterra summation equation 124

Wallis' formula 217
Weierstrass summation formula 366

z-transform 107-125,414
    table 132

QA 431 .K44 1991
Kelley, Walter G.
Difference equations